ANALYSIS OF NUCLEAR LEGISLATION

Volume I of this study deals with the international aspects of the regulation of nuclear trade, while volume II deals with national legislation.

This study is part of a series of analytical studies of the major aspects of nuclear energy legislation in force in OECD Member countries. The studies published to date are:

— Regulations governing Nuclear Installations and Radiation Protection (published in 1972);

— Nuclear Third Party Liability (first published in 1967 and published in a new edition in 1976);

— Regulations governing the Transport of Radioactive Materials (published in 1980);

— Regulatory and Institutional Framework for Nuclear Activities (volume I published in 1983; volume II published in 1984). This study brings up to date and expands a study published in 1969.

Also, a Description of Licensing Systems and Inspection of Nuclear Installations in OECD Countries was published in 1986.

*
**

Volume I has been prepared by the Secretariat with the collaboration of Jean Hébert. Its preparation has benefitted greatly from active participation by many experts from the international organisations concerned and from Member countries. The Secretariat wishes to express its appreciation of their kind assistance.

This study is based on information available to the Secretariat in February 1988 and neither the Secretariat nor national authorities assume any liability therefor.

**Subscription for one year
(2 issues and supplements)**
1989 Price: £17.60 FF150 $33 DM65

Covers legislative and regulatory developments, agreements and case law in the nuclear field in more than fifty countries. The Bulletin also reports on the regulatory work of international organisations competent in that area. Translations of the most important acts, regulations and agreements are published, as are legal studies and articles, signed by specialists in nuclear law. An analytical index, included in the subscription fee, is published every five issues and facilitates research of all information given in the Bulletin as from the first issue.

The Bulletin is also published in French.

THE REGULATION
OF NUCLEAR TRADE

NON-PROLIFERATION — SUPPLY — SAFETY

VOLUME I

INTERNATIONAL ASPEC⁻⁻

is to be
last date

Nuclear Legislation Series

NUCLEAR ENERGY AGENCY
ORGANISATION FOR ECONOMIC CO-OPERATION AND DEVELOPMENT

Pursuant to article 1 of the Convention signed in Paris on 14th December, 1960, and which came into force on 30th September, 1961, the Organisation for Economic Co-operation and Development (OECD) shall promote policies designed:

- to achieve the highest sustainable economic growth and employment and a rising standard of living in Member countries, while maintaining financial stability, and thus to contribute to the development of the world economy;
- to contribute to sound economic expansion in Member as well as non-member countries in the process of economic development; and
- to contribute to the expansion of world trade on a multilateral, non-discriminatory basis in accordance with international obligations.

The original Member countries of the OECD are Austria, Belgium, Canada, Denmark, France, the Federal Republic of Germany, Greece, Iceland, Ireland, Italy, Luxembourg, the Netherlands, Norway, Portugal, Spain, Sweden, Switzerland, Turkey, the United Kingdom and the United States. The following countries became Members subsequently through accession at the dates indicated hereafter: Japan (28th April, 1964), Finland (28th January, 1969), Australia (7th June, 1971) and New Zealand (29th May, 1973).

The Socialist Federal Republic of Yugoslavia takes part in some of the work of the OECD (agreement of 28th October, 1961).

The OECD Nuclear Energy Agency (NEA) was established in 1957 under the name of the OEEC European Nuclear Energy Agency. It received its present designation on 20th April, 1972, when Japan became its first non-European full Member. NEA membership today consists of all European Member countries of OECD as well as Australia, Canada, Japan and the United States. The commission of the European Communities takes part in the work of the Agency.

The primary objective of NEA is to promote co-operation between the governments of its participating countries in furthering the development of nuclear power as a safe, environmentally acceptable and economic energy source.

This is achieved by:

- *encouraging harmonisation of national, regulatory policies and practices, with particular reference to the safety of nuclear installations, protection of man against ionising radiation and preservation of the environment, radioactive waste management, and nuclear third party liability and insurance;*
- *assessing the contribution of nuclear power to the overall energy supply by keeping under review the technical and economic aspects of nuclear power growth and forecasting demand and supply for the different phases of the nuclear fuel cycle;*
- *developing exchanges of scientific and technical information particularly through participation in common services;*
- *setting up international research and development programmes and joint undertakings.*

In these and related tasks, NEA works in close collaboration with the International Atomic Energy Agency in Vienna, with which it has concluded a Co-operation Agreement, as well as with other international organisations in the nuclear field.

TABLE OF CONTENTS

Appendices 16 to 22

URANIUM — RESOURCES, PRODUCTION
AND DEMAND

Appendices 23 to 26

REACTORS —
NUCLEAR POWER PLANT IMPORTS

Appendices 27 to 31

NUCLEAR FUEL CYCLE — MATERIAL FLOW

FOREWORD

Nuclear trade has a special place in international economic relations largely due to both the strategic significance of transfers of fissile products and nuclear equipment, and to historical circumstances. Ever since man mastered the energy released by atomic fission, a cloud has weighed on the peaceful uses of this source of energy — the possibility of their being diverted to the production of atomic weapons. Fear that countries should be tempted in this direction, overtly or secretly, has resulted in multiple controls which have bestowed on nuclear trade very particular characteristics.

The 1968 Treaty on the Non-Proliferation of Nuclear Weapons (NPT) pinpointed the fundamental ambivalence of nuclear trade in its preamble which stressed the need to control the movements of nuclear material to prevent diversion, asserting at the same time that the advantages of the peaceful applications of nuclear technology should, without discrimination, be accessible to all the Parties. To a great extent, the law governing nuclear trade reflects the effort to reconcile both objectives. It is to be noted that, although the political aspects of horizontal nuclear-weapon proliferation have spawned a mass of literature, there have been relatively few recent studies on the actual regulation of nuclear trade despite the economic relevance of this sector.

The evolution of nuclear activities these past thirty years (in 1987, 417 reactors produced more than 16 per cent of the world's electricity, exceeding 23 per cent in the OECD area and 30 per cent in the Common Market) brought with it a corresponding growth in trade in that field. Starting with mining of ores and rounding off with disposal of radioactive waste, fuel cycle operations create widespread movements. Although uranium resources are more evenly distributed in the world than oil for instance, many countries with a nuclear power production programme must import materials to meet their needs. Moreover, few countries possess the capacity to enrich uranium, build reactors, manufacture or reprocess fuels — all the operations which constitute international nuclear trade. Also, the use of radioisotopes, particularly for medical purposes, gives rise to numerous transport operations.

However, while in many industrialised countries the continued development of nuclear power production programmes is considered a positive factor, examination of the law applicable to international transfers of nuclear material and equipment reveals that, as opposed to international commerce rules which are intended to promote and liberalise those activities, those in the nuclear field impose instead a series of restrictions. This apparent paradox confirms that preventing nuclear-weapon proliferation generally prevails over promoting economic relations in this area.

Therefore, unlike most other commodities, products for the nuclear industry do not benefit from international mechanisms helping the distribution of goods. Neither is there any homogeneous set of rules or general agreements governing the different economic aspects of nuclear trade. Instead, its regulation takes many forms: instruments of competent international organisations, commitments undertaken in bilateral or multilateral agreements, intergovernmental arrangements whose legal status is somewhat uncertain, resolutions with an even hazier scope adopted in international conferences and — not to be overlooked — national decisions of a political nature. Most of the texts record mutual obligations and establish controls ensuring that the goods exchanged have an exclusively peaceful vocation. This legal structure is capped by the International Atomic Energy Agency's Safeguards System which, in international law, is a unique example of acceptance of direct inspections on the territory of sovereign states. In short, while the law governing nuclear trade does have a certain specificity as regards its object and geopolitical context, the diversity of its sources is the aspect which is the most arresting.

The considerable influence wielded early on, and even today, by bilateral agreements for the supply of material, equipment and technology is another original feature of nuclear trade; it illustrates the prominent role of States in conducting nuclear business. In this way, supplier countries are able to impose prior conditions on the uses to be made of their exports. While adoption of the NPT in 1968 resulted in shifting the sometimes "unbalanced" dialogue between supplier and buyer countries to a multilateral framework, this has not undermined the importance of bilateral agreements in that field.

The entry into force of the NPT did not succeed in doing away with all concerns aroused by the proliferation risks associated with nuclear trade, and the seventies saw new developments in its regulation. Negotiations were held between supplier countries to specify conditions for applying safeguards and define a "code of good behaviour" to restrain the effects of an intensive economic competition. This policy of strengthening controls and placing embargoes on certain products and equipment disturbed importing countries, anxious to secure their supply sources and to avoid jeopardising legitimate desires to conduct their civil nuclear programmes successfully.

Compatibility of non-proliferation guarantees with security of supplies is a question which still dominates the law governing nuclear trade, and the numerous international negotiations it has initiated have found no entirely satisfactory solution as yet for all the interested parties. However, if we consider the unilateral character of co-operation agreements concluded at a certain time between exporting and importing countries, the more recent agreements for supply of materials and equipment reveal a tendency to greater flexibility and simpler controls as well as increased selectivity in the treatment accorded to importing countries, based on assurances the latter are expected to provide in that field. Finally, it would seem that nuclear trade law at its present stage of evolution is still seeking equilibrium; stronger, closer consultations at international level are needed to overcome its contradictions.

*
**

The importance of the non-proliferation problem in the regulation of nuclear trade should not, however, obscure its other features: statutory responsibilities of international organisations, mechanisms set up to secure fuel supplies, international industrial co-operation enterprises, technology transfers, physical protection of materials and equipment against risks of aggression, safe transport and insurance — all these topics are discussed in *Volume I* of this Study. On the other hand, no mention is made of the contract clauses in nuclear trade since,

as a rule, they are basically no different from those for other advanced technologies.

For consultation purposes, various documents which are important in the regulation of nuclear trade have been reproduced in full. In addition, the texts of some particularly representative bilateral agreements have been included. The Appendices to Volume I also contain information on the status of relevant international agreements, some basic economic data in this field as well as a selected bibliography.

Volume II of the Study deals with the national legislation of OECD countries with significant nuclear regulations and trade in that field and in essence, covers two aspects. The first concerns political and administrative controls over the import and export of "sensitive" products, namely fissile materials and large nuclear equipment as well as technology transfers. In most cases, this description is completed with the list of nuclear items whose export is restricted. The second concerns the licensing system governing trade in, as well as import and export of nuclear material to protect users and the public against the hazards created by its radioactive properties.

In most countries, there is in fact a twofold level of regulation: firstly, the nuclear texts properly speaking which regulate the different aspects of nuclear trade and secondly, the general regulations on foreign trade with, where necessary, special rules for nuclear material and equipment. As regards restrictions for the export of sensitive nuclear products, the measures may be taken in a specifically nuclear perspective or in the more general framework of items of strategic interest, or they may be parallel or combined.

Volume II also contains information on regulations concerning physical protection, industrial property and transport, as well as on multilateral and bilateral agreements involving nuclear trade. To facilitate research, the descriptions of national legislation follow a plan which is as uniform as possible, given the differences in the legal systems concerned.

CHAPTER ONE

THE ORIGINS OF NUCLEAR TRADE RULES AND EVOLUTION OF NON-PROLIFERATION POLICIES

A. INTRODUCTION

This chapter provides an overview of the stages of development of the law regulating trade in nuclear material, equipment and technologies and places this development within a historical, political and economic perspective. As implied by the title, trading relations in this area have, from the beginning, been inextricably linked to most exporting countries' concern to prevent the spread of nuclear weapons. The international rules applicable to nuclear trade are therefore largely determined by the potential interaction between the civil and military aspects of nuclear energy.

Historically, the dual nature of nuclear energy has raised fundamental political and legal problems. The fact that civilian atomic power was originally a spin-off from military programmes[1] and that some facilities used to supply nuclear fuel for power generation could also be used to produce nuclear material for nuclear explosives, has kept alive the notion that both facets of this form of energy cannot be wholly separated.

States which, during the Second World War, had been the most closely associated with building an atomic weapon were, in the post-war years, the best placed to capitalise on the experience acquired for the industrial development of atomic energy[2]. The United State's post-war decision to impose tight secrecy on atomic energy, even as regards its former allies, dampened, for at least a decade, any possibility of co-operation in the development of nuclear energy for peaceful purposes.

With the declassification of large amounts of nuclear information in 1955 at the first United Nations Conference on the Peaceful Uses of Atomic Energy, the way was now open to exploit this knowledge. Encouraged by growing public enthusiasm, Canada and Sweden joined the United States, the United Kingdom, the USSR and France in developing nuclear reactor models. By 1964, there were fifteen reactors in operation or completed[3], using either natural or enriched uranium. The sums thus invested were, already, considerable.

As an international market for the exchange of nuclear material developed, the United States, by virtue of its monopoly over enriched uranium supplies, was able to insist on verifying, through inspection, that the nuclear material transferred to foreign countries was not diverted to other than peaceful uses. Gradually, other industrialised countries adopted the same position, and thus, control over the non-proliferation of nuclear weapons entered on the scene, since access to nuclear material and equipment was possible only in exchange for assurance against their use for military purposes given by the recipient countries.

From the early post-war years to this day, nuclear export policies have vacillated between encouraging access to nuclear technology and restricting, for various reasons — mainly arising from the desire to avoid nuclear-weapon proliferation — the spread of that same technology. The development of nuclear trade regulations is itself a reflection of the difficulty of reconciling these two major interests. For purposes of analysis, the above period is broken down into several sections corresponding to the most significant phases of this development.

B. INTERNATIONALISATION AND SECRECY: 1945-1953

I. THE FRANCK REPORT AND THE DECLARATION ON ATOMIC ENERGY

The question of defining the post-war status of nuclear power arose in 1944-1945 when allied scientists responsible for the building of an atomic bomb, considered how to use for peaceful purposes the energy released by fission of the atom[4]. This concern led to the first proposals for an international control by a group of American nuclear scientists (the 1945 so-called Franck Report)[5] which urged their authorities to conclude an international treaty banning nuclear weapons and establishing a method for the control of nuclear energy based on a renunciation of certain sovereign rights.

This was followed in November 1945, by the "Agreed Declaration on Atomic Energy", issued by the leaders of the three countries that had co-operated in the development of the nuclear weapon (the United States, the United Kingdom and Canada). The Declaration reflected the opposing views of secrecy and internationalisation — on the one hand, it advocated a policy of secrecy, which would later be adopted in the United States' McMahon Act of 1946, and on the other it contained a proposal for the international control of all nuclear activities[6]. This ambivalence was to characterise the first stage in the development of atomic energy.

II. THE ACHESON — LILIENTHAL REPORT

Following this Declaration, the United Nations General Assembly adopted a proposal in January 1946 creating the United Nations Atomic Energy Commission (UNAEC) as a dependent body of the United Nations Security Council concerned with the development of atomic energy and the possible elimination of nuclear weapons. The United States' policy, to be presented before this Commission, was greatly influenced by a committee whose report in March 1946 to the Secretary of State was to become known as the *Acheson-Lilienthal Report*[7]. A comprehensive, international regime for control of atomic energy would be sought in exchange for the United States giving up its nuclear monopoly.

Since almost all industrial nuclear activities were classified as "dangerous", the Report proposed that internationalisation encompass all applications of nuclear power and that the competence of sovereign states with respect to the development of national nuclear programmes devolve to a world organisation. Because the Report advocated total nuclear disarmament and comprehensive international ownership and management of nuclear material, this international authority would not be charged with external inspections but was to be given a leadership role in the development of nuclear technology.

III. THE BARUCH PLAN

The Acheson-Lilienthal proposal to create a "supranational" body was, however, considered by the American government as being too idealistic to succeed. The notion that the United States' military atomic monopoly should be preserved was re-introduced in the new version of the proposal, now called the "Baruch Plan" which was presented in June 1946 to the UNAEC by the United States' Representative, Bernard M. Baruch. The Baruch Plan nevertheless maintained the internationalisation aspects of its predecessor in its proposal for the structure and role of the International Atomic Development Authority (IADA)[8].

This authority would be vested with complete managerial control of all atomic energy activities potentially dangerous to world security, with emphasis on the necessity for effective control, surveillance and licensing of national activities. In addition, no national governmental veto would be allowed so as not to interfere with IADA's control and inspection missions. The Plan, in fact, was intended to curtail, almost entirely, any national initiatives as concerns nuclear energy decisions. The new control and enforcement regime would thus have considered national definition of policies in nuclear matters as the exception rather than the rule.

The Soviet Union challenged the Baruch proposal by introducing a counter-proposal that nuclear weapons be prohibited before even discussing the question of control by an international organisation[9]. In the second meeting of the UNAEC, the Soviet Union made clear that it could not accept the no-veto proposal contained in the Baruch Plan and neither could it concede the necessity for major international intervention in the domestic nuclear sphere.

The Baruch Plan fell victim to the cold war rivalry developing between the United States and the USSR. The long negotiations to achieve international nuclear disarmament and world control over nuclear energy ended in a deadlock in 1948 and were followed by the eventual demise of the UNAEC and its official dismantling in 1952.

IV. THE McMAHON ACT[10]

The failure of the Baruch Plan may have confirmed the United States in its isolationist tendencies and convinced it of the necessity to protect its nuclear monopoly. Congress had already adopted, in 1946, the McMahon Act preventing the sharing of any knowledge, technology or materials which could be used to develop nuclear weapons by prescribing strict limitations on dissemination of restricted data, both domestically and abroad[11]. It called nevertheless for "dissemination of scientific and technical information relating to atomic energy". All questions relating to atomic energy were to be the responsibility of the United States Atomic Energy Commission (USAEC), which had just been created by the United States Congress. The Act also provided for the control of all classified information and severe penalties were prescribed for anyone guilty of passing information to a foreign government.

However, this policy of denial prevented neither the Soviet Union nor the United Kingdom from shortly thereafter achieving nuclear weapon capability[12]. At the same time, the United States resorted to massive nuclear testing and the build-up of its own nuclear arsenal.

The failure to contain the spread of nuclear weapons by blocking the flow of scientific and technical knowledge made it clear that a comprehensive regime to control the development and use of nuclear energy on a world-wide scale would be almost impossible to achieve. In fact only a weaker form of internationalisation would now be sought to permit the major nuclear powers to retain sovereignty over their national nuclear programmes while other States submitted voluntarily to limiting their use of nuclear technology to solely peaceful purposes.

C. OPENNESS AND INTERNATIONAL CO-OPERATION: 1953-1974

I. THE POLITICAL AND ECONOMIC CONTEXT

The loss of the American nuclear weapon monopoly in the early 1950s coupled with growing commercial interests for the development of civil nuclear energy, strongly influenced the elaboration of a new United States non-proliferation policy. In 1953, the nascent United States atomic energy industry consisted largely of government-owned laboratories and factories operated by private contractors. The 1946 Atomic Energy Act had placed nuclear activities in the United States under the ownership and control of the Atomic Energy Commission. Furthermore, the Commission had, in the light of the post-war situation, given priority to military applications to the detriment of civil uses. The manufacturers and operators of nuclear facilities and prototype naval reactors, on the other hand, were anxious to see a privatisation of nuclear activities through a modification of the McMahon Act[13], all the more so since the results of the United States' naval propulsion programme had indicated the possibility of adapting this technology to civil power production.

Other countries also chose to follow-up this possibility. By the end of the 1950s, viable alternatives to the American light-water reactor[14], particularly natural uranium-graphite moderated or heavy-water moderated reactor types, were seriously being pursued by Canada, the United Kingdom and France. Although the Federal Republic of Germany was not at the time a direct competitor in the reactor market, its industrial capability made it a highly sought-after partner. Its eventual decision, as that of Sweden, to opt for light-water reactors may have tipped the scales in favour of this reactor type instead of European reactors which the creation of EURATOM might have been expected to promote.

The establishment of a strong private nuclear industry in the United States was a prerequisite to the entering of foreign markets[15]. To do so, the Government took several steps to encourage nuclear development at home[16] and abroad[17]. Thus, the "Joint Program" under the EURATOM-United States Agreement of 8th November 1958, together with the sums credited by the Export-Import Bank promoted the export to Europe of light-water reactors fuelled with uranium enriched in the United States.

II. ATOMS FOR PEACE PLAN

It is in this context that President Eisenhower officially launched the Atoms for Peace Plan at the United Nations on 8th December 1953[18].

As has been seen, the basis of the Baruch Plan had been that the civilian and military applications of atomic energy could not be arbitrarily separated. The Plan had therefore proposed the full internationalisation of all potentially dangerous nuclear activities and that such an effort be linked with total nuclear disarmament. The United States now abandoned these principles and took the position that:

— civilian and military uses of atomic energy could be technically separated from each other through effective safeguards and inspections;
— as a result there would be no need to lay an embargo on the entirety of the fuel cycle, but only on certain sensitive nuclear facilities with control being exercised over the remainder;
— such control, though useful to achieve nuclear disarmament at a later stage was not to be understood as a disarmament strategy, for only non-nuclear weapon States (NNWS) would be subjected to it.

Clearly distinct from both disarmament and full internationalisation, the new regime would be limited to a general surveillance system. In parallel with this shift, the United States officially abandoned its earlier policy of denial of information as embodied in the 1946 McMahon Act, and launched a new strategy to promote the development of peaceful uses of nuclear energy under appropriate safeguards[19]. In particular, Section 123 of the 1954 Atomic Energy Act, by allowing flexibility in the negotiation of nuclear co-operation agreements, opened the way for an important transfer of United States nuclear technology by way of bilateral agreements[20].

The consequences of this shift have had a major impact on the history of international nuclear relations up to the present period. Atoms for Peace provided the basis for both the creation of various co-operation arrangements as well as for the growth of nuclear energy programmes throughout the world[21].

President Eisenhower's initiative also led in 1955 to the first United Nations-sponsored Geneva Conference on the Peaceful Uses of Atomic Energy. The broad international co-operation thus established was a precursor of the specialised intergovernmental organisations which were to be set up in the nuclear field, the first of which was created within the framework of the United Nations. The initial ambivalence pointed out above was transformed into a policy aiming to reconcile an opening towards international trade, with control requirements in the framework of both bilateral and multilateral co-operation.

III. NUCLEAR CO-OPERATION ORGANISATIONS

1. The International Atomic Energy Agency[22]

The Atoms for Peace proposal set forth an ambitious plan to reduce the threat of atomic weapons' development by directing nuclear explosive materials to an international pool of materials to be used solely for peaceful purposes — a far-reaching idea which will be echoed in more recent proposals for the internationalisation of certain parts of the fuel cycle.

However, the regime envisaged in the Atoms for Peace Plan differed in two important respects from the earlier Baruch Plan:

— instead of a comprehensive international regime granting a virtual monopoly in nuclear activities to a world agency, the new plan proposed a more classic form of co-operation whereby States would essentially retain their competence in nuclear affairs while agreeing to participate in an international control system;
— by contrast to the total nuclear disarmament envisaged in 1946, Atoms for Peace offered a more limited concept whereby nuclear-weapon States (NWS) would agree to deposit with a new world agency some of their fissionable materials, which had, till then, been earmarked for military uses.

At this stage, therefore, it was still being proposed that the agency retain some links with military applications of the atom albeit in a much less active manner than had been envisaged in earlier proposals. The link between peaceful and military applications would result from the agency's dual function as the custodian and supplier of nuclear material for civil purposes.

Of these two major functions, very little remained after the four years of negotiations leading to the adoption of the Statute of the International Atomic Energy Agency (IAEA) in 1956. The direct arms control function was dropped entirely in the final Statute, while the new agency's powers in respect of fuel storage and supply were also considerably reduced.

Nevertheless, the creation of the IAEA, a specialised institution in the United Nations family and based in Vienna, was the first realisation of the new policy of openness and co-operation in the nuclear field; this was to lead as well, in a more limited framework, to the creation of both EURATOM and the European Nuclear Energy Agency of the OEEC.

2. The European Atomic Energy Community[23]

The European Atomic Energy Community (EURATOM) was established by the Treaty of Rome in 1957, fulfilling the idea of a community to integrate the energy sector along the lines of the European Coal and Steel Community, which had been launched in 1955 at the Conference of Messina. This initiative was further motivated by energy supply considerations (the Suez Canal crisis of 1956) and also by the desire to assure the development of nuclear activities in the Community framework.

The EURATOM project contemplated the regrouping of the European nuclear industry and its development in close co-operation with the United States. This idea was evidenced in the Three Wise Men's report, "A Target for EURATOM"[24] published in 1957. The United States showed, from the outset, a very strong support for this European initiative.

In effect, the founding fathers of EURATOM hoped that the new organisation would restart the process of political integration by resolving the problem of nuclear weapons in Europe (a "no-weapons" clause, proposed by Jean Monnet, was eventually dropped from the final Treaty) and by providing an incentive in this direction through regrouping, under the aegis of the Community, the nuclear energy programmes of its Member States. The Treaty finally concluded did not go this far but a number of provisions which were retained reflected this initial intention, especially as concerns the Supply Agency and Community ownership of special fissionable material[25].

3. The European Nuclear Energy Agency[26]

The European Nuclear Energy Agency (ENEA), was created within the Organisation for European Economic Co-operation (OEEC) on 1st February 1958, one month after EURATOM. It was given wide statutory responsibilities, including the harmonisation of the scientific and industrial programmes of its Member countries; the establishment of joint undertakings; a security control function; the development of national legislation in the peaceful uses of nuclear energy, particularly in the field of nuclear third party liability[27].

Benefitting from a period favourable to international co-operation in this area, the objectives of the ENEA as concerns technical and industrial co-operation were very early given substance with the creation of three joint undertakings: EUROCHEMIC, in 1957; HAL-DEN, in 1958; and DRAGON, in 1959. Later these objectives were to evolve towards more traditional co-operation activities looking at all aspects of nuclear power, with emphasis on the technology and safety of nuclear energy.

The Agency remained essentially a European venture until 1972 with the joining of Japan, at which time the Agency's name was changed to the *OECD Nuclear Energy Agency (NEA)*. Eventually Australia, Canada, Finland and the United States successively became Members.

IV. USSR-USA CONSENSUS ON NON-PROLIFERATION

Although the period leading up to and immediately following the establishment of the three agencies was characterised by a high level of international co-operation, nevertheless great differences between the United States and the USSR, on the concept and modalities of international inspections persisted since the breakdown of the UNAEC negotiations. Competition between the superpowers in the peaceful uses of nuclear energy was evident as early as 1955 with both countries declaring that they would share nuclear technology with friendly nations. This resulted in an intensification, notably on the part of the United States, of nuclear energy aid through the use of bilateral agreements. This competition which favoured the promotion of civil nuclear energy throughout the world, led sometimes to unintended results, as witnessed by the case of the Soviet Union which, having agreed to co-operate in the construction of an enrichment plant in China, unwittingly facilitated the latter's development of its own atomic weapons[28].

The rupture between the Soviet Union and China at the end of the 1960s led to an important reassessment of the Soviet nuclear exports policy and had a fundamental influence on the evolution of Soviet attitudes towards the IAEA's safeguards function which it had originally opposed. In fact, both the United States and the USSR had, at one point, shared the view that internationalisation should be opposed and that nuclear aid and trade should follow a bilateral pattern. The difference was that the United States had progressively come to accept a multilateral safeguards system, while the USSR maintained its opposition to IAEA safeguards.

In these circumstances the plan for integrating bilateral control agreements within the IAEA safeguards regime was effectively blocked. A unified safeguards system was in fact not feasible until 1963 when the Soviet Union reversed its position in favour of international inspections and, as a result, the United States Atomic Energy Commission officially changed its policy to encourage the establishment and application of the Vienna Agency's safeguards[29]. An improvement of relations between the two countries after the Cuban missile crisis was most visibly demonstrated by the signing on 5th August 1963 of the Treaty Banning Nuclear Weapons Tests in the Atmosphere, in Outer Space and Under Water.

This policy shift laid the ground for placing national nuclear programmes and assistance in nuclear technology development provided by advanced countries, under the surveillance of a world authority. This new regime was achieved in two steps: the transfer of bilateral safeguards agreements to the competence of the IAEA (by 1967-1968 safeguards had become a major function of the IAEA) with a persistent weakness being the absence of automatic application to all nuclear transactions. This would finally be achieved in the framework of the negotiations on a new Treaty, the second step in the process.

V. ADOPTION OF THE TREATY ON THE NON-PROLIFERATION OF NUCLEAR WEAPONS

Prior to the adoption of the NPT, the bilateral agreements attributing control functions to the IAEA covered only part of the nuclear activities of the States concerned[30]. Article II of the Statute requires the Agency to ensure that any assistance provided under its supervision or control is not used to further any military purpose. The implementation of these safeguards is provided for by Article III.A.5 to three categories of cases:

— in connection with materials and equipment supplied by the Agency or placed under its supervision or control;
— at the request of the parties to any bilateral or multilateral arrangement; and
— at the request of a State, to any nuclear activity of that State.

In fact, the IAEA (like the NEA) did not play a major role as a supplier or, in a more general sense, in the field of nuclear trade. This in turn led to a transformation of the safeguards concept; whereas safeguards were originally perceived as a compulsory element of the Agency's supply of materials function, in practice, the other two cases mentioned above (voluntary decision of IAEA members to submit to safeguards either through bilateral or multilateral instruments, or individually) came to dominate.

The key then to a unified safeguards system was to tie in the provisions of the IAEA Statute to an international global agreement which would require that all the parties having adhered to such an agreement accept Agency safeguards. Two instruments were adopted with this objective in mind: the 1967 Tlatelolco Treaty for the Prohibition of Nuclear Weapons in Latin America and, more widely, the 1968 Treaty on the Non-Proliferation of Nuclear Weapons (NPT).

The NPT was opened for signature on 1st July 1968; among the sixty-two signatories are three nuclear-weapon States, the United Kingdom, the United States and the USSR. The Treaty divided countries into two categories to which the rights and obligations assigned are fundamentally different those already having the atomic bomb (the nuclear-weapon States — NWS) and those without (the non-nuclear weapon States — NNWS). The NNWS would receive the benefits of peaceful nuclear technology by renouncing nuclear weapons in the hopes of consolidating world peace. They pledged to allow international inspection of their nuclear plants in order to guarantee their compliance with the agreement. The NWS, on their part, undertook to allow equal access to nuclear technology for all and to pursue negotiations for ending the nuclear arms race and for general and complete disarmament.

Among the NWS, both France and and the People's Republic of China refused to sign the Treaty[31], as did some of the countries more advanced in nuclear technology such as Argentina, South Africa, India, Israel[32]. Despite these abstentions which compromised the universality of the NPT the Treaty entered into force in 1970[33], and the IAEA began negotiations with the Contracting Parties for the application of safeguards on their territories[34]. The first global regime for control of the peaceful uses of nuclear energy was therefore gradually put in place.

VI. LOSS OF CONSENSUS: INSECURITY OF SUPPLY AND INDUSTRIAL COMPETITION

Following the entry into force of the NPT, the prevailing trend of thought was that proliferation had been adequately blocked, without impeding the economic development of nuclear power which was growing with increased international trade in materials and equipment. The confidence engendered by the security system was short-lived and was to be shaken by events of the mid-1970s. The most important was the explosion in May 1974 of an Indian nuclear device made possible by the recuperation of (non-IAEA safeguarded) plutonium produced in a Canadian-supplied reactor. This explosion was characterised by official Indian announcements as a peaceful experiment[35], giving rise to questions as to the efficiency of the legal system which made that interpretation possible. In addition, the announcements of certain nuclear contracts involving sensitive installations, such as the French projects to supply reprocessing plants to South Korea and Pakistan and the Federal Republic of Germany's agreement with Brazil for the supply of installations covering the complete fuel cycle, raised serious concern on the part of some countries which were only somewhat allayed by the assurance that transfer of such equipment would be subject to IAEA safeguards.

In this context, much criticism was raised as to the effectiveness of the non-proliferation regime — notably as concerns the limited scope of the IAEA safeguards —

and as to the absence of coherent and transparent trading rules in this area.

Parallel to these developments, the United States Atomic Energy Commission's decision in 1974 to suspend new contracts for natural uranium enrichment services created uncertainties in many governments as to future supplies of enriched uranium[36]. The suspension lasted until September 1977. This situation involving uncertainties over uranium supplies[37] encouraged some countries that produce reactors for electricity generation, to expand or develop other indigenous fuel-cycle capacities. The Federal Republic of Germany had already concluded negotiations with the Netherlands and the United Kingdom in a joint venture for uranium enrichment: URENCO[38], and the European reprocessing plant EUROCHEMIC was commissioned. Encouraged by projections of higher electricity demands in Europe and increasing nuclear export market opportunities, the Federal Republic of Germany invested in the development of advanced reactors such as the high-temperature reactor, while acquiring the capacity to manufacture conventional nuclear reactors.

Several industrialised countries followed the same example, using licensing arrangements to further their ability to produce American-type reactor components principally in order to achieve a greater degree of autonomy. Such was the case for Italy, Japan, Spain and Sweden.

After the 1973 oil crisis, which demonstrated Europe's energy vulnerability, and which considerably increased the cost of its oil imports, France decided to launch an ambitious nuclear power production programme. Having earlier abandoned, not without regret, the development of the "indigenous" natural uranium gas-graphite reactor in order to concentrate on the Westinghouse PWR model, the French Government began a policy of diversification which was notably manifested by the creation of EURODIF[39] (marking, after URENCO, the failure of the EEC to construct a Community enrichment facility) as well as by the construction of a reprocessing facility at La Hague. Contrary to France, the United Kingdom chose to continue development of its advanced gas-cooled reactor (AGR), although this decision was not a commercially successful one. At the same time, the United Kingdom began to offer a wide range of services in the nuclear field.

Although in 1970s, the United States was still the undisputed leader in all areas of the nuclear cycle and in particular, in reactor exports[40], the situation gradually changed. France, the Federal Republic of Germany and Canada (which had developed its own heavy water reactor technology — CANDU), began to aggressively challenge this United States commercial monopoly. The search for new markets however, brought with it the temptation to offer sensitive fuel cycle technologies in addition to simple reactor sales. This tended to aggravate tensions and gave a political dimension to commercial rivalries.

Under these circumstances, the concern that some countries could thereby obtain "proliferating" nuclear facilities not subject to an effective control system would lead some countries (particularly the United States and Canada) to try to halt what they considered unrestrained competition for nuclear export markets, especially when the clients were found in regions they considered particularly unstable.

D. REINFORCEMENT OF CONTROLS: 1975-1985

The commercial competition among nuclear exporters as well as uncertainties concerning the effectiveness of non-proliferation policies gave rise to reconsideration of the basic principles of the NPT safeguards system. This reappraisal began with the challenge, by several supplier countries, of the NPT rule allowing the export of nuclear material and equipment under IAEA safeguards even when those safeguards did not encompass all activities of the fuel cycle in the recipient country.

These criticisms gave birth gradually to the idea that there should be no nuclear trade with non-nuclear weapon States without universal full-scope safeguards and that sensitive activities, such as reprocessing, should be either abandoned or reserved to nuclear-weapon States or dealt with under proper institutional arrangements. The discussions of the London Club were to focus on attempts to impose full-scope safeguards on a multilateral basis. On another level, the NPT's provisions on export controls required interpretation and clarification; this task was given to a Committee set up between the Signatories.

The adoption by the United States of the Nuclear Non-Proliferation Act in 1978 (NNPA) was to be on the other hand a unilateral attempt by one supplier country to tighten the rules of nuclear trade. The internationalisation of activities such as reprocessing or uranium enrichment led to a study by IAEA on regional nuclear fuel cycle centres[41] and then became the focal point of broader-scale discussions within the International Nuclear Fuel Cycle Evaluation (INFCE) exercise proposed by the United States.

The following section will look at these different attempts, some of which, by multiplying restrictions and controls beyond the NPT's provisions, resulted in affecting the support originally given by importing countries.

I. NUCLEAR EXPORTERS COMMITTEE (ZANGGER COMMITTEE)[42]

Following the entry into force of the NPT in 1970, this Committee (known as the "Zangger Committee", after its Chairman) was set up under the IAEA to interpret the safeguards clause (Article III.2) of the Treaty and to agree on common rules for its application in respect of exports requiring safeguards. Made up exclusively of NPT signatories (therefore excluding France), the members of the Committee[43] decided to establish and keep up to date, in step with technological developments, the list of materials and equipment subject to the Treaty, and sensitive from the viewpoint of nuclear weapon proliferation.

The Committee reached agreement on the definition of those items whose export would "trigger" the application of IAEA safeguards to the facility for which the items were being provided — the "Trigger List". It is important to note that this "Trigger List" is only for components, equipment and materials necessary for the nuclear fuel cycle. There was however, no ban on transfers of sensitive technologies such as enrichment and reprocessing. Moreover, safeguards were required only for the materials and equipment transferred.

Despite these limitations, the main result of this agreement within the Zangger Committee was to subject all future nuclear transactions — whether or not within the NPT — to a still limited but more stringent control system. The countries participating in the Zangger Committee resumed their negotiations after 1978.

II. NUCLEAR SUPPLIERS GROUP (LONDON CLUB)[44]

The Nuclear Suppliers Group (known as the London Club because of its early secret meetings in 1974 in London) originally consisted of seven countries (including France) and subsequently was expanded to fifteen members to include some importing countries. Supported by both the United States and the USSR, these negotiations dealt principally with two issues: obtaining from France an undertaking equivalent to that under Article III.2 of the NPT and seeking to achieve a consensus among exporting countries on a more restrictive export policy than that under the Treaty on two points. The first required full-scope safeguards for all nuclear activities in importing NNWS (including indigenously produced facilities) and the second called for an embargo as concerns the most sensitive parts of the fuel cycle, i.e. reprocessing and enrichment. Implementation of both points, however, was criticised as being counter to the principles of the NPT by various non-nuclear Signatory States, and gave rise to a certain amount of friction among the participants.

The voluntary agreement which was finally reached in 1976 was a sort of code of good behaviour. Known as the "London Club" Guidelines, it attempted to strengthen the existing mechanism of nuclear export control[45] by adoption of a list of materials and equipment (very similar to the Zangger Trigger List) whose export is subject to IAEA safeguards. Regarding the second point, the stricter proposals (subsequently incorporated in the United States legislation) did not achieve consensus; the compromise solution was to recommend particular caution for the export of "sensitive" materials (uranium enriched to more than 20 per cent and plutonium) and equipment.

III. THE UNITED STATES NUCLEAR NON-PROLIFERATION ACT OF 1978[46]

President Carter's attempts early in his administration to convince other nuclear suppliers to require full-scope safeguards for nuclear exports to non-nuclear weapon States and to embargo supply of sensitive nuclear technologies (i.e. those for enrichment and reprocessing) were not successful. Nevertheless, the United States in March 1978 enacted the Nuclear Non-Proliferation Act of 1978 which, among other things, required full-scope safeguards as a condition for United States supply, despite the absence of such a provision in existing agreements for nuclear co-operation. Moreover, the Act directed the President to renegotiate existing agreements to conform to the NNPA, including a set of provisions for United States post-export controls, or consent rights applying to transfers and reprocessing or enrichment of United States origin nuclear material.

The non-negotiable demands, retroactivity and lack of consultation stirred up strong controversy which was aggravated by a brief cutoff of United States enrichment exports to EURATOM States. Although most existing United States agreements included some provision for consent rights and safeguards, these did not conform with the stricter requirements of the new Act both for criteria for licensing United States exports and for post-export controls.

The enactment of the NNPA inevitably worsened the climate of uncertainty already existing among nuclear commercial entities involving the retransfer and reprocessing of United States origin fuel. Hostile to reprocessing and the use of plutonium, because these activities seemed to present too great a risk from the perspective of "proliferation", the United States Administration had suspended in 1977 its own civil reprocessing and fast breeder reactor programme and attempted to convince other countries to follow suit. Several Western European countries, which had already invested large sums in the development of reprocessing and fast breeder technol-

ogies felt themselves directly threatened by this initiative (which was soon after supported by Canada). Other countries, particularly in Latin America, perceived this as an attempt to maintain them in a position of dependency vis-à-vis the supplier countries. The overall negative reaction to the new United States policy was one of the factors in the decision to organise an international conference for the discussion of these problems.

IV. INTERNATIONAL NUCLEAR FUEL CYCLE EVALUATION (INFCE)[47]

INFCE was a proposal by the Administration of President Carter to create a forum to discuss different technical options as a background to finding an improved consensus on non-proliferation while allowing nuclear developments. The discussions, which began in 1978 and continued over a period of two years focused, at the insistence of the other countries, on the technical, economic and institutional aspects of nuclear trade and development.

The final conclusions of INFCE will be looked at in more detail, but it may be said here that if INFCE did not entirely satisfy all the parties concerned, it did help to resolve certain controversies, particularly as concerned plutonium. While INFCE questioned the assertion that there was sufficient uranium so as to make the use of plutonium as a fuel unnecessary, it also suggested that there was less need than had been expected to recycle plutonium in conventional reactors. The analysis also showed that there was interest and potential demand for fast breeder reactors and therefore a need for reprocessing to provide their fuel.

The INFCE Summary Report also recognised that "technical measures have a powerful influence on reducing the risk of theft, but only a limited influence on reducing the risk of proliferation" and that institutional measures are potentially even more important than technical measures. In other words INFCE confirmed that "technological fixes" could not be found which would allow the use of nuclear energy for peaceful purposes while at the same time preventing any military activities.

Nevertheless the dialogue initiated by INFCE played a significant role in restoring relationships among supplier countries. These countries have as a result co-operated in work organised by the IAEA as follow-up to the technical recommendations of the INFCE Report. This includes the Expert Group on International Spent Fuel Management as well as the Expert Group on International Plutonium Storage in the early eighties. This work, however, has not yet had concrete results at international level.

18

E. PRESENT CONTEXT

I.   STATUS OF NUCLEAR INDUSTRY

After a period of expansion marked by the preeminent position of the American nuclear suppliers[48], since the mid-1970s the nuclear industry has experienced, if not a real depression, at least in some countries a stagnation in new construction. The nature of this phenomenon and the reasons for it have been analysed in various publications[49]. Because of the long lead times in nuclear plant construction, one can predict, from the steadily declining rate of ordering of nuclear power plants in the West after Three Mile Island, (from thirty-three units/ year in the early 1970s to four or five units in the first half of the 1980s) that this situation will continue for some time. It may be foreseen that the Chernobyl accident is bound to have an additional negative impact, even if this accident did not dissuade several countries in particular France, Japan, the United Kingdom and even the USSR and other Eastern European countries from pursuing their electro-nuclear efforts.

The situation in this respect differs vastly from one industrialised country to another, while developing countries face both technical (adaptation of the electricity transportation system, for example) as well as financial obstacles to realise these nuclear projects. Recent OECD estimates for nuclear capacity and fuel cycle requirements are lower than previous estimates, even based on the assumptions that all projects currently underway will be completed, and that approximately 50 GWe of additional nuclear capacity will be ordered and construction begun between 1986 and 1995[50]. In consequence, the nuclear construction industry has reduced capacity and manpower to such an extent that questions have been raised as to whether a prolonged absence of orders in the short term could have a critical effect on its ability to respond to new orders if and when these occur. At the same time, the oversupply of oil has led to lower prices for traditional fossil fuels, and helped at least temporarily to erode the competitiveness of nuclear energy.

Clearly, this recession could have destabilising effects on trade relations among major nuclear suppliers. In particular, there is concern that some supplier States might be tempted to save their nuclear industries by offering export contracts which might include the transfer of technologically sensitive fuel cycle facilities. The fact that this concern has not materialised may favour the effectiveness of the measures studied above. But it may also be due to the still low electricity demand and other economic problems of developing countries. Indeed this situation has shifted the focus of competition among suppliers away from technological "sweeteners" towards a search for novel financing schemes and barter arrangements.

It can be noted that as a result of the reduction in nuclear power plant construction throughout the world, supplies of uranium and enrichment are abundant in comparison with the shortages predicted a few years ago.

The fact that enrichment services are now available from the Soviet Union and from European sources (EURODIF, URENCO) as well as from the United States, does not encourage the development of national enrichment facilities for purely economic reasons.

In the post-INFCE era of nuclear energy, there remains a certain hesitation as to how to manage the present situation. Discussions to clarify and reassert the goals of non-proliferation are now taking place in a climate marked by economic realities. These discussions concern both assurances of supply and the ways and means of strengthening the NPT system.

II.   COMMITTEE ON THE ASSURANCES OF SUPPLY (CAS)[51]

Uncertainties due to the supply of nuclear fuel, as well as to the negative impact of unilaterally imposed restrictions on retransfer and reprocessing provided an incentive to seek a consensus between importing and exporting countries. A Committee on the Assurances of Supply (CAS) open to all members of the IAEA, was therefore set up within the Agency in June 1980. It is mandated to consider the ways and means in which supplies of nuclear material, equipment and technology and fuel cycle services can be assured on a more predictable and long-term basis in accordance with mutually acceptable considerations of non-proliferation and the Agency's role and responsibilities in this area. The Committee, whose work is in abeyance, has attempted to reconcile differing viewpoints by seeking an agreement on supply principles which would be applicable without discrimination to all countries.

III.   NPT REVIEW CONFERENCES

Article VIII of the NPT provides for a Conference of Parties to the Treaty to take place at intervals of five years after the Treaty's entry into force. Since 1970, three Conferences have been held in order to review the implementation of the Treaty and progress made in meeting its objectives. The first two Conferences, particularly that of 1980, ended in political controversy surrounding the accusation that nuclear-weapon States had not lived up to their commitments to negotiate an ending to the arms race and that equipment and technology transfers necessary for the development of peaceful uses of nuclear energy were unnecessarily hampered.

The Third Review Conference, held in September 1985, produced a more promising outcome. The participants (80 out of the 130 Parties) reaffirmed their belief that the Treaty was "essential to international peace and security" and called upon the three nuclear-weapon States signatories of the Treaty to resume talks for a comprehensive ban on all nuclear testing. Although these countries were again criticised for not living up to the obligation concerning nuclear disarmament, the Conference nevertheless stressed the importance of safeguards and urged countries exporting nuclear material or technology to demand that the recipient country accept full-scope safeguards.

The Third Review Conference went one step further and called on those countries not parties to the Treaty and which are acknowledged as having potential nuclear weapon capacity to allow international inspections of all their installations in order to verify that they are being used for peaceful purposes only. These countries are not part of the "club" of countries which have developed the voluntary rules for nuclear exports. If these "emerging nuclear suppliers" were tempted to export nuclear material under lenient non-proliferation terms, the formation of such a system could result in the disruption of the NPT[52] and a loss of support by many countries. The Conference therefore urged these countries to accede to the NPT so as to reduce the risk of nuclear proliferation that unsafeguarded facilities represent.

In conclusion, the Conference which is scheduled for 1995 will have a determining influence on the future of the NPT regime because on that occasion, twenty-five years after the entry into force of the Treaty (Article X), the Parties will decide by majority whether to maintain the Treaty in effect for an indefinite period of time or whether to prolong it for one or several supplementary periods of fixed duration.

IV. THE UNITED NATIONS CONFERENCE FOR THE PROMOTION OF INTERNATIONAL CO-OPERATION IN THE PEACEFUL USES OF NUCLEAR ENERGY

In 1977, a Resolution of the United Nations General Assembly (32/50) approved the holding of the Conference (UNCPICPUNE) which took place in March-April 1987 in Geneva. In the tradition of the important nuclear conferences of the past, the major preoccupations and uncertainties confronting nuclear energy today were the main focus of these discussions: how to facilitate access to nuclear energy by developing countries, how to deal with horizontal and vertical proliferation, how to resolve the economic difficulties of nuclear energy programmes, and finally, how to guarantee the safety of nuclear installations in the aftermath of Chernobyl.

The work of the Conference[53] in differentiating technical and political aspects, was devoted in part to discussing the universal principles acceptable for international co-operation in the peaceful uses of nuclear energy. This Conference organised mainly on the initiative of developing countries ("Group of 77") did not succeed in surmounting the different points of view and interests on how to reconcile non-proliferation and the promotion of nuclear trade. The Conference did, nevertheless, provide an occasion to review international co-operation on the technical aspects of the use of nuclear energy, underlining on this occasion the essential role played in this field by the specialised intergovernmental organisations.

The conclusion to be drawn from this overview of the currents prevailing in the area of nuclear trade seems to be that only a political approach, based on a spirit of co-operation, is likely to provide a true guarantee of non-proliferation, in a climate favourable to international nuclear relations[54].

NOTES AND REFERENCES

1. The first official public communication on the "atomic" revolution was the report by Henry D. Smyth, *Atomic Energy for Military Purposes*, A General Account of the Scientific Research and Technical Development that went into the making of Atomic Bombs, published by the United States Government in August 1945. It was, in fact, the successful nuclear submarine project which started the United States off on its civilian nuclear programme. See B. Goldschmidt, *Le Complexe Atomique, Histoire Politique de l'Energie Nucléaire*, Paris, Fayard, 1980.

2. Collaboration for both military and commercial purposes was already foreseen as early as September 1944 when the President of the United States and the British Prime Minister signed the Hyde Park Aide-Memoire guaranteeing full collaboration between the two governments after Japan's defeat. With the death of Mr. Roosevelt shortly thereafter, this engagement was never implemented. M. Gowing and L. Arnold, *Independence and Deterrence: Britain and Atomic Energy, 1945-1952*, Volume 1: Policy Making, Macmillan, London, 1974.

3. Six were located in the United Kingdom, three in Italy, three in the United States, two in the USSR and one in France. See B. Goldschmidt, *op. cit.*

4. B. Goldschmidt, *Les Rivalités Atomiques, 1939-1966*, Paris, Fayard, 1967.

5. Text of Report reproduced in R. Jungk, *Brighter than a Thousand Suns*, New York, Harcourt Brace Janovich, Inc., 1958. See also Hewlett, R. and Anderson, O. Jr., Vol. I, *A History of the United States Atomic Energy Commission, The New World, 1939/1946*, Pennsylvania State University Press, 1962.

6. P. Szasz, *The Law and Practices of the International Atomic Energy Agency*, IAEA 1970, Legal Series No. 7. The Declaration acknowledged that the information required for the industrial development of nuclear energy was virtually the same as that needed for weapon production. It was agreed, therefore, that it was necessary to withhold this information until appropriate safeguards could be established to ensure that the information be used only for peaceful purposes.

7. *Report on the International Control of Atomic Energy*; the report was named after the Under-Secretary of State, Dean G. Acheson, Chairman of the Committee and David E. Lilienthal, President of the Tennessee Valley Authority and chairman of a subgroup of industrialists. P. Szasz, *op. cit.*

8. The IADA would be entrusted with all phases of the development and use of atomic energy. It would control or own all atomic energy activities dangerous to world security, and would control, license, and inspect all others. Its functions would include fostering the beneficial uses of atomic energy and conducting research and development in this field. Once the IADA was established, the production of all nuclear weapons would be halted and existing nuclear weapons would be destroyed. The IADA would possess all information associated with atomic energy. For more details, see statement by Bernard M. Baruch (14th June 1946) reprinted in *Nuclear Proliferation Factbook*, Congressional Research Service, Library of Congress, August 1985.

9. For a description of the negotiations and preparatory work leading to the creation of the International Atomic Energy Agency see J. Hall, *The IAEA: Origins and Early Years — Peace through scientific co-operation became an abiding purpose*, IAEA Bulletin, Vol. 2, 1987.

10. Public Law No. 79-585, 60 Stat. 755.

11. Restricted data was defined by the 1946 Act as: "all data concerning the manufacture and utilization of atomic weapons, the production of fissionable materials, or the use of fissionable material in the production of power, but shall not include any data which the Commission from time to time determines may be published without adversely affecting the common defense and security" [section 10(b)(1)]. See also W. Walker and M. Lönnroth, *Nuclear Power Struggles, Industrial Competition and Proliferation Control*, London, George Allen and Unwin, 1983.

12. The USSR tested its first nuclear device in 1949 and the United Kingdom in 1952. These countries were joined by France in 1960 and the People's Republic of China in 1964, which means that the "Club" of nuclear-weapon States is made up of the permanent members of the Security Council of the United Nations.

13. In 1954, Congress amended the 1946 Act to allow the development of commercial nuclear power and to facilitate United States nuclear exports and co-operation. The Atomic Energy Act therefore opened the way in the United States for large-scale declassification of information about nuclear power technology which led to the disclosure of previously classified information.

14. In the United States two types of light-water reactors were developed: the pressurised water reactor by Westinghouse and the boiling water reactor by General Electric. These two models will, in the 1960s-1970s dominate the export markets (excluding Eastern European countries) either through direct sales ("turnkey" contracts) or by licensing agreements with the national manufacturers.

15. See M. Gowing, *op. cit.* and P. Pringle and J. Spigelman, *The Nuclear Barons*, Holt, Rinehart and Winston, New York, 1981.

16. Among other things, the United States Government supported a large share of the research and development costs of nuclear power reactors, undertook a comprehensive power demonstration programme and also passed the Price Anderson Act of 1956 in order to limit the liability of the nuclear industry for the risks of nuclear accidents while assuring payment to victims of nuclear damage. See Wendy Allen, *Nuclear Reactors for Generating Electricity: US Development from 1946 to 1963*, Santa Monica, Rand Report R-2116-NSF, 1977.

17. In 1957, the Atomic Energy Commission authorised private parties to conduct most forms of nuclear transactions (outside of the Soviet Bloc), subject to specific licensing requirements on the export of equipment and materials [10 C.F.R. Part 810 (1978)]. The practical effect of this regulation was not only to permit normal commercial sales efforts but also to open the door to the development of licensing arrangements, and joint ventures between the U.S. and foreign firms.

18. The text of this proposal is reproduced in *Documents on Disarmament, 1945-1959*, Pub. No. 7008, United States Government Printing Office, 1960, Vol. I.

19. The Atomic Energy Act of 1954 (Pub. L. No. 83-703) revoking the Atomic Energy Act of 1946, implemented, as national legislation, the principles laid down in the Atoms for Peace proposal.

20. Section 123 of the Atomic Energy Act of 1954 as amended requires (1) that the terms, conditions, duration, nature and scope of the co-operation be set out in the Agreement; (2) a guarantee by the co-operating party that security safeguards and standards set forth in the Agreement will be maintained; (3) a guarantee that any material to be transferred thereunder will not be used for atomic weapons, or for the research on or development of atomic weapons or for any other military purpose; and (4) a guarantee by the co-operating party that any material transferred thereunder will not be retransferred except as permitted in the Agreement. For a further analysis of the Act, see Volume II of this Study on national regulations.

21. Goldschmidt and Kratzer, *Peaceful Nuclear Relations: A Study of the Creation and the Erosion of Confidence*, New York, Rockefeller Foundation, 1978 Report of the International Consultative Group on Nuclear Energy.

22. An analysis of the Agency's Statute can be found in chapter two of this Study. See also P. Szasz, The Law and Practices of the International Atomic Energy Agency, *op. cit.*

23. An analysis of the EURATOM Treaty is contained in chapter two of this Study.

24. The authors of this Report are Louis Armand (France), Franz Etzel (Federal Republic of Germany), and Francesco Giordani (Italy), NED No. 2291, May 1957.

25. See chapter three of this Study.

26. The ENEA's Statute and its Convention on the Establishment of a Security Control are analysed in chapter two of this Study.

27. P. Strohl, *La coopération internationale dans le domaine de l'énergie nucléaire — Europe et pays de l'OCDE*, Colloque de Nancy 21st-23rd May 1981, AFDI, Pédone, Paris. For an analysis of the relationship between the NEA, EURATOM and the IAEA, see P. Reyners, *L'Agence de l'OCDE pour l'Energie Nucléaire — ses relations avec l'AIEA et EURATOM*, in Annuaire Européen, Vol. XXXII, Martinus Nijhoff Publishers, Dordrecht, 1986.

28. D. Fischer and P. Szasz, *Safeguarding the Atom: A Critical Appraisal*, Stockholm International Peace Research Institute, 1985. The case of China, however, was an exception to the rule since Soviet export controls which combined political control with technical control, especially as regards the recuperation of irradiated fuel, were otherwise extremely efficient.

29. Walker and Lönroth, *op. cit.*

30. The IAEA's safeguards system is analysed in chapter four of this Study.

31. Although not a signatory, France announced in 1968 that it would abide by the Treaty as if it were a Party.

32. The motivations of these countries are naturally not the same. The developing countries mainly oppose the discriminating nature of the Treaty which they view as neo-colonialistic. Vice-Admiral Castro Madero, at the time President of the Atomic Energy Commission of Argentina, declared in September 1982 before the VIIIth FOR-ATOM Congress that "moral principles shared by all should not be invoked to protect the political and economic interest of a minority" and called for "the elimination of industrial hegemony which hides behind the noble ideal of non-proliferation". Quotation from article by Simone Courteix, *Le contrôle de la prolifération des armes nucléaires*, (The Control of Nuclear Arms Proliferation), Numéro spécial sur le désarmement, Revue de droit de McGill, Vol. 28, No. 3, July 1983.

33. The list of signatures, ratifications and accessions to the NPT appears in appendix 1 to this Study.

34. It should be noted that the IAEA is not itself a Party to the NPT, the Treaty having been negotiated outside the Vienna Agency. It does not have, therefore, the power to carry out directly the obligations undertaken by the NPT Parties. The latter are nevertheless obliged to conclude safeguards agreements with the IAEA to have their non-proliferation commitments verified. See L. W. Herron, *A Lawyer's Point of View of Safeguards and Non-Proliferation*, IAEA Bulletin, Vol. 24, No. 3, September 1982.

35. It should be noted that in this respect, the United States and the USSR had initiated in the 1960s experimental programmes on the use of nuclear explosive devices for peaceful purposes and in so doing had opened a sort of Pandora's Box, an argument which India did not fail to invoke. A Treaty between both countries banning underground testing for peaceful purposes was signed on 28th May 1976 but has yet to come into force. See Lawrence Scheinman, *The International Atomic Energy Agency and World Nuclear Order*, published by Resources for the Future, Washington, 1987.

36. Prior to this announcement, countries which had concluded co-operation agreements with the United States could have their own uranium enriched in United States plants. See B. Goldschmidt, *Le complexe atomique* (The Atomic Complex), *op. cit.*. See also J. Simpson, A. McGrew, Ed., *The International Nuclear Non-Proliferation System, Challenges and Choices*, Macmillan, London, 1984.

37. Over the years, the evolution of the uranium market was very contrasted. After a period of rapid growth in the 1940s and '50s, the mining industry experienced a slump when the reduced military demand occurring in the 1960s was not compensated by a corresponding increase in civilian demand. Measures taken by certain States and uranium producers in order to maintain a viable national mining industry created important distortions in the market. As a result, the implementing legislation of Section 16 of the United States Private Ownership of Special Materials Act of 1964 prohibited completely, then partially, until 1977 the enrichment in the United States of foreign uranium when destined for American users. One should also note the cancellation by the United States in 1959 of Canadian uranium purchasing contracts and the policy

decision in 1971 to reduce uranium stocks through enrichment contracts with the USAEC. The reaction of the main uranium producers outside the United States was supposedly the creation of a cartel in 1972, combined with a stock policy — particularly in Canada — which contributed to increased prices. This increase would be accelerated by a turnabout in 1973 of the enrichment policy of the United States (long-term contracts) at a time when the oil crisis obliged many electricity producers to undertake important nuclear programmes. The slowdown of these programmes in the 1980s, coupled with a greater flexibility in enrichment contracts, has produced a new phase of depression in the uranium markets.

For the evolution of United States regulations in this field, see Richard M. Stein, *La nouvelle législation américaine relative à la propriété privée des matières nucléaires*, in Aspects du droit de l'énergie atomique, Vol. 2, CNRS, Paris, 1967.

38. The Treaty of Almelo signed in 1970 among the three countries provided for the completion of several uranium enrichment ultra-centrifuge facilities.

39. Joint venture set up by France in 1973 for uranium enrichment by gas diffusion in which Italy, Spain, Belgium and Sweden also participated. Sweden withdrew its participation in 1974. Iran was also briefly associated with this venture.

40. See Walker and Lönroth, *op. cit.*

41. Regional Nuclear Fuel Cycle Centres: 1977 Report of the IAEA Study Project.

42. For a more detailed analysis, see chapter five of this Study.

43. Although not a body of the IAEA, the Zangger Committee meets in Vienna, with the British Embassy acting as Secretariat. See Ashok Kapur, *International Nuclear Proliferation — Multilateral Diplomacy and Regional Aspects*, Praeger Publishers.

44. For a more detailed analysis, see chapter five of this Study.

45. Another mechanism for the control of nuclear exports is the one exercised by the Co-ordinating Committee for Multilateral Export Controls (COCOM) which co-ordinates national export controls to Eastern Europe for NATO countries (except Iceland) plus France and Japan. For more details, see chapter five of this Study.

46. Public Law No. 95-242. For a more detailed analysis, see Volume II of this Study on national regulations. Adoption of the NNPA was preceded by two amendments (Symington in 1976 and Glenn in 1977) of the 1961 Foreign Assistance Act providing for stricter export conditions, with a threat of suspension of American economic and military assistance. On these questions, see Lawrence Scheinmann, *op. cit.*

47. See chapter five of this Study.

48. America's lead in the area of light water reactor technology and enrichment services resulted, by 1974, in the construction by American companies of over 70 of the reactors throughout the world and their holding of two-thirds of the export market. In the period 1968-1973, the revenues realised by the United States for the exportation of equipment surpasses the total revenues of all the other nuclear exporters. In 1976, more than 80 of all reactor orders are of American origin. This domination which naturally has an impact in the area of regulation, will in the period which follows, decrease considerably due to the internal crisis in the American nuclear programme. This crisis is illustrated by the fact that in the last ten years, there have been no new reactor orders in the United States, with numerous cancellations having been experienced. The nuclear industry is likely to find a new equilibrium as a result, even if other countries may also feel the effects of the recession; see Joseph A. Camilleri in *The State and Nuclear Power — Conflict and Control in the Western World*, Wheatsheaf Books Ltd., 1984.

49. P. Lellouche and R. Lester, *The Crisis of Nuclear Energy*, The *Washington Quarterly*, Summer 1979; Walker and Lönnroth, Nuclear Power Struggles, *op. cit.*

50. *Nuclear Energy and Its Fuel Cycle, Prospects to 2025* (Yellow Book), 1987 Report, OECD/NEA.

51. For more details, see chapter five of this Study.

52. The bombing by Israel in 1981 of the Nuclear Research Centre in Tamuz, Irak, was certainly an event which was capable of shaking confidence in the Treaty since the aggression was committed by a non-Party to the NPT against a Party which had submitted its installations to IAEA safeguards. As the Director General of the Agency declared "a country non Party to the NPT was not convinced by either our findings or our capacity to efficiently carry out our responsibilities as concerns safeguards". Far from weakening the NPT, this event, on the contrary, gave rise to a reaction of defence on the part of the international community and led to an unequivocal condemnation both within the IAEA and the United Nations, as Simone Courteix has remarked, *op. cit.*

53. Report of the United Nations Conference, July 1987, A/Conf.108/7.

54. W. G. Vitzthum, *World Nuclear Order and Equality of States* in Law and the State, Vol. 25, Institut für Wissenschaftliche Zusammenarbeit, Tübingen, 1982.

CHAPTER TWO

ROLE OF INTERNATIONAL ORGANISATIONS IN THE FIELD OF NUCLEAR TRADE

A. INTERGOVERNMENTAL ORGANISATIONS

I. INTERNATIONAL ATOMIC ENERGY AGENCY (IAEA)

In his famous speech, "Atoms for Peace", delivered on 8th December 1953 before the General Assembly of the United Nations, the President of the United States, Dwight D. Eisenhower, proposed that the countries most advanced in nuclear technology should make available part of their stocks of uranium and special fissionable materials and entrust an international agency, to be set up, with the task of using these materials in peaceful installations in such a way as to satisfy the needs of humanity. Negotiations were then conducted, first between the United States and the USSR, then with other States, which led to the adoption, on 23rd October 1956, of the Statute of the International Atomic Energy Agency; the Statute entered into force on 29th June 1957.

The *creation* of the IAEA is thus not the result of a Resolution of the General Assembly of the United Nations but of a Statute which has the characteristics of an international treaty. Consequently, the Agency does not strictly speaking form part of the United Nations' "system" and, in particular, its activities are not subject to the Security Council's right of veto. Nevertheless, the Agency is required, under its Statute, to establish close links with the United Nations Organisation and its specialised agencies and organs, in particular by submitting reports (Article VI.J).

An international interstate organisation based on the principle of sovereign equality, the Agency enjoys, on the territory of each of its members, the legal capacity and privileges and immunities required to carry out its functions. The Agency's headquarters are in Vienna (Austria), and its members today number 113.

The *objectives* of the Agency, laid down in Article II of its Statute, are to accelerate and enlarge the contribution of atomic energy to peace, health and prosperity throughout the world and to ensure, as far as it is able, that assistance provided by it or at its request or under its supervision or control is not used in such a way as to further any military purpose.

The *organs* of the Agency are the General Conference, the Board of Governors, and the Secretariat, headed by a Director General.

i) *General Conference*

The General Conference (Article V) consists of representatives of all the Member States. It meets every year in regular session, and special sessions may be convened at the request of the Board of Governors or of a majority of members.

The Conference may discuss any questions within the scope of the Statute or relating to the powers and functions of the Agency's organs and may make recommendations concerning such questions to the membership of the Agency and/or the Board of Governors.

ii) *Board of Governors*

The composition of the Board (Article VI) reflects a compromise between the concern of the countries most advanced in nuclear technology to retain an important influence, and the desire to ensure an equitable representation of the various regions of the world.

The members representing the "most advanced" countries are designated by the outgoing Board. The equitable representation of the less advanced countries is ensured by members elected by the General Conference.

The Board has authority to carry out the functions of the Agency and it drafts an annual report on the affairs and projects approved by the Agency, for submission to the General Conference.

iii) *Director General*

The Director General, appointed by the Board of Governors with the approval of the General Conference for a term of four years, is responsible, subject to the authority and control of the Board, for the appointment, organisation and functioning of the staff (Article VII).

The Statute entrusts the IAEA with various tasks of a general nature such as:

— to foster the exchange of scientific and technical information on peaceful uses of atomic energy (Article III.A.3);
— to encourage the exchange and training of scientists and experts in this field (Article III.A.4);
— to establish standards of safety for protection of health and minimisation of danger to life and property (Article III.A.6);
— to encourage and assist, throughout the world, the development and practical application of atomic energy for peaceful uses (Article III.A.1).

When performing this latter task, which concerns the subject-matter of this study, the Agency endeavours to reconcile the objective of the non-proliferation of nuclear weapons with the need for a regular supply of nuclear material and access to nuclear fuel cycle services.

Article III.A.1 specifies that the Agency may, if requested to do so, act as an intermediary to secure the performance or supply, by one of its members for another, of services, materials, equipment, or facilities; the following paragraph specifies that it should ensure the supply of materials, services, equipment or facilities required to produce electricity, taking into account the needs of the underdeveloped regions of the world.

The Statute also provides that members may make available to the Agency such quantities of special fissionable materials as they deem advisable and on such terms as shall be agreed with the Agency (Article IX). These functions, though not involving any obligation for members, reflect the initial idea of President Eisenhower. In practice, the amounts of materials made available to the Agency or, through its intermediary, to other members, in the form of gift, sale, loan or lease, have remained fairly modest. They have, however, been sufficient to satisfy needs in the field of research and training and, in three cases, to supply power reactors.

The IAEA is responsible for establishing and applying measures designed to ensure that special fissionable and other materials, services, equipment, facilities and information made available by the Agency or at its request or under its supervision or control are not used in such a way as to further any military purpose, and to extend the application of these *safeguards*, at the request of the parties, to any bilateral or multilateral arrangement, or, at the request of a State, to any of that State's activities (Article III.A.5). It was thought at the start that the IAEA's responsibilities regarding safeguards would be triggered according to the evolution of its role as a supplier of materials and its capacity to follow through projects (Article XI of the Statute) and, to a lesser degree, by the control duty States would entrust it with, either bilaterally or multilaterally. Since the adoption of the Non-Proliferation Treaty, this latter task has been one of the Agency's main activities and plays a major role as a framework for the development of nuclear trade in the world.

Recently, the Agency set up an intergovernmental Committee on Assurances of Supply, which is working on reconciling the objectives of non-proliferation with the need for regular supply of nuclear material and access to the nuclear fuel cycle services (see chapter five of this Study).

II. ORGANISATION FOR ECONOMIC CO-OPERATION AND DEVELOPMENT — NUCLEAR ENERGY AGENCY

In the interests of clarity, the activities of this Organisation which are relevant from the viewpoint of the regulation of nuclear trade, will be dealt with separately from those of its Nuclear Energy Agency (NEA).

1. Organisation for Economic Co-operation and Development (OECD)

The activities of the OECD (formerly the Organisation for European Economic Co-operation — OEEC) relating directly to the regulatory aspects of the use of nuclear energy are, for the most part, conducted by its Nuclear Energy Agency[1] (see section 2 below). However, for some years, the Member countries of the OECD have actively sought to conclude agreements to limit export subsidies from Governments, and particularly export credits, which tend to distort international trade. Thus, as regards the export of nuclear power plants, the OECD countries have concluded an arrangement relating to officially supported export credits in this field, in the context of a comprehensive Consensus on export credits. This Consensus and a Sector Understanding are analysed in the following paragraphs.

a) *Arrangement on Guidelines for Officially Supported Export Credits*[2]

The OECD has traditionally played an active role in limiting export credit practices since 1955, the year in which its predecessor, the OEEC, adopted rules designed to eliminate certain of these practices. These rules, prohibiting specific measures concerning export subsidies, were incorporated into the General Agreement on Tariffs and Trade (GATT) when, in 1961, the OEEC became the OECD.

However, the economic disturbances of the early 1970s, in particular the oil shock of 1973, revived fears that a trade war, started by export credit practices supported by Governments anxious to defend the trading positions of their countries, could, after all, break out.

In order to prevent such a situation from developing, the OECD countries adopted in May 1974 a "Trade Pledge" in terms of which they undertook not to engage in sterile competition by means of official subsidies to export credits. In 1976, these countries agreed on a Consensus on converging policies in the field of export credits establishing, for a maximum period, the minimum payments and interest rates applicable with respect to all export credits of two years or more. Subsequently, the countries participating in the Consensus decided to prepare a document based on these principles, entitled Arrangement on Guidelines for Officially Supported Export Credits[3].

Under this Arrangement, at least 15 per cent of the contract price must be paid in cash; reimbursement must be spread over a period not exceeding eight and a half years (which may be extended to ten years for the poorest countries) and the minimum interest rates, called "matrix rates", are automatically adjusted every year in January and July for credits lasting five, eight and a half, and ten years. The Arrangement, which does not apply to agricultural commodities, nuclear power plants, aircraft, ships or military equipment, was subsequently completed by the adoption of Sector Understandings in some of these fields, and particularly that of nuclear power plants.

b) Sector Understanding on Export Credits for Nuclear Power Plants[4]

This Understanding entered into force on 10th August 1984 after several years of negotiations. It applies to new contracts for the export of complete nuclear power plants or parts thereof.

The maximum repayment term for an export credit is fifteen years, and at least 15 per cent of the cost of the project must be reimbursed in cash. The minimum interest rates are established on the basis of the matrix rates of the said Arrangement applicable to export credits with the longest repayment terms destined for "relatively rich", "intermediate" and "relatively poor" countries respectively.

The participants agreed to notify each other in advance of any derogation from the conditions laid down in the Understanding and to undertake an annual review of its provisions.

2. OECD Nuclear Energy Agency (NEA)

Before examining the statutory responsibilities of this Agency as regards international nuclear trade, a brief description will be given below of the circumstances of its creation and its objectives.

a) Creation of the Agency

Studies undertaken in the mid-1950s within the Organisation for European Economic Co-operation on the problem of Europe's energy supply led, in December 1957, to the creation of a European Nuclear Energy Agency (ENEA) (Decision of the Council of the OEEC of 20th December 1957). Under this Decision, which entered into force on 1st February 1958, ENEA was given the status of a department of the OEEC specialised in nuclear energy (without, therefore, its own legal personality). The Agency, composed of the seventeen Member countries of the Organisation participating in the Decision creating the Agency, was given the task of furthering and developing co-operation, among all the countries of Western Europe, with regard to the use of nuclear energy for peaceful purposes.

Following the admission of non-European countries at the beginning of the 1970s, the Agency was, in 1972, renamed the OECD Nuclear Energy Agency (NEA), and now comprises most industrialised market-economy countries[5]. The tasks entrusted to it are implemented, under the authority of the OECD Council, by the Steering Committee for Nuclear Energy, assisted by an international staff.

b) Objectives laid down in the initial Statute

Since the Agency was originally conceived with a view to promoting the development of the production and use of nuclear energy in a European context, the Statute it was given at its creation contains a number of relevant provisions from the viewpoint of trade in this field.

— In the first place, the Agency was given the task of promoting the "confrontation and harmonization of programmes and projects of participating countries relating to research and industry" in the field of the peaceful uses of nuclear energy.
— It also had the task of promoting the formation of joint undertakings, endeavouring to secure the participation of the greatest possible number of countries.
— Other provisions specified that the Agency had to help ensure that joint undertakings as well as participating countries were regularly supplied with the raw materials necessary for implementation of their nuclear programmes, in particular by the conclusion of supply agreements.
— The Agency was also instructed to examine jointly with the other competent Committees of the Organisation, measures to liberalise international trade in products involved in the production and use of nuclear energy for peaceful purposes.

In order to ensure that the implementation of the two objectives of promoting the creation of joint undertakings and helping ensure the supply of raw materials did not lead to the proliferation of nuclear weapons, the Statute provided for the setting up of a *security control*.

Despite these fairly extensive responsibilities, the implementation of the objectives in question has, in reality, remained somewhat limited, except for the creation of joint undertakings. The beginnings of the Agency were marked by the creation of three large joint undertakings of industrial and scientific co-operation: the EUROCHEMIC Company, in December 1957, the HALDEN Project in June 1958, and the DRAGON Project in March 1959 (see section B below). Studies were made of other similar projects, but led to no further undertakings.

The other functions of the Agency relating to trade have hardly been implemented since Member countries generally preferred at the time to give priority to their national programmes. On the other hand, the Agency's work has become increasingly technical, scientific and regulatory in nature, particularly in the spheres of nuclear safety, radioactive waste management, radiation protection and nuclear law.

c) Adapting the Statute to the needs of the Member countries

In order to take account of developments in the needs of the Member countries, the NEA Statute was *revised* in 1978 (Decision of the Council of 5th April 1978). The new Statute still refers to the Agency's general objective of furthering the development of the production and uses of nuclear energy [Article 1(b)] but, as already indicated, emphasis was henceforth placed on the more traditional activities of technical and regulatory co-operation in the field of the safe application of nuclear energy. Thus, some of the above-mentioned tasks, relating for example to the supply of raw materials, were abolished or made less important. Since that time, the provisions of the Agency's Statute relating to international nuclear trade have been as follows:

— The Agency retains the task of promoting the formation of joint undertakings (Article 5).

No other joint undertakings have been formed since the three mentioned above, and only the HALDEN Project continues to function at the present time. However, a large number of research and development projects, more modest in scope and not therefore formally joint undertakings, but which meet to a greater extent the needs of Member countries, has been set up in recent years under the aegis of the NEA.

— It is also responsible for contributing to the elimination of obstacles to international trade or to development of the nuclear industry, insofar as consistent with the obligation to prevent the proliferation of nuclear weapons [Article 8(a)(v)].

No significant use has yet been made of the Agency's powers in this field.

— The Agency is required also to encourage measures to ensure the most efficient use of patented inventions in the field of nuclear energy [Article 8(a)(iv)].

So far, this provision has not been implemented.

— Lastly, the Agency has a duty to promote technical and economic studies and undertake consultations on the programmes and projects of participating countries relating to the development of research and industry for peaceful uses (Article 4). (To this end, the Steering Committee of the Agency may examine the programmes and projects of participating countries).

This last provision is less stringent than the equivalent one in the previous Statute. It is implemented by the Agency in the form of studies conducted within the framework of the *Committee for Technical and Economic Studies on Nuclear Energy Development and the Fuel Cycle (FCC)*. The work of this Committee consists in particular of assessing the supply and demand of uranium and the services linked to the fuel cycle and forecasting the development of nuclear energy. These studies give rise to various technical and statistical publications such as the "Red Book" which deals with uranium resources, production and demand, and the "Yellow Book" dealing with nuclear fuel cycle services over the long term (published in collaboration with the IAEA). The work conducted by the FCC is of direct relevance to the formulation of Member countries' energy plans.

d) Security control

A security control — provided for under Article 6 of the NEA Statute in order to prevent the proliferation of nuclear explosive devices resulting from the operation of joint undertakings or the materials, equipment and services made available by the Agency or placed under its supervision — was set up by the Convention of 20th December 1957 on the Establishment of a Security Control in the Field of Nuclear Energy[6] (which entered into force on 22nd July 1959), and operated until 1976. The technical means of control included the examination of installations, the keeping of accounts of source materials and special fissionable materials, on-site inspection by international inspectors designated by the Agency and, in the event of failure to comply with the obligations inherent in the scheme, the possibility of sanctions if voted by a two-thirds majority of the members of the Steering Committee for Nuclear Energy. Control was carried out, under the authority of the Steering Committee, by a Control Bureau, assisted by an international Secretariat (provided by the Agency), headed by a Director of Control. The Bureau was responsible for working out the security regulations applicable to the installations subject to control and for drafting the agreements necessary for their implementation. Four such security regulations were prepared. The first of these concerned the HALDEN and DRAGON joint undertakings (1961); the second was applicable to installations taking charge of the nuclear material from the DRAGON and HALDEN undertakings (1966); the third dealt with plants for the chemical reprocessing of irradiated fuels — i.e. the EUROCHEMIC plant (1967); the last regulation was applicable to installations using nuclear material recovered from or obtained in the three Agency joint undertakings (1969), and it replaced the second regulation.

In fact, this security control, relating essentially to the three joint undertakings of the Agency, fairly rapidly came to duplicate the work of the equivalent systems of the IAEA and EURATOM which were applied on a wider geographic scale and with distinctly superior logistical resources. Due to the enlargement of the European Communities in 1973 and the growing number of NEA Member countries having concluded safeguards agreements, discussions were undertaken with the Commission of the European Communities and the IAEA with a view to concluding arrangements to avoid duplicating controls over installations subject to NEA control. However, as this solution did not seem likely to reduce such duplication substantially, the Steering Committee for Nuclear Energy finally decided, on 14th October 1976, to authorise the Director of Control to suspend the application of the control regulations until further notice. It also instructed the Director of Control to keep the situation under review and, in consultation with the Chairman of the Control Bureau, to initiate reconsideration of the matter in the event of a significant change in the circumstances which had led to the suspension of the control. So far, this has not been the case.

III. EUROPEAN COMMUNITIES — EURATOM — SUPPLY AGENCY

1. European Communities — Historical background

Formal negotiations leading to the two Treaties signed in Rome on 25th March 1957 were conducted at a Conference in Messina in 1955. The first of these Treaties set up the European Economic Community (EEC) which covered goods and services as a whole. The second Treaty, setting up the European Atomic Energy Community (EURATOM), took up the idea of sectoral integration already adopted with respect to the European Coal and Steel Community (ECSC), in the hope that, applied to an industry then in its infancy, such integration could be achieved more extensively and more quickly than that envisaged by the EEC Treaty.

Since 1957, the European Communities have developed along the following lines:

— Merger of the institutions of the three Communities which were originally distinct

The Treaty signed in Brussels on 8th April 1965 set up one Council and one Commission for the three Communities and merged their staff. Although these bodies exercised their powers under the conditions envisaged in each of the three founding Treaties respectively, the independent nature of the EURATOM Treaty has in practice been somewhat blurred.

— Extension of the powers of the Assembly (European Parliament)

Wider budgetary powers were given to the Assembly — which decided, on 26th April 1962 to take the title of "European Parliament" — by Treaties of 22nd April 1970 and 22nd July 1975 and by the Declarations dated 4th March 1979 and 30th January 1982.

— Adhesion of new Member States to the European Communities

The first enlargement of the Community took place when the United Kingdom, Ireland, Denmark and Nor-way signed the Treaty of Adhesion in Brussels on 22nd January 1972 (Norway did not, however, ratify this Treaty). On 28th May 1979, Greece also joined the Communities, as did Spain and Portugal, by an Agreement dated 12th June 1985.

— Single European Act

A need was felt to give fresh impetus to the work undertaken, not only to extend the common economic policies and pursue new objectives but also to work progressively towards economic and monetary union (putting into place a wide Community market) and to formulate and implement a European foreign policy. To this end, a new Treaty, the *Single European Act*, was signed in Luxembourg on 17th February 1986 and the Hague on 28th February 1986; ratified already by eleven Contracting Parties, it should enter into force shortly.

2. Objectives of EURATOM

Under the Treaty of Rome setting up this Community, EURATOM was given the task of contributing to the raising of the standard of living in the Member States and to the development of relations with other countries by creating the conditions necessary for the speedy establishment and growth of nuclear industries.

In addition to duties not of direct relevance to this study, Article 2 provides that EURATOM shall:

— ensure that all users in the Community receive a regular and equitable supply of ores and nuclear fuels;
— make certain, by appropriate supervision, that nuclear materials are not diverted to purposes other than those for which they are intended;
— exercise the right of ownership conferred upon it with respect to special fissile materials;
— ensure wide commercial outlets and access to the best technical facilities by the creation of a common market in specialised materials and equipment, by the free movement of capital for investment in the field of nuclear energy and by freedom of employment for specialists within the Community.

One of the sections of the Single European Act deals with European co-operation with respect to foreign policy. The importance assumed by questions of non-proliferation and physical protection in international trade in nuclear materials and, to a lesser extent in equipment and technology transfers, is one reason why the Member States of the European Communities are anxious to formulate and implement jointly a common foreign policy.

3. The Institutions of the European Communities

Leaving aside the institutions fulfilling a role that is budgetary, judicial, supervisory (European Parliament, Court of Justice of the European Communities, Court of Auditors) or advisory (Economic and Social Committee, Scientific and Technical Committee of EURATOM), Community action is directed in tandem by the Council and Commission.

The *Council* is comprised of representatives of the Member States, each Government delegating to it one of its members. The office of President is held for a term of six months by each member of the Council in turn. Since 1974, the European Council acts as the Council when it deals with Community problems.

The Council ensures co-ordination of the general economic policies of Member States and has power to take decisions. It delegates power to the Commission to implement the rules it establishes or, in specific cases, may retain such powers for itself. Its decisions are prepared by the Committee of Permanent Representatives to the European Communities (COREPER), set up by the 1965 Treaty which merged the Community institutions.

The Commission: its members are appointed by common accord of the Governments of the Member States from among their nationals, this joint nomination ensuring their independence from these States. It functions on the principle of collective responsibility. Nevertheless, the Commissioners share out the task of following the various activities of the Communities, and the Commission may, subject always to control by it and the principle of collective responsibility, delegate certain tasks of implementation.

The Commission is responsible for ensuring that the provisions of the Community treaties and Community law derived therefrom are applied. It exercises the powers conferred on it by the Council. Moreover, it has its own power of decision and participates in the shaping of measures taken by the Council and the Parliament by means of proposals (right of initiative), and formulates recommendations or opinions in the fields defined by the treaties.

4. EURATOM Supply Agency

Since implementation of the provisions of Chapter VI of the EURATOM Treaty on supply requires the conclusion of contracts with the producers and users of ores, source materials or special fissile materials, the drafters of the Treaty preferred to entrust this commercial, rather than administrative task, to a body separate from the Commission itself, though placed under its control: the Supply Agency. Under Article 52 of the Treaty, the Agency is given a right of option on materials as well as an exclusive right to conclude contracts relating to supply within the Community. The Agency's role is described in chapter three of this Study.

The Agency is governed by the provisions of Section I of Chapter VI of the Treaty and by its Statutes, adopted by the Council on 8th November 1958 (and amended on 8th March 1973).

The Agency was given legal personality and financial autonomy. It enjoys, in each of the Member States, the most extensive legal capacity accorded to legal persons under public and private law and may, in particular, acquire and dispose of property, conclude all types of contracts and carry out all commercial acts. Only Member States in fact contribute to its capital.

The Agency operates on a non-profit-making basis and is recognised as an institution of public interest status. The majority share of its capital must be retained by the Community and the Member States. The Agency operates under the control of the Commission, which issues directives to it, has a right of veto over its decisions, and acts as the judge of any act of the Agency referred to it by any person concerned. The Commission appoints the Director-General and the Deputy Director-General of the Agency.

An *Advisory Committee to the Agency* was set up under the Statutes. The members of this Committee are appointed by the Council, acting on a proposal from the Member States, from among representatives of producers, users and highly qualified experts.

The task of the Committee is to act as a link between the Agency on the one hand and users and sectors concerned on the other. Empowered, under certain conditions, to issue opinions, it must be consulted by the Director-General of the Agency before certain important decisions are taken.

IV. COUNCIL FOR MUTUAL ECONOMIC ASSISTANCE

Depending on the document, this intergovernmental international organisation is referred to as SEV (Soviet Ekonomitcheskoï Vzaïmopomochtchi), CAEM (French acronym), CMEA or COMECON (English acronyms).

The Council was established in January 1949 following a conference of representatives from the USSR and several Eastern European countries. However, following the adhesion as from 1962 of socialist countries from other continents, the regional character of the CMEA stems today from the economic and political system adopted by its members[7] rather than from their geographic location. The current Charter dates from December 1959. Its Statutes were amended in 1962, 1974 and 1979 with a view to increasing the economic integration of Member countries.

The main objectives of CMEA are, by uniting and co-ordinating the efforts of Member countries, to help improve economic, technical and scientific co-operation and develop socialist economic integration, accelerate economic and technical progress, and gradually approximate economic development levels.

The Sessions of CMEA constitute the supreme organ of the Council, and lay down the main guidelines for its activities. Thus, the 41st Extraordinary Session (January 1986) adopted a new "complex programme of scientific and technical progress in Member countries of CMEA till the year 2000", in which priority is given to five sectors, including nuclear power.

The main executive organ is the *Executive Committee* composed of representatives from Member countries. It directs the Council's activities, especially the co-ordination of development, specialisation and co-operation plans for production, and systematically monitors the implementation by Member countries of the obligations flowing from the Council's recommendations as adopted by these countries. The Executive Committee is assisted by committees on co-operation in the field of planning and on scientific and technological co-operation, including the Commission on the Peaceful Uses of Atomic Energy. Furthermore, multilateral agreements signed under the auspices of the CMEA have also created bodies with tasks in the field of co-operation and co-ordination. These include the Interstate Council for the Co-ordination of Co-operation in the Production of Nuclear Equipment, and the Group of Socialist Nuclear-Exporter Countries, whose objective seems to be to harmonize their control and security policies with respect to the work of the IAEA Committee on Assurances of Supply (CAS) (see chapter five of this Study).

In the above-mentioned "complex programme" of scientific and technical progress, the development of nuclear energy is given the role of a locomotive for the economy as a whole, leading to an improved electricity supply and a restructuring of the energy sector by reducing consumption of organic fuels. Hence, projects for concerted action, in the context of "integrated close co-operation", for the construction of power plants equipped with water-cooled and moderated VVER 440 and VVER 1000 reactors made possible by the standardization of equipment, and relating to the various stages of the fuel cycle, and the interest shown in dual-purpose power plants (producing both electricity and heat) as well as in fast-breeder and high-temperature reactors[8].

V. GATT — UNCTAD

Both these aspects of international co-operation in the field of trade are simply mentioned *pour mémoire* as they do not play a significant role in nuclear trade.

The General Agreement on Tariffs and Trade (GATT) entered into force on 1st January 1948 and was subsequently amended on several occasions. GATT which is not a formally constituted organisation, is based in Geneva and provides the framework for wide-scale international negotiations on trade. Based on the most favoured nation clause and aimed at eliminating quantitative or tariff restrictions in international trade, the Agreement nevertheless makes certain exceptions, in particular regarding security interests. In the absence of specific arrangements on nuclear trade, it may be noted that Article XXI of the Agreement provides that it does not affect the right of a Contracting Party to take any action it considers necessary for the protection of its essential security interests, referring in particular to "fissionable materials or the materials from which they are derived".

Following a recommendation of the 1964 Geneva International Conference on Trade Problems, the *United Nations Conference on Trade and Development* (UNCTAD) was established by a Resolution of the United Nations General Assembly (Resolution 1995 of 30th December 1964) as a subsidiary organ of the Assembly. The UNCTAD Secretariat is also based in Geneva. Compared to GATT, UNCTAD deals more particularly with promoting international trade by means of preferred treatment accorded by industrialised countries to developing countries. Until now, UNCTAD's work has had no direct bearing on trade in nuclear material and equipment.

B. JOINT UNDERTAKINGS AND SIMILAR ENTERPRISES

I. JOINT UNDERTAKINGS OF THE OECD NUCLEAR ENERGY AGENCY

As has been seen, the Statute of the OECD/NEA provides for the formation of joint undertakings for the production and uses of nuclear energy for peaceful purposes (Article 5).

The rules relating to such undertakings are flexible:

— The participation of all Member countries is not required. However, the agreements concluded for the creation of joint undertakings must provide that Member countries or groups of Member countries not taking part may subsequently accede to the joint undertakings or benefit from the result of their activities, at least when such joint undertakings have been established on the initiative or with the assistance of the Agency.

— It is for the countries declaring their intention to set up a joint undertaking to carry out, within the Agency but at their own cost, the necessary work, through working parties or study groups. The Steering Committee for Nuclear Energy, the body which directs the Agency, must be informed of the progress and conclusions of the work.

— When joint undertakings have been established on the initiative or with the assistance of the Agency, the agreements concerned may assign any functions to the Steering Committee, or a restricted group thereof composed of representatives of the countries taking part in the undertaking. The undertaking reports each year to the Steering Committee and, where appropriate, to the restricted group on the progress of its work.

— The Steering Committee considers such problems of general interest as may be raised by the operation of joint undertakings, with a view to proposing any necessary measures to the Governments.

1. Eurochemic[9]

The rules governing the European Company for the Chemical Processing of Irradiated Fuels (EURO-CHEMIC) are to be found in the first place in a Convention signed on 20th December 1957 by the Governments of twelve of the Agency's Member countries and, in the second place, in a Statute for a commercial company with a share capital, approved by these Governments, but signed by the shareholders, of whom certain were private electricity companies.

The international character of EUROCHEMIC resulted from the fact that the Company was regulated essentially by the above-mentioned Convention and Statute, and only in a subsidiary capacity by the law of the country of the registered place of business (Belgium). In consequence, EUROCHEMIC was given various privileges and immunities under the Convention, particularly of a fiscal nature. Control was exercised by a Special Group of the Steering Committee for Nuclear Energy (made up of representatives of the participating countries) which, in particular, had to approve any important decisions by the Company.

The objects of the Company were:

— to conduct research activities concerning the reprocessing of irradiated nuclear fuel, and to train specialists in this field;
— to build and operate a reprocessing plant.

This plant was constructed at Mol in Belgium. Inaugurated in July 1968, the plant will have processed some 220 tonnes of irradiated fuel in all.

In spite of positive results achieved at the technical level, it was decided on commercial grounds, in 1974, to cease industrial operations. A Convention concluded in 1978 between EUROCHEMIC and the Belgian Government, subsequently completed by two Protocols, provides for the transfer of the installations to a Belgian company (Belgoprocess) and establishes a programme of work in particular for the conditioning and storage of low and medium-level waste. In 1985, Belgoprocess took over the site, but the decision to dismantle the plant has not yet been taken. As for EUROCHEMIC, it has been in liquidation since 1982.

2. OECD Halden Reactor Project

Created by an Agreement of 11th June 1958 between various Member countries and EURATOM (which ceased to participate in 1964), the objective of the HAL-DEN undertaking was to construct and operate jointly a heavy boiling water reactor located in Halden, Norway. This reactor was commissioned in 1959; it is owned by the Norwegian Institutt for Energiteknikk and is operated under its responsibility.

As from 1967, the programme moved towards using the reactor as an advanced experimental installation for testing the effects on fuel of a high rate of combustion, the computerisation of control and instrumentation, and man/machine relations. Under the terms of the current Agreement between Norway, Denmark, Finland, the Federal Republic of Germany, Italy, Japan, Sweden, the United Kingdom, the United States and the Netherlands (as Associates), this programme is to be continued at least until 31st December 1990.

3. High-Temperature Gas-Cooled Reactor Project (Dragon)

An Agreement between eleven Member countries of the Agency and EURATOM was signed on 23rd March 1959 for the construction and operation of a high-temperature gas-cooled reactor (the DRAGON Project) at Winfrith in the United Kingdom. This Agreement did not confer separate international legal personality on the undertaking, the Project being managed by the United Kingdom Atomic Energy Authority.

The reactor was inaugurated in October 1964. However, since there ceased to be any interest either in the host country — the United Kingdom — or in the other participating countries, in proceeding from the prototype to the industrialisation stage, it was decided to end the project in 1976. Winding up was completed in 1978[10].

II.   EURATOM JOINT UNDERTAKINGS

Article 45 of the EURATOM Treaty stipulates that "Undertakings which are of fundamental importance to the development of the nuclear industry in the Community may be established as Joint Undertakings ...".

Under the rules on joint undertakings laid down in the EURATOM Treaty, the organs of the Community are given much wider supervisory powers than those exercised by the NEA over its joint undertakings.

In the years following the signature of the Treaty, however, the Member States were unable to agree on the desirability of constructing, under these rules, the installation of "fundamental importance" which an isotopic separation plant would have represented. In practice, the status of joint undertaking has been granted to companies set up by various electricity producers in a single Member State — the Federal Republic of Germany: KRB, KWL, KWO in 1963 and 1964 — or by electricity producers in two Member States, for example France and Belgium: SENA and SEMO (1961 and 1974), to construct and operate nuclear power plants. Joint undertakings participate in international trade in nuclear material and equipment in the capacity of customers only. Although falling outside the scope of this Study, since of an exclusively scientific vocation, mention may be made of the Joint European Torus (JET) joint undertaking as being that which best reflects the objectives of Chapter V of the EURATOM Treaty which deals with joint undertakings.

The collaboration involved in constructing and operating plants for isotopic separation by gaseous diffusion (EURODIF) or by centrifuging (URENCO/CENTEC), or in co-ordinating investments in the field of reprocessing (United Reprocessors GmbH), or lastly in constructing and operating a prototype fast-breeder reactor on an industrial scale (NERSA), falls outside the scope of Chapter V of the EURATOM Treaty.

III. URENCO/CENTEC[11]

Although falling outside the framework of international or Community organisations, URENCO/CENTEC is nevertheless an industrial joint undertaking in the nuclear field. It was created by the Treaty of Almelo, signed in March 1970 by the Federal Republic of Germany, the Netherlands and the United Kingdom (and entering into force in July 1971). The objective of this Agreement is to develop the gas centrifuge process for enrichment of uranium, and to construct and operate plants in accordance with this process.

Policy decisions, in particular those concerning security control (safeguards), the protection of the confidentiality of information, and safety are taken by the three Governments acting unanimously.

Their intention was to promote construction and operation of plants on a commercial basis by two companies created in 1971, one in the Federal Republic of Germany, CENTEC GmbH, the other in the United Kingdom, URENCO Ltd., even though most or all of the capital of the Dutch and British shareholders belongs to the State.

These companies were set up in compliance with the legislation of the country in which their registered place of business is located, and they remain subject to such legislation. Each national partner holds one-third of the capital of the two companies taken together, but the partner of the country in which the registered place of business is located is the majority shareholder in the company established in that country, and is responsible for its management.

The partners have endeavoured to combine high-level integration under joint management in the field of the marketing of enrichment services, the management of contracts, the co-ordination of production plans and implementation of a joint R & D programme with a decentralisation of decisions of a technical nature or relating to investments.

IV. EURODIF

EURODIF is a shareholding company, under French law, with a Supervisory Committee and Board of Directors, created in November 1973 with the following object: study, research and development in the field of isotopic separation as well as the completion and operation of gaseous diffusion plants and the marketing of uranium (see note 39, chapter one of this Study). It built and operates the gaseous diffusion enrichment plant at Tricastin in France.

It is thus not an international company nor an NEA or EURATOM joint undertaking. However, a Convention was concluded on 20th March 1980 between Belgium, France, Italy and Spain, which provides that the Parties undertake to ensure that the Tricastin installations cannot be used by a non-nuclear-weapon State to manufacture or otherwise acquire nuclear weapons or other nuclear explosive devices, nor accede to the control of such weapons or devices.

The Convention stipulates that all non-nuclear-weapon States which, as a consequence of the activities of EURODIF, have access to or receive raw or special fissile materials, must, if they have not already done so, first have adopted suitable control measures in compliance with IAEA procedures and in the light of the international obligations undertaken by each of them; these measures provide for control by EURATOM, verified by the IAEA for such of these States which are members of the European Community.

C. NON-GOVERNMENTAL ORGANISATIONS

I. THE URANIUM INSTITUTE

After allegations that certain non-American producers of uranium had formed a cartel, sharing out the market and supporting agreed prices, there was, in the 1970s, a lively debate in the United States giving rise to Congressional inquiries and legal proceedings. This situation led to sixteen mining companies creating, in June 1975, a body entitled the Uranium Institute which, by publishing its Memorandum of Association, objectives and activities, would be above suspicion. The Articles of Association were amended in January 1976 to allow consumers to join the Uranium Institute.

The Uranium Institute is an international industrial association based in London. Its membership of sixty-eight companies comprises uranium producers and consumers, and associate members, drawn from eighteen countries and the European Communities.

The Institute, a company limited by guarantee without share capital, is governed by English company law. Its objectives are:

— to promote the use of uranium for peaceful purposes;
— to conduct research into uranium requirements, uranium resources and uranium production;
— to consult for these purposes with governments and other agencies;
— to provide a forum for the exchange of information on these matters.

The Institute has two full (or Ordinary) classes of membership — Producer Ordinary Members and Consumer Ordinary Members — with exact equality of rights. Both sides play an equal part in the running of the Institute, the chair alternating every two years between the uranium producers and the electricity utilities. There is also an Associate Membership class designed to encourage participation from other organisations with closely related activities (for example, fuel cycle companies). Associate members play a full part in the work of the Institute, other than being excluded from the Council of Management and election to officerships. The Council of Management is responsible for the running of the Institute, assisted by an Executive Committee. The Institute is wholly financed from members' subscriptions, with Associate members paying a reduced fee.

The work programme of the Institute is organised through three specialist committees comprised of representatives of interested member companies.

— *The Supply and Demand Committee* studies the balance of the uranium market. At approximately two-yearly intervals since 1979, it has published assessments of the medium-term supply and demand outlook. Its most recent report was published in December 1986.

— *The Committee on International Trade in Uranium* monitors economic and political factors affecting trade in nuclear fuel. In early years it made a number of submissions to the INFCE Conference (International Nuclear Fuel Cycle Evaluation), and in 1985 it submitted its view to the Non-Proliferation Treaty Review Conference. Current concerns of the Committee include a description of uranium price-reporting systems, bilateral nuclear trade agreements, and protectionism in the uranium industry.

— *The Nuclear Energy and Public Acceptance Committee* supports member companies addressing public concerns. It does so by pooling and analysis of members' public relations experience, and by publishing specialist studies of controversial issues. Among recent and planned publications are studies of waste management, uranium mill tailings, and the financing of plant decommissioning.

The Institute also organises an annual symposium open to the public. Topics covered at the symposium include uranium supply and demand, the fuel cycle industry, more general energy policy considerations, and the industry's public relations. The proceedings of the symposium are published.

II. ORGANISATION OF NUCLEAR ENERGY PRODUCERS (OPEN)

On 14th January 1974, a number of Belgian, French, German, Spanish and Swiss electricity producers formed the OPEN (Organisation des Producteurs d'Energie Nucléaire) with the legal status of a "Groupement d'Intérêt Economique" (GIE)[12]. The objectives of OPEN include:

— studying ways of satisfying their needs with respect to the separation of uranium isotopes, using, in particular, enrichment plants located in Europe;
— undertaking measures designed to obtain optimum safety and economy for their nuclear fuel supplies;
— seeking for and promoting all measures capable of improving, through the collaboration of its members, procedures for managing nuclear fuels.

To this end, the GIE is empowered:

— to study the detailed provisions of separation contracts and all provisions likely to stabilize this market or facilitate the financing of separation plants;
— to establish procedures for supplying electricity to a separation plant using the gaseous diffusion process, for example involving an enrichment exchange

in addition to electricity supply, and draw up the reciprocal commitments of those members participating in this supply;

— to study detailed rules for sharing nuclear fuels and exchanging energy, so as to make supply contracts more flexible and reduce the technical risks involved in constructing and operating power plants.

Membership of the GIE is restricted to European electricity producers. The admission of new members is subject to the unanimous consent of existing ones.

Working parties have been formed to study questions such as plutonium transfers.

OPEN has also drafted an agreement, opened for signature by its members in September 1982, defining the conditions on which each signatory undertakes to make certain quantities of natural uranium available to the other signatories, as well as laying down the rules for loans and repayments resulting from these commitments.

III. EUROPEAN ATOMIC FORUM (FORATOM)

Founded in Paris in 1960, the European Atomic Forum Association is a non-governmental international organisation. It acts as a consultative body to the IAEA.

Fourteen national fora (of which nine are in Member States of the European Community and six in other Western European countries) are members of FORATOM. The members of these national fora in turn comprise electricity producers, suppliers of installations and parts, research institutes, consulting engineers, etc.

The object of FORATOM is to contribute to the economic development of the peaceful uses of nuclear energy. To this end, the Association defines and studies the problems facing the nuclear industry, proposes solutions to governments and supports measures designed to improve public understanding of the issues related to the production of nuclear energy.

NOTES AND REFERENCES

1. There is also, within the OECD, an International Energy Agency (IEA), which is given the task under its Statute of promoting secure oil supplies and of taking appropriate measures to cope with needs in the event of an emergency. The IEA encourages the rapid introduction of alternative energy sources by promoting research and development of activities in the energy field, including controlled thermonuclear fusion and nuclear safety.

2. Most of the information in this section is taken from J. E. Ray, *The OECD 'Consensus' on Export Credits*, The World Economy, Vol. 9, No. 3, September 1986, Basil Blackwell for the Trade Policy Research Centre, London.

3. The current Arrangement, which took effect as from April 1978, is not an official act of the OECD Council. It is an Agreement concluded directly between the participants comprising all the OECD Member countries with the exception of Iceland and Turkey.

4. The Understanding has been published in *The Export Credit Financing Systems in OECD Member countries*, OECD, Paris, 1987.

5. NEA Member countries are as follows: Australia, Austria, Belgium, Canada, Denmark, Finland, France, the Federal Republic of Germany, Greece, Iceland, Ireland, Italy, Japan, Luxembourg, the Netherlands, Norway, Portugal, Spain, Sweden, Switzerland, Turkey, the United Kingdom and the United States. The Statute provides that the Commission of the European Communities may also participate in the work of the NEA; the International Atomic Energy Agency (IAEA) is linked to the Agency by a Co-operation Agreement.

6. The Convention of 20th December 1957 also created a *European Nuclear Energy Tribunal*, competent to hear claims from private undertakings or governments against control measures taken by the Agency. The Convention also provides that the Tribunal shall be competent to hear any other question relating to joint action by OECD Member countries in the field of nuclear energy. This provision has been used to make the Tribunal competent to decide on disputes relating to the interpretation and implementation of the Convention creating the EURO-CHEMIC Company, as well as the Paris Convention and the Brussels Supplementary Convention on nuclear third party liability. In fact, however, no dispute has yet been referred to the Tribunal.

7. USSR, Bulgaria, Hungary, Poland, Romania and Czechoslovakia, joined by Albania (from 1949 to 1961), GDR (1950), Mongolia (1962), Cuba (1972), and Vietnam (1978). Yugoslavia (1964) is entitled to participate in certain CMEA bodies, and a collaboration agreement was concluded with Finland in 1973. *Source: CAEM, 25 ans,* ed. by Progrès, USSR 1974. See also *Intégration économique socialiste*. Académie des Sciences de l'URSS "Sciences Sociales d'Aujourd'hui", Collection "Problèmes du monde contemporain" (18), March 1973, No. 1.

8. According to a CMEA publication, the co-operation of CMEA countries with respect to the peaceful uses of nuclear energy originates in an Agreement concluded in 1955 by certain of them with the USSR. Subsequent agreements have provided for the construction of research institutes, power plants using Soviet equipment, and the supply of artificial radionuclides. It is doubtless under bilateral agreements that the USSR acquires the uranium ore mined in the various CMEA countries (except Romania), supplies the nuclear fuel needed for the reactors and takes back the irradiated fuel, thus relieving the customer country of any obligation to transport and dispose of the waste, in return for which the customer gives up any right to the materials recovered by reprocessing. The most industrialised Member countries, other than the USSR, were subsequently called on to produce some of the equipment for these power plants, especially after a multilateral specialisation and co-operation agreement for the production and mutual supply of equipment for nuclear power plants for the period 1981-1990, signed in June 1979. In particular, Czechoslovakia manufactures steam generators, pumps and pipes. More recently, Czechoslovakia has apparently undertaken to construct VVER 440 reactors, under licence.

Juridically separate from CMEA, but taking part in the meetings of its Commissions, are joint undertakings, involving all or some of the Member countries. Thus, in the nuclear field, there is "Interatominstrument", created on 22nd February 1972, which organises scientific, technical, industrial and commercial co-operation in the construction of measuring and monitoring equipment designed for research or for nuclear installations, as well as co-operation in the use of artificial radionuclides. This body has production subsidiaries in Bulgaria, Poland and the USSR, *op. cit.*, supra.

The international technical plant "Interatomenergo", the creation of which was decided upon at the XXVIIth Session of CMEA and whose Statute, approved by the Executive Committee, was signed in December 1973, is responsible for co-operation in the manufacture of equipment, apparatus, spare parts and materials for nuclear power plants, for specialisation and co-operation in this field and in that of the training of maintenance staff, and for helping to organise start-up and other work. This body ensures direct collaboration between the technical Ministries, and multilateral co-operation between the production groups and undertakings of the countries having signed its Statute (Bulgaria, Hungary, GDR, Poland, Romania, USSR, Czechoslovakia and Yugoslavia) [CAEM, 25 ans, *op. cit.*, p. 32].

Mention should also be made of the construction, in the USSR, of jointly owned power plants. Under an agreement approved by the Executive Committee on 29th March 1979, Czechoslovakia, Poland and Hungary are paying half the construction costs of the Khmelnitskaya power plant in the Ukraine. These countries then receive supplies of the electricity produced. Similarly, Bulgaria is partly financing the construction of the Konstantinovska power plant, near Nikolaev, and the high-voltage line which, passing through Romania, will supply it with its share of the electricity produced.

9. *International Technical Co-operation Enterprises. Legal aspects — Evaluation — Prospects.* Round Table of 17th April 1985. SFDI, OECD/NEA, IEA and ESA. Paris 1986, pages 10 to 28.

10. For further details, see *Europe's Nuclear Power Experiment* (History of the OECD DRAGON Project), by E. N. Shaw, OECD/NEA, Pergamon Press, 1983.

11. See note 9, *supra*, pages 35 to 41.

12. For the moment, GIE has a legal structure peculiar to French law (Ordinance No. 87.821 of 23rd September 1967). With a place, under this law, between a company and an association, and made up only of two or more natural or legal persons, for a fixed period, its object is to allow them to put into effect all means designed to facilitate or develop the economic activity of its members, and to improve or increase the results of this activity.

CHAPTER THREE

THE LAW OF THE EUROPEAN COMMUNITIES GOVERNING INTERNATIONAL TRANSFERS OF NUCLEAR MATERIAL, EQUIPMENT AND TECHNOLOGY

A. INTRODUCTION

It goes without saying that Member States of the European Communities are obliged to comply with Community law. The main provisions of this law relating to the international transfer of nuclear material, equipment and technology may also be of concern to nationals of third countries, either in the case of transfers between them and a Member State of the European Communities, or of exports to the Community followed by a retransfer between Member States.

A major aspect of the Community is the free movement of goods, persons and capital between Member States. However, the goal of the founding Treaties was to go beyond the setting up of a simple customs union and to promote co-operation between Member States and their nationals, leading to the adoption of common policies and even to joint measures. Section B, below, will deal with this fundamental aim of the founding Treaties, especially the EURATOM Treaty.

In this same spirit, but linked to the specific question of the development of nuclear energy, the EURATOM Treaty introduced special rules for the supply of nuclear materials (Chapter VI) and the ownership of special fissile materials (Chapter VIII), and set up safeguards (Chapter VII) regulating certain exchanges between Member States and also exchanges between such States and third countries. Since safeguards systems are already analysed in chapter IV of this Study, the following paragraphs will describe the EURATOM Treaty provisions relating to supplies and ownership.

In addition, during the thirty years since the signature of the EURATOM Treaty, problems relating to the adaptation of the Treaty to developments in the international transfers of materials have arisen, and these will also be examined.

B. CO-OPERATION BETWEEN MEMBER STATES IN THE FIELD OF NUCLEAR TRADE

The purpose of Chapter IX of the EURATOM Treaty, entitled "The nuclear common market", was to anticipate the gradual removal of obstacles to intra-Community trade and the rolling-back of customs barriers to borders with third countries envisaged under the EEC Treaty, with regard to the nuclear materials in List A1 and the equipment and materials in Lists A2 and B, annexed to the EURATOM Treaty (see *annex I* to this chapter). By Agreement dated 22nd December 1958, the Common Customs Tariff was applied to these goods and products.

Subject to the transitional period envisaged for new Member States, the nuclear Common Market has become a reality, whether the goods considered be among those listed in Chapter IX of the EURATOM

Treaty or whether simply covered by the more general provisions of the EEC Treaty. It can be seen from the reservations included in the policy declarations made by certain Member States in the framework of the work of the London Club or the Zangger Committee, or from the Declaration of Common Policy notified to the IAEA in the name of the European Community on 22nd March 1985 (see chapter five of this Study), that the Member States intend to preserve the benefits which the Common Market represents for them.

In other chapters, the EURATOM Treaty goes beyond the stage of customs union to promote active co-operation between Member States and their enterprises in the nuclear field. Thus, the Commission is given the task of promoting nuclear research in Member States, supplemented by the implementation of a Community research and training programme (Chapter I of the Treaty). Consequently, and in keeping with the spirit of the EURATOM Treaty, Chapter II gives Member States and "persons or enterprises" carrying on nuclear activities within their territory, the right to information owned by the Community or to which it has access (Articles 12 and 13). The Commission must also endeavour to secure for such persons or enterprises, the communication of "information which is of use to the Community in the attainment of its objectives" (Chapter II, Section II of the EURATOM Treaty).

Conditions may nevertheless be imposed under the Treaty (and then further specified by Community contractual practice) with regard to such communication or exchanges of information, for the purpose of protecting the legitimate interests of Community undertakings and, in particular, of contractual partners in research undertaken by the Community — hence, for example, restrictions on dissemination and use in the form of classification procedures (Articles 13 and 15). Security regulations (Section III of the said Chapter of the EURATOM Treaty) may also be applied to information the disclosure of which is liable to harm the defence interests of one or more Member States.

The fact that the Commission is given the option of imposing such restrictions suggests that it should be possible to reconcile Community provisions on the dissemination of information with recent policies to limit the transfer of certain sensitive technologies. However, the guiding spirit of the EURATOM Treaty remains the promotion of technology transfers between Member States and their undertakings, the above-mentioned restrictions constituting exceptions to the rule. Furthermore, in terms of the second paragraph of Article 29 of the Treaty, the Commission may authorise a Member State, a person or an undertaking to conclude agreements for the exchange of scientific or industrial information with a third State, an international organisation or a national of a third State, on such conditions as it considers appropriate.

Another task of the Community is to facilitate investment and encourage ventures for the establishment of the basic installations necessary for the development of nuclear energy (Article 2). As a means of encouraging the establishment of basic installations, such as enrichment or reprocessing plants, the EURATOM Treaty introduced rules for the creation of "joint undertakings" (see chapter two of this Study). In fact, in the years following the signature of the EURATOM Treaty, Member States were unable to agree on the desirability of establishing such installations in the Community and, except for the special case of JET, the status of joint undertaking has only been used for the establishment of power reactors.

C. OWNERSHIP OF SPECIAL FISSILE MATERIALS AND RULES ON SUPPLY

I. OWNERSHIP

At the time the EURATOM Treaty was being negotiated, the United States Atomic Energy Act had already provided that the United States Atomic Energy Commission should retain ownership of special fissile materials, users being allowed to possess them under licence only. To facilitate the supply to Community users of special fissile materials produced, at the time, almost exclusively by the United States, it seemed advisable to provide for a "parallel" system in the Treaty, notwithstanding the resulting interference with the rules of economic liberalism to which certain Member States were attached. The result was a compromise — the special regime was kept for materials likely to be diverted without control.

Article 86 of the Treaty lays down the principle that special fissile materials shall be the property of the Community. This right extends to all such materials produced or imported by a Member State, a person or an undertaking; they are also subject to safeguards by the Commission. Because of the link thus established between the Community's right of ownership and safeguards, this right of ownership does not extend to materials "intended to meet defence requirements which are in the course of being specially processed for this purpose or which, after being so processed, are, in accordance with an operational plan, placed or stored in a military establishment"; Article 84, paragraph 3 provides that control is not extended to such materials. Another exception to Article 86 is expressly provided for in Article 75(c) in respect of material which has been processed, converted or shaped inside the Community and is then returned either to the original international organisation or national of a third State or to any other consignee likewise outside the Community, although this material is subject to safeguards while in the territories of the Member States. A third exception results from the Additional Agreement to the Co-operation Agreement between EURATOM and the United States of 11th June 1960 and the amendments, dated 21st-22nd April 1962, to this Additional Agreement and the Co-operation Agreement, under which the United States accepted to lease special nuclear material to the Community. The United States retained ownership of the material, but it was specified that the Community was entitled to exercise over it the powers provided for under Chapter VIII of the EURATOM Treaty.

Indeed, the rules regulating Community ownership of special fissile materials represent a dismemberment of the traditional law of ownership, insofar as its economic content has been removed[1]. However, the Commission, with approval from the Court of Justice of the European Communities (Judgment of 1st December 1971, Case 7/71, and deliberations 1/78 of 14th November 1978), has found in these provisions an additional basis for exercising its powers in the sphere of safeguards[2], supply, and negotiating international agreements dealing with these questions. This no doubt explains why the simplified Article 90 procedure was not used to revise Chapter VIII of the EURATOM Treaty when, under the Private Ownerhsip of Nuclear Material Act of 1964, the United States gave up the system of public ownership it had used until then — this might have been followed by the Communities in order to retain the initial parallelism between United States domestic legislation and the EURATOM Treaty on this question. In any event, the Community system of ownership could not be abolished pursuant to Article 90 since the exercise of the right of ownership constitutes one of its tasks under the Treaty [Article 2(f)].

The non-economic nature of the system of Community ownership is expressed by the Supply Agency's obligation to keep a "Special Fissile Materials Financial Account" (Article 89), in which variations in the value of these materials are expressed in such a way as not to give rise to any loss or gain to the Community. The risk of any loss or gain is borne by the holder alone, and balances are payable forthwith upon the request of the creditor.

Holders — i.e. Member States, persons or undertakings — which have properly come into possession of special fissile materials, have, in respect them, an "unlimited right of use and consumption" (Article 87). This expression should be taken to mean the economic rights and right of use and abuse conferred under the usual law of ownership. Any loss or gain resulting from variations in value are borne by or accrue to the holders. The Treaty does not contain any definition of the expression "properly come into their possession", but it is generally agreed that this implicitly involves compliance with the provisions relating to supply in Chapter VI.

II. COMMUNITY PROVISIONS RELATING TO SUPPLY, AND THEIR EVOLUTION

Those involved in negotiating the EURATOM Treaty envisaged a major and rapid development of nuclear energy. They were also aware that such a development might have led to competition among Member States for access to uranium resources, which were limited at that time, something which would hardly be in keeping with the co-operation it was hoped to introduce. Furthermore, the United States, a producer of uranium ore but, above all, the only country capable at that time of providing enrichment services, wished to conclude a contract with a sole partner in the form of a co-operation agreement with the Community, to replace the agreements previously concluded with the various Member States.

The development of the nuclear industry and, consequently, of the market in nuclear materials, was significantly different from that forecast and gave rise to distortions, of varying importance depending on the period, Member State or category of nuclear materials considered, between the letter or spirit of Chapter VI and its application. What is more, the need to amend the Community supply system had been foreseen, as shown by the simplified Chapter VI review procedure (Article 76). Various attempts to revise Chapter VI have not yet, however, made it possible to adapt the EURATOM Treaty to a situation characterised by the importance of co-operation agreements with supplier countries followed, more recently, by the increased importance of safeguards and physical protection concerns relating to supply.

1. Chapter VI of the EURATOM Treaty

Under Article 2 of the EURATOM Treaty, the Community is given the task of ensuring that "all users in the Community receive a regular and equitable supply of ores and nuclear fuels". Chapter VI provides that this is to be achieved by applying the principle of equal access to sources of supply, which means that resources inside or outside the Community must be distributed on the basis of requirements for implementing nuclear programmes. The Supply Agency may not discriminate between users on grounds of the use which they intend to make of the supplies requested, and "all practices designed to secure a privileged position for certain users", including pricing practices which could have this effect, are prohibited [Articles 52.2 to 62.2(b) and (c), and Article 68]. Under Chapter VI, the regularity of supply is to be ensured by a system for redistributing materials implemented by means of the exercise by the Supply Agency of the rights given to it, as well as by implementation of a common supply policy.

The Agency has:

— a right of option relating, depending on the category of materials involved, either to the right of ownership, or to the right to use and consume ores, source materials and special fissile materials produced in the territories of Member States. This right is exercised on the basis of the needs of users as compared to the offers which producers are obliged to make [Articles 52.2(b) and 57 to 63]
— an exclusive right to conclude contracts relating to the supply of nuclear materials coming from inside the Community or from outside [Articles 52.2(b) and 64 to 66].

Under Articles 68 and 69 of the Treaty, the Commission may re-establish prices at a level compatible with the principle of equal access, and the Council may fix prices. In the event of a shortage of resources, the Agency is empowered to share out supplies proportionately (Article 60) or suggest an equalisation of prices.

Other aspects of the common supply policy (Section V of Chapter VI of the Treaty), on the other hand, essentially retain the character of guidelines (except for the last paragraph of Article 70). The Commission may subsidise prospecting activities, make recommendations to Member States in this sphere and establish commercial or emergency stocks.

A more detailed examination of Chapter VI shows, however, that its "dirigiste" approach has been softened by certain concessions to liberal concepts, political realities or the programming of industrial operations.

Thus, the Supply Agency is required, under the Treaty, to meet all orders by users unless prevented from so doing by legal or material obstacles (Article 61), and an exception is made to the Agency's right of option when the producer uses the materials for his "own requirements" or, on certain conditions, makes them available to another undertaking in the Community (Article 62). Similar concessions can be seen also in the price fixing mechanism (Article 67), and in the right of users to conclude directly contracts relating to supplies from outside the Community should the Agency not be in a position to satisfy their orders (Article 66). Moreover, in the absence of orders from users, the Agency may choose not to exercise its right of option (Article 59). This makes it possible for producers to export ores or source materials, subject to authorisation from the Commission (but special fissile materials may be exported only through the Agency) [Articles 59 and 62.1(c)].

Provision is made for exceptions to the rule of non-discrimination on grounds of use in the event of use which is unlawful either under public international law (using materials for military purposes prohibited by certain States), or under national regulations made for reasons of public policy or public health (Articles 52 and 195). Exceptions may also be made because of the need for the Community to ensure compliance with the conditions (specific use clauses for example) imposed by suppliers outside the Community under co-operation agreements concluded with third countries. None of these restrictions, however, should result in a compartmentalisation of the Community market by national markets.

Industrial realities are also taken into account by allowing Community producers to choose the stage of production at which to offer the product to the Agency (Article 58). Similarly, and subject to transparency with regard to the Commission, account is taken of the connection between undertakings, especially for processing, conversion and shaping operations performed outside the Community with provision for return to the original undertaking, or inside the Community on behalf of operators from Member States or of nationals from third countries (Articles 58, 62 and 75). Exemptions from the provisions of Chapter VI are also allowed with respect to deliveries which are part of a larger agreement (Article 73) or deliveries of "small quantities" (Article 74).

2. Application of Chapter VI of the EURATOM Treaty

There is general agreement that Chapter VI of the Treaty has been applied pragmatically, especially as regards the exercise by the Supply Agency of its right of option and right to conclude supply contracts.

This situation is the result of two factors. First, contrary to forecasts, the state of the ores and source materials market has, since signature of the Treaty, been characterised by an abundance of supply. Secondly, in a form which has varied with time, the legislation of certain third countries, especially the United States, has had an effect on the conditions of implementation of the Treaty. In practice, the Supply Agency only began to perform its functions on 1st June 1960[3], on the tacit understanding that it would make limited use of its right of option and right to conclude contracts.

As regards the right of option, the Agency Rules of 5th May 1960[4] provide that "producers" shall be "considered" to have fulfilled their offer obligation to the Agency pursuant to Article 57.2 paragraph two of the Treaty, by submitting their tenders, at the invitation of the Agency, during the procedure of balancing demand against supply. While the Rules refer to the Agency's right to decide whether and over what quantities it will exercise its right of option, this right has not, in practice, been exercised.

With regard to the exclusive right to conclude contracts, a distinction should be made between its application to ores and source materials on the one hand, and special fissile materials on the other.

In respect of *ores and source materials*, the above-mentioned Rules, as amended by a Regulation of 15th July 1975, introduced a so-called simplified procedure under which, should the Commission find a clear surplus of supply over demand, the parties concerned are authorised to negotiate directly and to sign supply contracts which are then deemed to have been concluded by the Agency.

However, such contracts must:
— satisfy the "general conditions" communicated by the Agency on 30th November 1960[5], which simply list the essential elements of all supply contracts;
— be communicated to the Agency, which then has eight days to notify any justified objection. Apparently, such an objection could arise only from unlawful use as defined in Article 52 *in fine* of the Treaty, it being provided that the notification of the contract must indicate the intended destination of the materials.

Note should also be taken of the definitions of the concentration of ores in Regulation No. 9 of the Council of 2nd February 1960[6] and of the "small quantities" exempt from the application of Chapter VI[7] subject to a quarterly report being sent to the Agency on transfer operations involving such "small quantities", in which, in particular, the intended use of the materials must be specified.

However, the application of the Treaty by "legal fiction" has, ostensibly at least, been affected by the amendment of the Regulation of the Agency of 15th July 1975[8]. Since the entry into force of this amendment, users have continued to negotiate directly with producers, but contracts have had to be signed by the Agency.

However, with certain exceptions, whether in the case of notification or of signature by the Agency of a contract directly negotiated between the undertakings concerned, the Agency no longer has an active commercial role as provided for in the Treaty. It does, on the other hand, carry out duties of a more administrative nature such as keeping a permanent watch on markets or advising operators[9]. It also ensures that transactions in this sphere conform to Community law.

Certain users have created their own purchasing service. In other Member States (for example the Federal Republic of Germany), intermediary companies have been created to play the commercial role which the Agency should have had by reason of its exclusive right to conclude contracts. Initially, such "intermediaries" simply communicated to the Agency contracts concluded with users. Subsequently, however, they agreed to communicate also contracts concluded with producers, and this has allowed the Agency to have a better, though still incomplete, idea of stocks in the Community.

It should, however, be pointed out that following the 1975 amendment, users may, outside the procedure for balancing supply and demand, place orders with the Agency, which must satisfy them on the best conditions possible given the state of the market. Indeed, certain users find it is to their advantage to ask the Agency to procure nuclear materials for them. More frequently still, users ask the Agency for advice when concluding or implementing supply contracts, especially as regards obtaining prior consent and export licences from third countries. This latter aspect of the assistance given by the Agency relates, in particular, to operations involving special fissile materials.

As concerns *special fissile materials*, Chapter VI has been applied more strictly than with regard to source materials. For one thing, contracts have always been signed by the Agency even if, in actual fact, they had been negotiated by producers or users (or "intermediaries") alone. For, although it could have been argued that isotope separation, carried out under contract on source material supplied by the customer, was a conversion operation to which, as provided by Article 75, Chapter VI did not apply, and only requiring notification of the commitment concerned, the Commission never-

theless opted for the interpretation that such an operation results in a change in nature and category. It will be noted that this interpretation of the Treaty is in line with the United States concept of the operation in question.

3. Influence of the domestic law of third states and of international agreements

The influence of the domestic legislation of third countries on the application of Chapter VI, particularly marked in the case of the transfer of special fissile materials, results from certain provisions of such legislation being included in co-operation agreements concluded between these countries and the Community. This, in turn, was due to the fact that the Community was relatively dependent on such countries, with the United States having a monopoly on enrichment services until 1973, and countries such as Canada and Australia possessing considerable mineral resources.

Under the United States Atomic Energy Act of 1954, special nuclear material could be exported only within the framework of a co-operation agreement (enrichment under contract with a foreign country being considered as an export). The EURATOM/United States Co-operation Agreement of 8th November 1958[10] quickly became the only co-operation agreement regulating supplies between the United States and the Member States of the Community. Under Article 1 C of an amendment, dated 21st-22nd May 1961, to the Additional Agreement of 11th June 1960 to the said Agreement, contracts — and especially enrichment contracts — were to be concluded between the Agency and the United States Atomic Energy Commission (USAEC). That the designation of the Agency as exclusive partner was later changed (amendment of 20th September 1972) to the more flexible "authorised persons", does not change the fact that, for many years, the Agency's exclusive right to conclude contracts was enforced at least as much by this Agreement as by the actual force of the EURATOM Treaty. Furthermore, these Agreements give the United States Government the right to follow up what happens to materials supplied to the Community (use defined in joint programmes, restrictions on retransfers outside the Community, etc.). It is true that, in negotiating this Agreement, the Additional Agreement of 11th June 1960 and subsequent amendments, the Commission suceeded in preserving fundamental principles of Community law such as non-discrimination between Community users, freedom of intra-Community transfers, etc. This does not alter the fact that the exception to the provisions of Chapter VI concerning "conditions imposed by suppliers outside the Community" (see Articles 52 and 64) became the rule, resulting in the control of supply being shared between the Commission and the Government of a third state.

For a description of EURATOM/United States relations since the passing of the United States Nuclear Non-Proliferation Act (1978), see chapter six of this Study.

"Conditions imposed by suppliers outside the Community" also affect the working of Chapter VI under Agreements concluded by the Community with Australia and Canada (see chapter six of this Study). These Agreements are based on the principle that the prior consent of Australia or Canada is required for transfers of nuclear materials originating in these countries to third states. Australia and Canada have, however, agreed to waive this requirement in favour of less strict procedures (notification by the Community, for example, when it retransfers materials subject to the Agreements to certain third states). Canada draws up the list of such third states itself, while in the case of Australia, the third state must have a transfer agreement with it and, even in this event, Australia has reserved the right to warn the Community not to undertake the proposed transfer. Similarly, the reprocessing of materials subject to the Agreement is allowed only if the materials recovered are to be used in compliance with a fuel cycle programme agreement between the Parties (Australia) or described to the other Party (Canada).

4. Reorientation of Supply Agency activities, and draft amendments to Chapter VI

The importance assumed by the co-operation agreements with the main Community suppliers of source materials and/or enrichment services, together with the increasingly strict non-proliferation and physical protection conditions attached, by these countries, to the supply of such materials or services, have led to a reorientation of the activities of the Agency. Rather than concluding contracts, the Agency now concentrates on advising the Commission when such agreements are being negotiated and on finding an overall solution to any implementation difficulties encountered, and participates at the same time in work organised under the auspices of the IAEA such as CAS (see chapter five of this Study).

Moreover, emphasis is now placed on the assistance afforded Community undertakings to obtain authorisation for the supply of highly enriched uranium for research reactors, reprocessing and retransfer to third states. Assistance is also given to convert enrichment contracts of the old type (fixed commitment, for example), concluded with the United States Department of Energy, to the more recent Utility Service Contract model established in 1984.

A more systematic approach to the supply problem was tried in the form of a revision of Chapter VI of the Treaty. Following proceedings before the Court of Justice of the European Communities of 14th November 1978[11] which brought to light the discrepancy between a literal or teleological interpretation of the Treaty and its application in practice, the Commission prepared, in December 1982, a proposal for revision[12], including the changes made following various opinions and consultations[13] and, in the version dated 30th November 1984[14], may be described as follows:

— substitution of the principle of the unity of the Community market for that of equal access, considered an insufficient incentive for investment in the mining industry or the fuel cycle. Consequently, the elimination of all restrictions on transfers of materials between Member States or on imports from outside the Community would be accompanied by conditions and criteria to be drawn up by the Community institutions, after agreement at government level, to take account of Member States' duties with regard to non-proliferation. In addition, all exports of materials from the Community would have to be subject to authorisation by the Commission, in conformity with the non-proliferation and safeguards policy of the Community;
— provision is made for "solidarity measures". They, in fact, constitute an extension of those currently in force, and, like them, need a decision taken by the Council by majority vote to become effective;
— the Agency would lose its rights of option and right to conclude contracts, and would verify that all supply contracts — which must, together with certain items of information, be notified to it — comply with the provisions of the new Chapter VI. It would also have to ensure market transparency and, at the request of Member States or undertakings, or if provision were so made in an international agreement, provide assistance with regard to the negotiation or conclusion of supply contracts.

Without waiting for the revision procedure to be completed, the Commission proposed to the Council in 1985, regulations[15] laying down the conditions for transfers between Member States and imports from outside the Community, as well as the criteria for export authorisations. The Council has not yet taken any decision on these proposals. As regards these proposals, *annex II* to this chapter contains extracts from an information note circulated by the Spokesman's Group of the Commission. Meanwhile, on 20th November 1984, the Foreign Ministers of the Member States of the Community adopted a Declaration of Common Policy (see chapter five of this Study).

Under this Declaration of Common Policy[16], transfers between Member States of "sensitive" fissile materials (plutonium or uranium enriched to more than 20 per cent) require a certificate from the consignee specifying the final destination, the quantities, the approximate date of delivery, the timetable for utilisation, the form in which the delivery is to take place and the allocation of the material to one of the uses authorised under the Declaration. The accuracy of this information is certified by the Member State to which the consignee belongs. Retransfers to a third state require the mutual agreement of the Member State that has separated the plutonium or enriched the uranium, and of the Member State desiring the retransfer, without prejudice to any other rights of prior consent that may exist. Operations involving "small quantities", however, are exempt from these requirements.

NOTES AND REFERENCES

1. See *Le régime juridique des matières nucléaires dans le Traité instituant EURATOM: propriété, approvisionnement, contrôle*, by J. P. Delahousse, in Aspects du droit de l'énergie atomique, Volume 2, Paris, 1967, p. 475 *et seq*. See also *Le régime de propriété dans le Traité d'EURATOM*, by V. G. Vedel, in Annuaire français de droit international, 1957, p. 586 *et seq*.

2. Sanctions for infringements of safeguards obligations include, in particular, the total or partial withdrawal of special fissile materials, following a decision by the Commission, which decision is enforceable (Articles 83.1 and 2, and 164 of the Treaty).

3. Decision of the Commission of 5th May 1960 (OJEC of 11th May 1960).

4. Regulation determining the manner in which demand is to be balanced against the supply of ores, source materials and special fissile materials (OJEC of 11th May 1960).

5. OJEC of 30th November 1960.

6. OJEC of 20th February 1960, as corrected by OJEC of 7th March 1960.

7. The definition of this exemption, first established by the Regulation of the Commission of 29th November 1961 (OJEC of 19th December 1961) was considerably widened by Regulation No. 17/66 of 29th November 1966 (OJEC of 28th December 1966), concerning transfers, imports and exports of 1 tonne of uranium and/or thorium content per operation and up to 5 tonnes per annum, per user and for each of these materials. For U-235, U-233 or Pu, exemptions were granted for transfers or imports of up to 200 grammes per operation and 1 000 grammes per annum and per user (subject, where appropriate, to the terms of co-operation agreements relating to such imports). The exemption relating to special fissile materials was extended to exports by Regulation No. 3137/74 of 12th December 1974 (OJEC of 13th December 1974).

8. OJEC L 193/37 of 25th July 1975.

9. See recent Annual Reports of the Supply Agency.

10. Co-operation Agreement between EURATOM and the United States dated 8th November 1958 (OJEC 309/51 of 19th March 1959).

    Amendment to the Co-operation Agreement dated 21st-22nd May 1962 (OJEC 2038/62 of 8th August 1962).

    Additional Agreement dated 11th June 1960 (OJEC 668/61 of 20th April 1961).

    Amendment to the Additional Agreement dated 21st-22nd May 1962 (OJEC 2045/62 of 8th August 1962).

    Amendment to the Additional Agreement dated 22nd-27th August 1963 (OJEC 88/64 of 21st October 1964).

    Amendment to the Additional Agreement dated 20th September 1972 (OJEC L 139/24 of 22nd May 1974).

11. *A propos d'une délibération de la Cour de Justice des Communautés Européennes — le régime des matières nucléaires et la capacité de la Communauté de conclure des accords internationaux* by Jean-Pierre Puissochet, in Annuaire Français de Droit International, XXIV, 1978, pp. 977-988.

12. OJEC C 330/4 of 16th December 1982.

13. For example, the opinion of the Supply Agency's Advisory Committee dated 26th October 1983: the Committee noted that in practice the Agency's monopoly had not been exercised; it recommended solutions based on organisation. Failing a formal revision, the Committee envisaged proceeding by interpretation, the purpose being to avoid discriminating between operators and to superimpose Community constraints and national constraints, while liberalising intra-Community trade.

14. Court (84)606 Final/2.

15. OJEC C 29/10 of 31st January 1985.

16. It should be noted that the purpose of the said draft regulations is to implement, within the framework of the Community, the provisions of the Declaration of Common Policy and that failing their adoption, this Declaration does not officially form part of Community law.

*
**

EURATOM, analyse et commentaires du Traité, by Messrs. Errera, Symon, Van der Meulen and Vernaeve, les Editions de la librairie encyclopédique, S.P.R.L., Brussels, 1958.

ANNEX I

LIST OF GOODS AND PRODUCTS SUBJECT TO THE PROVISIONS OF CHAPTER IX ON THE NUCLEAR COMMON MARKET

LIST A[1]

Uranium ores containing more than 5 per cent by weight of natural uranium.

Pitchblende containing more than 5 per cent by weight of natural uranium.

Uranium oxide.

Inorganic compounds of natural uranium other than uranium oxide and uranium hexafluoride.

Organic compounds or natural uranium.

Crude or processed natural uranium.

Alloys containing plutonium.

Organic or inorganic compounds of uranium enriched in organic or inorganic compounds of uranium 235.

Organic or inorganic compounds of uranium 233.

Thorium enriched in uranium 233.

Organic or inorganic compounds of plutonium.

Uranium enriched in plutonium.

Uranium enriched in uranium 235.

Alloys containing uranium enriched in uranium 235 or uranium 233.

Plutonium.

Uranium 233.

Uranium hexafluoride.

Monazite.

Thorium ores containing more than 20 per cent by weight of thorium.

Urano-thorianite containing more than 20 per cent of thorium.

Crude or processed thorium.

Thorium oxide.

Inorganic compounds of thorium other than thorium oxide.

Organic compounds of thorium.

LIST A[2]

Deuterium and its compounds (including heavy water) in which the ratio of the number of deuterium atoms to normal hydrogen atoms exceeds 1:5 000.

Heavy paraffin in which the ratio of the number of deuterium atoms to normal hydrogen atoms exceeds 1:5 000.

Mixtures and solutions in which the ratio of the number of deuterium atoms to normal hydrogen atoms exceeds 1:5 000.

Nuclear reactors.

Equipment for the separation of uranium isotopes by gaseous diffusion or other methods.

Equipment for the production of deuterium, its compounds (including heavy water) and derivates, and mixtures or solutions containing deuterium in which the ratio of the number of deuterium atoms to normal hydrogen atoms exceeds 1:5 000:

— equipment operating by the electrolysis of water;
— equipment operating by the distillation of water, liquid hydrogen, etc.;
— equipment operating by isotope exchange between hydrogen sulphide and water by means of a change of temperature;
— equipment operating by other techniques.

Equipment specially designed for the chemical processing of radioactive material:

— equipment for the separation of irradiated fuel:
  • by chemical process (solvents, precipitation, ion exchange, etc.);
  • by physical process (fractional distillation, etc.);
— waste-processing equipment;
— fuel-recycling equipment.

Vehicles specially designed for the transport of highly radioactive substances:

— railway and tramway goods vans, goods wagons and trucks of any gauge;
— motor lorries;
— motorised works trucks for the handling of goods;
— trailers and semi-trailers and other non-motorised vehicles.

Containers with lead radiation shielding for the transport or storage of radioactive material.

Artificial radioactive isotopes and their inorganic or organic compounds.

Remote-controlled mechanical manipulators specially designed for handling highly radioactive substances:

— mechanical handling gear, fixed or mobile, but not being capable of being operated manually.

LIST B

Lithium ores and concentrates.

Nuclear-grade metals:

— crude beryllium;
— crude bismuth;
— crude niobium (colombium);
— crude zirconium (hafnium-free);
— crude lithium;
— crude aluminium;
— crude calcium;
— crude magnesium.

Boron trifluoride.

Anhydrous hydrofluoric acid.

Chlorine trifluoride.

Bromine trifluoride.

Lithium hydroxide.

Lithium fluoride.

Lithium chloride.

Lithium hydride.

Lithium carbonate.

Nuclear-grade beryllium oxide.

Refractory bricks of nuclear-grade beryllium oxide.

Other refractory products of nuclear-grade beryllium oxide.

Artificial graphite in the form of blocks or bars in which the boron content is less than or equal to 1 part per million and in which the total microscopic thermal neutron absorption cross-section is less than or equal to 5 millibarns.

Artificially separated stable isotopes.

Electromagnetic ion separators, including mass spectographs and mass spectrometers.

Reactor simulators (special analog computers).

Remote-controlled mechanical manipulators:

• hand-controlled (i.e., operated manually like a tool).

Liquid-metal pumps.

High-vacuum pumps.

Heat exchangers specially designed for nuclear power stations.

Radiation detection instruments (and spare parts) of one of the following types, specially designed, or adaptable, for the detection or measurement of nuclear radiation, such as alpha and beta particles, gamma rays, neutrons and protons:

- geiger counter tubes and proportional counters;
- detection or measuring instruments incorporating Geiger-Muller tubes or proportional counters;
- ionization chambers;
- instruments incorporating ionization chambers;
- radiation detection or measuring equipment for mineral prospecting and for reactor, air, water and soil monitoring;
- neutron detector tubes using boron, boron trifluoride, hydrogen or a fissile element;
- detection or measuring instruments incorporating neutron detector tubes using boron, boron trifluoride, hydrogen or a fissile element;
- scintillation crystals, mounted or in a metal casing (solid scintillators);
- detection or measuring instruments incorporating liquid, solid or gaseous scintillators;
- amplifiers specially designed for nuclear measurements, including linear amplifiers, preamplifiers, distributed amplifiers and pulse height analysers;
- coincidence devices for use with radiation detectors;
- electroscopes and electrometers, including dosimeters (but excluding instruments intended for instruction purposes, simple metal leaf electroscopes, dosimeters specially designed for use with medical X-ray equipment and electrostatic measuring instruments);
- instruments capable of measuring a current of less than 1 picoampere;
- photomultiplier tubes with a photocathode which gives a current of at least 10 microamperes per lumen and in which the average amplification is greater than 10^5, and any other types of electric multiplier activated by positive ions;
- scalers and electronic integrating meters for the detection of radiation.

Cyclotrons, Van de Graaff or Cockroft-Walton electrostatic generators, linear accelerators and other machines capable of imparting an energy greater than 1 MeV to nuclear particles.

Magnets specially designed and constructed for the above-mentioned machines and equipment (cyclotrons, etc.).

Accelerating and focusing tubes of the type used in mass spectrometers and mass spectrographs.

Intense electronic sources of positive ions intended for use with particle accelerators, mass spectrometers and similar devices.

Anti-radiation plate glass:

- cast or rolled plate glass (including wired or flashed glass) in squares or rectangles, surface-ground or polished but not further worked;
- cast or rolled plate glass (whether or not ground or polished) cut to shape other than square or rectangular, or curved or otherwise worked (for example, bevelled or engraved);
- safety glass, consisting of toughened or laminated glass, shaped or not.

Airtight clothing affording protection against radiation or radioactive contamination:

- made of plastic;
- made of rubber;
- made of impregnated or coated fabric:
  - for men;
  - for women.

Diphenyl (when it is in fact the aromatic hydrocarbon $C^6H^5C^6H^5$).

Triphenyl.

NOTE D'INFORMATION DU GROUPE DU PORTE-PAROLE DE LA COMMISSION

(31 octobre 1984)

APPROVISIONNEMENT NUCLÉAIRE — CHAPITRE VI : LA COMMISSION AJUSTE SA PROPOSITION AFIN DE FAIRE JOUER PLEINEMENT LE «MARCHÉ COMMUN» DES MATIÈRES NUCLÉAIRES A UTILISATION PACIFIQUE

(Extraits)

Conformément à l'esprit et à la lettre du Traité EURATOM, la Commission européenne avait proposé en décembre 1982 aux Gouvernements des Dix*, à l'initiative du Vice-Président DAVIGNON, d'adapter aux circonstances des années'80, le régime d'approvisionnement des matières nucléaires à utilisation pacifique prévu par ce Traité. En termes pratiques, la Commission propose un régime qui réalise effectivement l'unité du «marché commun nucléaire» en vue de promouvoir la solidarité entre pays qui disposent d'une vraie dimension industrielle dans le secteur et leurs partenaires. Cela favoriserait les débouchés de l'industrie nucléaire des premiers et offrirait la possibilité aux seconds de ne pas devoir se lancer dans des investissements lourds, sans rapport avec leur programme nucléaire limité. Cette solidarité implique par ailleurs une attitude commune vis-à-vis des pays tiers dont certains ont une tendance à assortir leurs exportations de conditions contraires à l'unité du marché communautaire.

La proposition de la Commission a donné lieu entretemps à des débats approfondis qui se sont déroulés de manière continue pendant les 22 derniers mois. Toutes les institutions communautaires, autorités nationales et milieux intéressés ont participé à ce débat.

A la lumière de ce large débat, la Commission a décidé d'apporter plusieurs modifications et éclaircissements à sa proposition initiale. Ces amendements, tout en simplifiant davantage encore le nouveau régime d'approvisionnement nucléaire envisagé, ne compromettent pas la réalisation de l'objectif de la révision, à savoir, mettre la Communauté en mesure de veiller efficacement à l'approvisionnement régulier et équitable de tous les utilisateurs en minerais et combustibles nucléaires.

La Commission estime que le moment est venu d'engager la phase finale des travaux : les débats, estime-t-elle, ne peuvent se prolonger indéfiniment. La proposition amendée constitue aux yeux de la Commission la base d'un accord au Conseil : il sera mis fin ainsi à l'incertitude qui entoure cette question dans un secteur de pointe qui présente une importance vitale pour la Communauté, incertitude qui entraîne des violations du droit injustifiables.

. . . .

POINTS ESSENTIELS DE LA PROPOSITION AJUSTÉE

En ce qui concerne le champ d'application du nouveau Chapitre VI la Commission n'amende pas sa proposition initiale mais confirme dans sa communication que le Chapitre VI révisé s'appliquerait exclusivement à l'approvisionnement «à des fins civiles non explosives». La Commission souligne également que le nouveau régime tout en tenant dûment compte de la diversité de statut des Etats membres au regard de l'armement nucléaire, devrait mettre fin à l'incohérence et aux discriminations qui caractérisent la situation actuelle dans laquelle une proportion croissante (et différente suivant les Etats membres) des contrats d'approvisionnement civils échappent aux règles du système d'approvisionnement communautaire.

* Voir note d'information P-76 du Porte-Parole de la Commission ; décembre 1982.

Note by the Secretariat : Text available in French only.

Il s'agit de remplacer un régime inadapté, contesté et largement inappliqué par un système qui sera appliqué mais prendra pleinement en compte les transformations profondes de l'industrie et du commerce nucléaires comme aussi des conditions politiques dont les opérateurs industriels doivent tenir compte.

Quant à l'unité du marché et aux exportations, l'examen par le Conseil de la proposition de révision de la Commission de décembre 1982 a mis en lumière la nécessité — si l'on veut maintenir un système d'approvisionnement communautaire — d'un minimum de convergences de politiques de non-prolifération des Etats membres dans la mesure où celles-ci concernent les transferts de matières nucléaires entre Etats. La Commission a modifié sa proposition initiale sur ces deux chapitres. Les amendements proposés permettent d'éviter toute ambiguïté quant à l'exclusivité de la compétence des Etats membres en matière de non-prolifération et de réconcilier les impératifs des politiques de non-prolifération des Etats membres avec les nécessités d'un système d'approvisionnement communautaire régulier et équitable.

Pour ce qui regarde le contrôle de l'application du nouveau Chapitre VI, la Commission, modifiant son option initiale, propose un système allégé comportant une simple notification de l'existence des contrats complétée par certains éléments d'information importants permettant à l'Agence d'approvisionnement de s'acquitter de ses responsabilités de nature économique. Toutefois, la position de la Commission sur ce point demeure ouverte à une solution comportant la communication intégrale des contrats négociés avec des entreprises ressortissant de pays tiers dans le but de renforcer le «bargaining power» des contractants européens.

Enfin, si, comme le propose la Commission dans la présente communication de révision amendée, on s'en tient à un système de notification des contrats, il n'y a plus d'obstacle à ce que l'Agence puisse prêter, sur demande des opérateurs qui le souhaitent, son assistance à la négociation et à la conclusion de contrats d'approvisionnement. La Commission a élargi le rôle de l'Agence sur ce point, par rapport à la proposition de décembre 1982.

CHAPTER FOUR

PREVENTION OF NUCLEAR-WEAPON PROLIFERATION
(SECURITY CONTROL)

A. INTRODUCTION

Following changes in its legislation in 1951 and, above all, in 1954 (Atomic Energy Act), the Government of the United States agreed to supply other countries, within the framework of co-operation agreements, with nuclear information, equipment and materials. It decided to make these supplies subject to "safeguards", to ensure that the recipient country would not use the materials and equipment obtained for military purposes, or transfer them to third parties without the approval of the United States. To check compliance with these obligations, a security control technique was drawn up which served as a model for security control systems established subsequently.

It rapidly became clear that it was more in keeping with the national sovereignty of countries receiving such transfers to delegate the exercise of these controls to international organisations rather than, for example, having installations inspected by civil servants from a foreign country. This led to powers in the field of security control being given to the IAEA (safeguards in the terminology of this Agency) under Article XII of its Statute, to EURATOM, under Chapter VII of the Treaty of Rome, and lastly to the European Nuclear Energy Agency under Article 8 of its Statute and the Convention of 20th December 1957. It will be recalled, however, that the exercise of control by this latter Agency was suspended in 1976 to avoid duplication with the controls exercised by the IAEA and EURATOM (see chapter two of this Study). Consideration will therefore be given here only to those systems of control currently in force[1].

The Treaty on the Non-Proliferation of Nuclear Weapons of 1st July 1968 and, to a lesser extent, the Treaty of Tlatelolco of 14th February 1967 for the Prohibition of Nuclear Weapons in Latin America[2] followed, in a second stage, by policy declarations by certain States consequent to the work of the Zangger Committee and the "London Club" (see chapter five of this Study) led to a development of the system of safeguards implemented by the IAEA and an approximation of the EURATOM safeguards system with that of the IAEA. The section dealing with EURATOM will therefore be restricted to a description of the system resulting from this approximation, with references to features of the original system only when necessary for a proper understanding of the current situation.

B. THE IAEA SAFEGUARDS

I. THE SYSTEM PRIOR TO THE NPT

While the EURATOM safeguards system applicable prior to Regulation 3227/76 of 19th October 1976 is no longer in force, the safeguards regime applied originally by the IAEA remains applicable to certain nuclear installations in a few States.

Article III.A.5 of the IAEA Statute provides that the Agency is authorised "To establish and administer safeguards designed to ensure that special fissionable and other materials, services, equipment, facilities, and information made available by the Agency or at its request or under its supervision or control are not used in such a way as to further any military purpose; and to apply safeguards, at the request of the parties, to any bilateral or multilateral arrangement, or, at the request of a State, to any of that State's activities in the field of atomic energy". Article XII of the Statute outlines the control mechanism and provides for sanctions if a State subject to safeguards breaches the obligations resulting therefrom.

The dissuasive value of the safeguards system set up by the IAEA consists essentially in the alert given to the international community following any breach by a State of its obligations. For, should the Board of Governors of the IAEA be informed by the Director General of any non-compliance reported by the inspectors, it informs all Member States of the non-compliance and refers the matter to the Security Council and General Assembly of the United Nations (Article XII.C). It is only in cases where assistance has been provided by the Agency that the Board of Governors must call upon the recipient State or States to take the necessary measures to remedy the non-compliance with the safeguards agreement. In the event of failure of the recipient State or States to comply with this requirement within a reasonable time, the Board may either direct curtailment or suspension of the assistance being provided and call for the return of materials and equipment made available, or suspend the non-complying member from the exercise of the privileges and rights of membership of the Agency.

It can be seen from these provisions that the object of control is either a State receiving assistance from the Agency, or receiving supplies from another State when both States have agreed with the Agency that safeguards would be exercised by the Agency over such supplies, or else a State which has agreed to submit to Agency control over certain of its activities, although in practice, most of the obligations involved weigh upon the undertakings concerned in the State which is being controlled. In the EURATOM system set up by the Treaty of Rome, on the other hand, it is "persons or undertakings" which are concerned directly.

Another aspect of the original IAEA system was the contractual element involved in bringing it into play. For the IAEA safeguards to apply, the State controlled had to request, and obtain, the assistance of the Agency (Project Agreement), or agree with the supplier State and the Agency on a transfer of the exercise of control to the Agency (Safeguards Agreement) or else reach agreement with the Agency on the voluntary submission of certain of these activities. The original EURATOM system, on the other hand, was regulatory in nature, linked to the presence of nuclear materials subject to control on the territory of the European Community.

The scope of IAEA safeguards, at least when the power of control stems from assistance provided by the Agency, is particularly wide since it extends to services, equipment, facilities and information and also allows the Agency to require "the observance of any health and safety measures prescribed by the Agency" (Article XII.A.2). The original EURATOM safeguards, on the other hand, applied only to "ores, source materials and special fissile materials".

However, this difference was not so great in practice since the IAEA system, introduced in 1961, applied in fact only to nuclear materials even though it was later extended to large reactor units (1964), processing plants (1966) and then to nuclear materials in conversion and fabrication plants. In its final version, the initial safeguards system, reviewed in 1965, is defined in document *INFCIRC/66/Rev. 2* (1968) (see *annexes I and II* to this chapter). According to the Annual Report of the Agency for 1986, safeguards continue to be applied in compliance with this document in nine countries, either following project agreements or voluntary-offer agreements. The gradual reduction in the number of States subject to the original regulations is due to the substitution of safeguards agreements concluded in implementation of the NPT for agreements regulated by INFCIRC/66/Rev.2.

Transfers of nuclear materials subject to safeguards in accordance with document INFCIRC/66/Rev.2, and performed outside of the territory of the State where they are subject to these safeguards, are authorised subject to verification by the Agency that at least one of the following conditions is met:

i) When the nuclear material subject to safeguards has been returned to the State that originally supplied it, the material ceases to be subject to safeguards if it was till then so subject only by reason of such supply and if it was not improved while under safeguards and if, in addition, any special fissionable material that was produced in the nuclear material has been separated out, or safeguards with respect to such produced material have been terminated.

ii) When the nuclear material subject to safeguards is being transferred, under an arrangement or agreement approved by the Agency, for the purpose of processing, reprocessing after irradiation, testing, research or development, within the State concerned or to any other Member State or to an international organisation, the safeguards are suspended provided that the quantity of nuclear material does not exceed any of the limits laid down in document INFCIRC/66/Rev.2, paragraph 24 (for example, one effective kilogramme of special fissionable material). Safeguards with respect to nuclear material in irradiated fuel which is transferred for the purpose of reprocessing may also be suspended if the State or States concerned have, with the agreement of the Agency, placed under safeguards substitute nuclear material for a period laid down in paragraph 25.

iii) Arrangements have been made by the Agency to safeguard the material in the State to which it is being transferred.

iv) Lastly, except for material subject to safeguards pursuant to a Project Agreement, the material must be subject, in the State to which it is being transferred, to safeguards other than those of the Agency but nevertheless generally consistent with such safeguards and accepted by the Agency.

In the case of transfers to any one of these States, the obligations assumed by the supplier State, whether pursuant to the NPT or to commitments undertaken in accordance with the work of the Nuclear Exporters'

Committee ("Zangger Committee") or the Nuclear Suppliers Group ("London Club") (see chapter five of this Study), are not *ipso facto* satisfied unless the supplies are intended for a facility subject to the provisions of INFCIRC/66/Rev.2 or fall within its scope by virtue of a Project Agreement or voluntary submission.

II. SAFEGUARDS SYSTEM UNDER THE NPT

The 1968 Treaty on the Non-Proliferation of Nuclear Weapons provides that each non-nuclear-weapon State Party to the Treaty undertakes to accept the IAEA "safeguards" to allow the Agency to verify fulfilment of the obligations assumed by the State in question under the Treaty (Article III.1).

Furthermore, each State Party to the Treaty (and particularly the nuclear-weapon States: the United States, the United Kingdom and the USSR) undertakes not to provide

i) source or special fissionable material; or
ii) equipment or material specially designed or prepared for the processing, use or production of special fissionable material,

to any non-nuclear-weapon State whatsoever, i.e. whether or not Party to the Treaty, unless the source or special fissionable material is made subject to the IAEA safeguards (Article III.2).

This responsibility, entrusted to the IAEA under the Treaty, obliged the Agency to remodel its safeguards system. It is true that, as in the initial system, the subjects of control remain States, and the system remains a contractual one in that only those States having ratified the Treaty are bound and, above all, because Article III expressly provides for the conclusion of safeguards agreements with the Vienna Agency. But in other ways, the new system moved closer to the initial regulatory character of the EURATOM system. First, because the scope of the safeguards is limited to source or special fissionable materials and, unlike the Statute of the Agency (if not in practice the initial system defined by INFCIRC/66), does not cover services, equipment, facilities and information. The similarity between the two systems is even greater if the territorial aspect is considered, since the above-mentioned nuclear materials are subject to safeguards in all peaceful nuclear activities within the territory of a non-nuclear-weapon State, under its jurisdiction or carried out under its control anywhere (Article III.1 *in fine*).

The number of safeguards agreements to be concluded and the equality of treatment that it was necessary to ensure all non-nuclear-weapon States, led to a high degree of standardization of safeguards agreements. This trend towards standardization is reflected in the brochure published by the IAEA, commonly called the "Blue Book" (*INFCIRC/153* "The Structure and Content of Agreements between the Agency and States required in connection with the NPT" — 1983 revised edition). This has reduced the contractual aspect of the initial IAEA system, and restricted this aspect to the signing of the safeguards agreement and, above all, to adapting the rules to each specific situation by the "Subsidiary Arrangements" provided for under these agreements. Because of these subsidiary arrangements, the examination of a concrete case of application of safeguards, such as the transfer of nuclear materials from State A to State B, requires a knowledge of the safeguards agreements concluded by these States with the Agency. Details of these agreements are given in the IAEA publication entitled "Agreements Registered with the International Atomic Energy Agency, Ninth Edition, Legal Series No. 3, 1985"; this information may be updated by referring to subsequent Annual Reports of the Agency[3].

For a better understanding of the safeguards procedures applicable to the international transfer of nuclear materials, a brief description is necessary of how control is exercised when the materials are in a fixed installation, this being the most usual situation for which the system was designed.

The safeguards system is essentially based on the keeping of accounts of materials, backed up by containment and inspection provisions.

Material accounting is in principle a national system of accounting and control applying to all materials subject to safeguards. As the NPT (Articles III.3 and IV) provides that the implementation of safeguards must "avoid hampering the economic or technological development of the Parties", material flow is measured only at strategic points called "key measurement points" at the edges of "material balance areas". Accounts are kept for each "material balance area" (MBA), as defined by the safeguards agreement, using, as far as possible, existing containment and surveillance measures such that every transfer into or out of a material balance area can be recorded. This, together with design information and an initial inventory, should allow the State and the Agency to know, at any given moment, the quantity, location and movement of these materials. The definition of each area and its exact size depends on the accuracy with which it is possible to draw up the relevant material balance. It may be a facility, part of a facility sufficiently autonomous by reason of its technological function and containment or, on the other hand, several facilities located on different sites but technologically integrated. However, general accounts of nuclear materials are not sufficient since the isotopic composition varies depending on the operations carried out in the facility (a reactor, for example). That is why the book inventory is determined with the help of the physical inventory using records of operations corroborated by means such as the calibration of equipment, sample analysis, etc. These book and physical inventory changes must be established in respect of "batches" of nuclear material and may be adjusted or corrected in the light of the results of measurements or analyses. Periodic inventories also make it possible to check that the record system is reflecting the real situation and detect any losses. The Agency has set up an International Safeguards Information System (ISIS) to process these reports.

As a rule, safeguards agreements are accompanied by Subsidiary Arrangements agreed between the IAEA and the State concerned. These Arrangements define containment and surveillance provisions such as locks, seals, cameras, non-destructive analyses, etc. of such a type as to ensure verification that the content of the MBA, known to the Agency by means of the records system, has not been altered and to control facility activities involving movements of nuclear material. In order both to optimise use of its means of inspection and to "avoid hampering economic development" by placing too heavy a burden on operators, the Agency has made great efforts to perfect effective containment and surveillance devices.

The IAEA is informed of the results of the national system of accounting by means of periodic accounting reports (inventory changes and inventories) and, in the event of losses, or suspicion thereof, or of changes in containment, by means of special reports. It may also ask the State concerned for amplification or clarification of reports. Moreover, it is responsible for checking the results obtained by the national system, in particular by means of independent observation procedures. It has a laboratory for conducting analyses for safeguards purposes. The design information relating to facilities and accounting data are processed and evaluated with the help of ISIS.

Above all, the Agency is empowered to carry out inspections and has, for this purpose, a body of trained inspectors. Inspections are either ad hoc, to check the information contained in the initial report and identify any subsequent changes, or carried out to ensure that reports are consistent with records, verify the location, quantity and composition of the materials, and check the information on possible causes of material unaccounted for in the event of uncertainties in the book inventory. Lastly, special inspections may be carried out to verify the information contained in special reports or whenever the explanations supplied by the State concerned are not satisfactory. The rules (frequency and intensity) relating to routine inspections are laid down in the Subsidiary Arrangements.

By means of both the system for accounting for materials and the containment and surveillance measures verified by inspections, the system set up is designed to be able to detect any diversion of a significant quantity of nuclear materials, i.e. a large enough quantity to enable a nuclear weapon to be manufactured. Such detection must occur within the time required to transform the materials into nuclear weapons or other nuclear explosive devices.

Clearly, in such a system, an international transfer of materials subject to safeguards constitutes merely a special case of material flow out of an MBA or, when the recipient State is also Party to the Treaty as a non-nuclear-weapon State, a transfer between MBAs. Document INFCIRC/153 (the relevant extracts are given in *annex III* to this chapter) specifies in this respect that the safeguards agreement should provide that the State concerned is responsible, under the safeguards system:

— in the case of *import*, from the time that such responsibility ceases to lie with the exporting State, and no later than the time at which the nuclear material reaches its destination;
— in the case of *export*, up to the time at which the recipient State assumes such responsibility, and no later than the time at which the nuclear material reaches its destination.

Furthermore, the safeguards agreement must stipulate that any intended transfer out of the State concerned must be notified to the Agency after conclusion of the contractual arrangements leading to the transfer and at least two weeks before the nuclear material is to be prepared for shipping. This notification must make it possible for the Agency to identify the material concerned, verify its quantity and composition and, if necessary, affix seals.

It is for the States concerned to make arrangements to determine the point at which the transfer of responsibility will take place. No State shall be deemed to be responsible for nuclear material merely by reason of the fact that the material is in transit on or over its territory or territorial waters, or that it is being transported under its flag or in its aircraft.

In cases where the nuclear material will not be subject to Agency safeguards in the recipient State, the exporting State shall make arrangements for the Agency to receive, within three months of the time when the recipient State accepts responsibility for the material, confirmation by the recipient State of the transfer.

A special report must be made if circumstances or any unusual incident (such as significant delay) leads the shipper State to believe that there is or may have been loss of nuclear material.

These provisions do not, however, apply to a single transfer, or successive shipments to the same State within a period of three months, of less than one effective kilogramme of material[4].

III. SAFEGUARDS SYSTEM UNDER THE TLATELOLCO TREATY

On 14th February 1967, prior to the signature of the NPT, a majority of Latin American States concluded a Treaty prohibiting nuclear weapons in this region of the world. Article 12 of this Treaty, which entered into force on 22nd April 1968, provides for the establishment of a control system to verify, in particular, that no activity for military purposes, prohibited by Article 1, using imported nuclear material is undertaken on the territory of the Contracting Parties. Rather than setting up a separate control system, Article 13 requires the Parties to negotiate agreements with the IAEA for the application of its safeguards to their nuclear activities.

The Treaty itself is supplemented by two Protocols which refer to States other than those Party to the Treaty. Protocol No. I concerns extracontinental States nevertheless having *de juro* or *de facto* responsibility for territories within the geographical limits established by the Treaty. Under this Protocol such States undertake to apply the provisions of the Treaty to the territories concerned. Protocol No. II concerns the nuclear-weapon States which are invited, on ratification, to observe the denuclearised status of Latin America and to withhold from using nuclear weapons against Parties to the Treaty. (The list of Contracting Parties to the Treaty and the Protocols is reproduced in *appendix 5* to the Study.)

Two Latin American States, Colombia and Panama, have signed but not yet ratified the NPT; they are, on the other hand, bound by the Tlatelolco Treaty. In both countries, the IAEA exercises control over all nuclear material on national territory, in compliance with agreements concluded with the Agency pursuant to the Tlatelolco Treaty alone (Colombia: Agreement entering into force on 22nd December 1982, INFCIRC/306; Panama: Agreement entering into force on 23rd March 1984, INFCIRC/316). For other States bound by both the Tlatelolco Treaty and the NPT, the safeguards agreements cover the two Treaties (Bolivia, Costa Rica, Dominican Republic, El Salvador, Equador, Guatemala, Haiti, Honduras, Jamaica, Mexico, Nicaragua, Paraguay, Peru, Suriname, Uruguay and Venezuela). The other Latin American countries (Argentina, Brazil, Chili and Cuba), are governed by the provisions of INFCIRC/66/Rev.2 alone, under Project Agreements, voluntary submission or tripartite transfer agreements (for example, the Agreement between Brazil, the Federal Republic of Germany and the IAEA).

IV.   THE RAROTONGA TREATY

The South Pacific Nuclear Free Zone Treaty was adopted on 6th August 1985 at Rarotonga (Cook Islands)[5]. It entered into force on 11th December 1986. The Treaty is supplemented by three Protocols signed in Suva on 8th August 1986. The Protocols are addressed to the nuclear-weapon States calling on them to refrain from using a nuclear weapon against a Party to the Treaty and from conducting nuclear explosive tests in the nuclear-free zone (see INFCIRC/333 and Addendum 1).

The Treaty provides for the establishment in the South Pacific of a nuclear-weapon free zone. The Parties have pledged not to develop or supply nuclear weapons and to supply nuclear material and equipment for peaceful uses only on condition they are subject to IAEA safeguards under the Non-Proliferation Treaty. (The status of signatories and ratifications is given in *appendix 6* to the Study.)

C.   EURATOM SAFEGUARDS SYSTEM

I.   GENERAL

One of the objectives of EURATOM (cf. chapter two of this Study) is to ensure that nuclear material within the Community is not diverted for purposes other than those for which it was intended. Consequently, Chapter VII of the EURATOM Treaty, and its implementing regulations of 1959, set up a system of safeguards (since the Treaty was revised following the adhesion of the United Kingdom to the European Communities, the term "security control" was replaced by "safeguards"; consequently, this latter term has been used throughout the Study). This control is used either to verify that materials are not diverted from their intended use as declared by the users and, in this case, it does not apply to materials intended to meet defence needs, or is used to verify compliance with particular obligations the Community has undertaken in an agreement concluded with a third state, such as an obligation to use materials for non-military purposes. However, the scope of the EURATOM safeguards is particularly wide since it applies to ores, source materials and special fissile materials, except for materials intended for use by France or the United Kingdom for defence needs, whenever these are located on Community territory. Infringements may be punished by the withdrawal of the source or special fissile materials, in accordance with decisions taken by the Commission which are enforceable.

The non-nuclear-weapon Member States of the European Community wished to sign the NPT but did not wish to have to comply with two sets of obligations arising from the simultaneous application of parallel safeguards systems, that of the IAEA pursuant to the NPT, and that of EURATOM under the EURATOM Treaty.

Since the NPT (Article III.4) allows the possibility of joint safeguards agreements with the IAEA, the European Community, Belgium, Denmark, the Federal Republic of Germany, Ireland, Italy, Luxembourg and the Netherlands therefore signed, on 5th April 1973, an Agreement with the Vienna Agency in implementation of Article III, paragraphs 1 and 4 of the NPT, often referred to as a "Verification Agreement"[6]. In the case of Denmark (except for the Faroe Islands), this Agreement replaced a previous NPT-type bilateral agreement. Greece and Portugal, for their part, acceded to the Agreement of 5th April 1973 on 17th December 1981 and 1st July 1986, respectively.

Under this Agreement, the IAEA, the European Community and the above-mentioned States co-operate in facilitating the implementation of safeguards with a view to avoiding unecessary duplication of control measures and taking due account of the nuclear common market which provides for the free movement of nuclear materials among all Community Member States. These objectives are achieved by the application by the Community of its safeguards system, it being understood that the Community provides the Agency with the information necessary for Agency control. The IAEA retains the right to proceed with measures, observations or independent inspections to clarify the results obtained by the Community safeguards system, but in accordance with procedures defined in the Agreement.

The nuclear common market is taken into account under the Verification Agreement in the case of the transfer of nuclear material subject to the Agreement. On the one hand, the territories of signatory States are considered as constituting a single zone for the application of safeguards, and no notification by the Commission to the Agency of transfers between such States is required. Transfers out of the territory of these States or into such territory must, on the other hand, be notified by the Commission (Article 12, the provisions of which are developed in Articles 91 to 97). The safeguards are terminated when the recipient State has assumed responsibility for the materials. The aim of notification is to allow the Agency to keep records of each transfer and, where appropriate, to reapply safeguards, and to proceed, if necessary, with an ad hoc inspection to identify the materials transferred and verify their quantity and composition.

Furthermore, provision was made for adhesion to the Agreement by non-nuclear-weapon States Party to the NPT which subsequently became Members of the Community. Lastly, provision was also made in the Verification Agreement for the special case of France, a Member of the Community but not a signatory of the NPT, by cancelling the option of ad hoc inspection of transfers by signatory States to France, subject to the Commission's confirming to the IAEA that the materials did in fact arrive on French territory (see the relevant extracts from the Agreement in *annex IV* to this chapter). These latter provisions also apply to the United Kingdom which is a nuclear-weapon State but which, unlike France, is

a signatory of the NPT and which became a full Member of the Communities some time before signature of the Verification Agreement.

In these circumstances, an updating of the EURATOM regulations relating to safeguards appeared necessary, to adapt them to the NPT-type system defined by INFCIRC/153 ("Blue Book"). Furthermore, there was a need to specify the particular safeguards procedures applicable to the two nuclear-weapon Member States: the United Kingdom (signatory of the NPT) and France (a non-signatory) which, by voluntary submission, had made or were proposing to make certain of their facilities or nuclear materials subject to the IAEA system of safeguards. That is why previous Regulations (No. 7 and No. 8) were repealed and replaced by the "codification" carried out by *Regulation No. 3227/76* of the Commission, dated 19th October 1976 (OJEC L 363 — 31st December 1976) (see relevant extracts from this Regulation in *annex V* to this chapter).

The new Regulation gives the same importance as the initial system to the declarations of the basic characteristics and programmes of activities of installations, but reflects the influence of the IAEA system by providing that "particular safeguard provisions" are to be laid down by individual decisions of the Commission after consultation with the undertaking and Member State concerned. There is also an approximation with the IAEA system in that accounting is by "MBA" and "batch".

As concerns more specifically transfers, imports and exports, Part III of the Regulation provides for notification to the Commission, after conclusion of the contractual arrangements leading to the transfer but in time to reach the Commission eight working days before the material is to be prepared for shipment:

— of any export of source or special fissile materials to a non-Member State;
— of any export from a Member State party to the Verification Agreement to a Member State not party to the Agreement (i.e. the United Kingdom and France);
— of any export from the United Kingdom to a Member State party to the Agreement,

whenever the consignment exceeds one "effective kilogramme", or even when it is less than this amount if notification is required under the safeguards provisions in question. This is the case for installations habitually transferring large quantities of materials to the same State.

Similar provisions apply to imports, in respect of which notification must reach the Commission at least five working days before the material is unpacked.

Forms for the advance notification of exports and imports of nuclear materials are reproduced in Annexes V and VI to the Regulation. Any change of date in the preparation for shipment, in the shipment or in the unpacking of nuclear materials as compared to the dates given in the notifications must be communicated without delay to the Commission, with an indication of the revised dates, if known (Article 28).

Moreover, a special report must be sent to the Commission by the "persons or undertakings" concerned if, following exceptional circumstances or an incident, they have received information that nuclear materials have been or appear to be lost, particularly when there has been a considerable delay during transfer (Article 27).

The Regulation ensures that transfers are covered exhaustively by requiring notification not only by "the persons or undertakings" obliged to declare the basic technical characteristics, but also by all other exporters or importers of nuclear materials (Article 26). Although the safeguards in principle cover only source and special fissile materials, uranium and thorium producers exporting ores to non-Member States must also notify the Commission by means of a declaration of ore shipments/exports in accordance with the model contained in Annex VII to the Regulation (see *annex V* to this chapter).

Carriers or persons temporarily storing source or special fissile materials during shipment may accept them or hand them over only against a duly signed and dated receipt. This receipt must be kept for at least one year. However, documents compiled in accordance with regulations in force on the territory of the Member State in which such undertakings operate may take the place of the said receipt provided that they contain all the required information (Articles 32 and 33).

Finally, "intermediaries" must keep all documents relating to the supply of nuclear materials under contract for at least one year after the expiry of the contract. Such documents must mention the names of the Contracting Parties, the date of the contract, and the quantity, nature, form and composition together with the origin and destination of the materials (Article 34).

II. SPECIAL CASE OF NUCLEAR-WEAPON MEMBER STATES

Part V of Regulation No. 3227/76 contains special provisions applicable to nuclear-weapon Member States, i.e. the United Kingdom and France.

In compliance with Article 84 of the EURATOM Treaty, the provisions of this Regulation do not apply:

— to installations or parts of installations which have been assigned to meet defence requirements and which are situated on the territory of a Member State not party to the Agreement of 5th April 1973; or
— to nuclear materials which have been assigned to meet defence requirements by such a Member State.

They do, on the other hand, apply in principle to installations or materials intended for civilian use. However, it proved necessary to clarify a number of special cases such as materials or installations, or parts of installations "which are liable to be assigned to meet defence requirements", and installations which either alternatively or simultaneously use or process nuclear materials assigned or liable to be assigned to meet defence requirements and "civil" nuclear materials. The problems of safeguards raised by these special cases must be regulated by procedures defined by the Commission in agreement with the Member State concerned. The provisions of the Regulation governing declarations of the "basic technical characteristics" and the formulation of "particular safeguard provisions" apply:

— first, to installations or parts of installations which at certain times are operated exclusively with nuclear materials liable to be assigned to meet defence requirements but at other times are operated exclusively with civil nuclear materials;
— secondly, to installations or parts of installations which produce, treat, separate, reprocess or use in any other way simultaneously both civil nuclear materials and nuclear materials assigned or liable to be assigned to meet defence requirements, with, in this case, exceptions being allowed on grounds of national security, in particular when a restriction of access is involved.

Moreover, the civil nuclear materials held in such installations or parts of installations remain subject to the other provisions of the Regulation concerning in particular the programme of activities, the accounting system and — of direct concern to this Study — transfers, imports and exports.

D. VOLUNTARY SUBMISSION TO IAEA SAFEGUARDS BY NUCLEAR WEAPON STATES

At different dates and using different methods, four of the five[7] States recognised as having nuclear weapons have, by agreement with the IAEA, made certain of their activities, facilities or nuclear materials subject to the safeguards applied by the Agency.

Under an Agreement which entered into force on 9th December 1980 (INFCIRC/288), the United States made all its nuclear activities except those directly related to defence subject to safeguards. The United States draws up a list of "civil" installations, including certain facilities owned by the State; the IAEA chooses those with regard to which it applies its safeguards, avoiding, in agreement with the United States Government, creating any discrimination between commercial enterprises. Provision is also made for a rotation system of installations subject to control and for limiting inspections to one-third only of the facilities on the Agency list.

Another nuclear-weapon State which is not a member of the European Communities is the USSR, which signed a voluntary submission Agreement dated 21st February 1985 (INFCIRC/327), which entered into force on 10th June 1985. This Agreement contains provisions similar in principle to those in the United States/IAEA Agreement. In the event of an international transfer, the USSR communicates to the IAEA the information specified in a letter of 10th July 1974 (INFCIRC/207) and, where appropriate, drafts a special report (Articles 13 and 89).

In its final declaration (21st September 1985), the Third NPT Review Conference urgently requested the People's Republic of China also to conclude a safeguards agreement with the IAEA, even though this State is not Party to the NPT.

Both members of the European Communities which are nuclear-weapon States (the United Kingdom and France) have signed safeguards agreements with the IAEA.

The United Kingdom, which has ratified the NPT, concluded an Agreement with the IAEA and the European Community on 6th September 1976, under which the source and special fissile materials in the installations or parts of installations contained in a list drawn up by the United Kingdom and including all its civilian installations are submitted to the safeguards applied by both the Community and the Vienna Agency, the only exceptions allowed being those based on national security

considerations. Using measures and observations distinct from those of the Commission of the European Communities, the Agency verifies that such materials are not withdrawn from civilian activities.

The Agreement concluded between France, the European Community and the IAEA on 27th July 1978 (INFCIRC/290) adopts the same general layout as the above-mentioned Agreements for controls applied by both the European Community and the Vienna Agency. However, its scope is more restricted than that of the Agreement concluded by the United Kingdom, since France reserves the right to designate not only the installations, but also the source and special fissile materials which it agrees to submit to control.

It will have been noted that in the case of the United States and the USSR, the safeguards exercised over certain of their nuclear installations by the IAEA is the only control system applied in these countries. In the United Kingdom and France, on the other hand, over and above the "common law" safeguards applied, with the exceptions related to defence, by the Commission under the EURATOM Treaty and Regulation No. 3227/76, the IAEA also, to a greater or lesser extent, applies control. This latter control is conceived as a verification of the effectiveness of the initial control exercised by the Commission of the European Communities.

It may also be noted that, on 11th July 1974, the three nuclear-weapon States which have signed the NPT (United States, USSR and United Kingdom) jointly informed the IAEA of their decision to supply it with information concerning exports and imports of nuclear materials in excess of one effective kilogramme which they intend to carry out or have received (INFCIRC/207 — 26th July 1974). France made a similar declaration on 16th February 1984 (INFCIRC/207 Add. 1).

NOTES AND REFERENCES

1. No mention will be made either of the weapons control system set up under the Protocols amending and supplementing the Treaty of Brussels of 23rd October 1954 since this system has not been implemented.

2. The text of these two Treaties is reproduced in the IAEA publication entitled "International Treaties relating to Nuclear Control and Disarmament", Legal Series No. 9, IAEA 1975.

3. The following paragraphs are taken from the IAEA Annual Report for 1986:

   392. At the end of 1986 there were 164 safeguards agreements in force with 96 States, compared to 163 agreements with 96 States at the end of 1985, a project agreement concluded with Thailand for the supply of nuclear fuel for a research reactor having entered into force in September.

   393. A safeguards agreement was concluded with Albania covering all its nuclear material and facilities. Safeguards agreements pursuant to NPT were concluded with Belize, Brunei Darussalam, Equatorial Guinea and Tuvalu; these agreements have not yet entered into force.

   394. In July, Portugal acceded to the safeguards agreement of 5th April 1973 between the non-nuclear-weapon States of the European Community, EURATOM and the Agency.

   395. Negotiations commenced for the conclusion of a safeguards agreement pursuant to the offer by China to place some of its civilian nuclear installations under Agency safeguards. When this agreement enters into force there will be safeguards agreements in force with all five nuclear-weapon States.

   396. Discussion of a trilateral safeguards agreement between Spain, EURATOM and the Agency was initiated.

   397. During 1986, four non-nuclear-weapon States became Party to NPT: Malawi in February, Colombia in April, the Yemen Arab Republic in May, and Trinidad and Tobago in October. The total number of States party to the Treaty at the end of 1986 was 134, including three nuclear-weapon States. [Note by the Secretariat: Spain acceded to the NPT on 5th November 1987.]

   398. At 31st December 1986, 46 of the 131 non-nuclear-weapon States party to NPT had not complied, within the prescribed time limit, with their obligations under Article III.4 of the Treaty regarding the conclusion of the relevant safeguards agreement with the Agency. However, with the exception of Viet Nam, none of these States has, as far as the Agency is aware, significant nuclear activities.

   399. Safeguards were actually applied in 41 non-nuclear-weapon States under agreements pursuant to NPT or to NPT and the Tlatelolco Treaty, and in one non-nuclear-weapon State pursuant to the Tlatelolco Treaty.

   400. Thirty-nine safeguards agreements based on INFCIRC/66/Rev.2 were in force with the following nine non-nuclear-weapon States not party to either NPT or the Tlatelolco Treaty: Argentina, Brazil, Chile, Cuba, India, Israel, Pakistan, South Africa and Spain. Safeguards were actually applied in eight of the nine States pursuant to these agreements. Also, safeguards were applied pursuant to INFCIRC/66/Rev.2-type agreements in Viet Nam and the Democratic People's Republic of Korea, both party to NPT. The Agency also applies safeguards to nuclear facilities in Taiwan, China.

   401. In five of the nine States referred to in the first sentence of the preceding paragraph, unsafeguarded facilities of significance for safeguards were known to be in operation or under construction.

   402. All five nuclear-weapon States have complete unsafeguarded nuclear fuel cycles. Voluntary-offer agreements were in force with four of these States during the whole of 1986. In accordance with the relevant agreements, certain facilities in the four States were designated by the Agency for inspection and were inspected. In addition, in one of these States safeguards were applied in 1986 to some facilities in accordance with INFCIRC/66/Rev.2-type agreements.

   403. On 31st December 1986, there were in non-nuclear-weapon States 485 nuclear facilities under safeguards or containing safeguarded nuclear material (474 in 1985); there were also 414 locations outside facilities containing small amounts of safeguarded material (413 in 1985) and two non-nuclear installations (two in 1985). There were also nine facilities in nuclear-weapon States under Agency safeguards pursuant either to voluntary-offer agreements or to safeguards transfer agreements (ten in 1985).

   404. At the end of 1986, the nuclear material under Agency safeguards, including that covered by voluntary-offer agreements with nuclear-weapon States, amounted to 8.4 t (7.9 t in 1985) of separated plutonium (in safeguards, "separated plutonium" includes recycled plutonium in fuel elements until their discharge from the reactor), 194.5 t (156.2 t) of plutonium contained in irradiated fuel, 13.2 t (12.3 t) of high-enriched uranium (HEU), 27 911 t (24 546 t) of low-enriched uranium (LEU) and 47 402 t (43 044 t) of source material. The greater part of this material was in those non-nuclear-weapon States party to NPT where safeguards are being applied to all peaceful nuclear activities. Non-nuclear material under Agency safeguards included 1 470 t (1 432 t) of heavy water.

4. See chapter nine "Definitions" of this Study. Account may have to be taken of exemptions to and terminations of safeguards.

5. The text of the Treaty is reproduced in Nuclear Law Bulletin No. 35.

6. The text of this Agreement is contained in IAEA document INFCIRC/193, of 14th September 1973. It entered into force on 21st February 1977. See extracts in annex IV to this chapter.

7. At the IAEA General Conference of September 1985, the Delegate from China officially announced that the Chinese Government had decided to place voluntarily, in due course, some of its civilian nuclear installations under Agency safeguards.

*
**

"L'Accord entre EURATOM et l'AIEA en application du Traité sur la non-prolifération des armes nucléaires", by Jan Gijssels, Annuaire français de droit international, 1-837 *et seq.*, 1972.

ANNEX I

THE AGENCY'S SAFEGUARDS SYSTEM
(1965, AS PROVISIONALLY EXTENDED IN 1966 AND 1968)

(24th September 1968)

Since 1961, the Safeguards System has developed as follows:

System		Set forth in document
Nature	Name	
The first system	The Agency's Safeguards System (1961)	INFCIRC/26
The 1961 system as extended to cover large reactor facilities	The Agency's Safeguards System (1961, as extended in 1964)	INFCIRC/26 and Add. 1
The revised system	The Agency's Safeguards System (1965)	INFCIRC/66
The revised system with additional provisions for reprocessing plants	The Agency's Safeguards System (1965 as provisionally extended in 1966)	INFCIRC/66/Rev. 1
The revised system with further additional provisions for safeguarded nuclear material in conversion plants and fabrication plants	The Agency's Safeguards System (1965, as provisionally extended in 1966 and 1968)	INFCIRC/66/Rev. 2

ANNEX II

EXTRACTS FROM THE INFCIRC/66/Rev. 2 PROVISIONS

(24th September 1968)

C. SUSPENSION OF SAFEGUARDS

24.   Safeguards with respect to *nuclear material* may be suspended while the material is transferred, under an arrangement or agreement approved by the Agency, for the purpose of processing, reprocessing, testing, research or development, within the State concerned or to any other Member State or to an international organisation, provided that the quantities of *nuclear material* with respect to which safeguards are thus suspended in a State may not at any time exceed:

   a)   1 *effective kilogram* of special fissionable material;
   b)   10 metric tons in total of natural uranium and depleted uranium with an *enrichment* above 0.0005 (0.5 per cent);
   c)   20 metric tons of depleted uranium with an *enrichment* of 0.005 (0.5 per cent) or below; and
   d)   20 metric tons of thorium.

25.   Safeguards with respect to *nuclear material* in irradiated fuel which is transferred for the purpose of reprocessing may also be suspended if the State or States concerned have, with the agreement of the Agency, placed under safeguards substitute *nuclear material* in accordance with paragraph 26(d) for the period of suspension. In addition, safeguards with respect to plutonium contained in irradiated fuel which is transferred for the purpose of reprocessing may be suspended for a period not to exceed six months if the State or States concerned have, with the agreement of the Agency, placed under safeguards a quantity of uranium whose *enrichment* in the isotopic uranium 235 is not less than 0.9 (90 per cent) and the uranium 235 content of which is equal in weight to such plutonium. Upon expiration of the said six months or the completion of reprocessing, whichever is earlier, safeguards shall, with the agreement of the Agency, be applied to such plutonium and shall cease to apply to the uranium substituted therefor.

D. TERMINATION OF SAFEGUARDS

26.   *Nuclear material* shall no longer by subject to safeguards after:

   a)   It has been returned to the State that originally supplied it (whether directly or through the Agency), if it was subject to safeguards only by reason of such supply and if:

      1.   It was not *improved* while under safeguards; or
      2.   Any special fissionable material that was produced in it under safeguards has been separated out, or safeguards with respect to such produced material have been terminated; or

....

   e)   It has been transferred out of the State under paragraph 28(d), provided that such material shall again be subject to safeguards if it is returned to the State in which the Agency had safeguarded it.

....

E. TRANSFER OF SAFEGUARDED NUCLEAR MATERIAL OUT OF THE STATE

....

28.   No safeguarded *nuclear material* shall be transferred outside the jurisdiction of the State in which it is being safeguarded until the Agency has satisfied itself that one or more of the following conditions apply:

a) The material is being returned, under the conditions specified in paragraph 26(a), to the State that originally supplied it; or

b) The material is being transferred subject to the provisions of paragraph 24 or 25; or

c) Arrangements have been made by the Agency to safeguard the material in accordance with this document in the State to which it is being transferred; or

d) The material was not subject to safeguards pursuant to a *project agreement* and will be subject, in the State to which it is being transferred, to safeguards other than those of the Agency but generally consistent with such safeguards and accepted by the Agency.

ANNEX III

EXTRACTS FROM THE INFCIRC/153 (CORRECTED) PROVISIONS, 1983 EDITION

(Blue Book)

THE STRUCTURE AND CONTENT OF AGREEMENTS BETWEEN THE AGENCY AND STATES REQUIRED IN CONNECTION WITH THE TREATY ON THE NON-PROLIFERATION OF NUCLEAR WEAPONS

TRANSFER OF NUCLEAR MATERIAL OUT OF THE STATE

12.   The Agreement should provide, with respect to *nuclear material* subject to safeguards thereunder, for notification of transfers of such material out of the State, in accordance with the provisions set out in paragraphs 92-94 below. The Agency shall terminate safeguards under the Agreement on *nuclear material* when the recipient State has assumed responsibility therefor, as provided for in paragraph 91. The Agency shall maintain records indicating each transfer and, where applicable, the re-application of safeguards to the transferred *nuclear material*.

....

Special reports

68.   The Agreement should provide that the State shall make special reports without delay:

   a)   If any unusual incident or circumstances lead the State to believe that there is or may have been loss of *nuclear material* that exceeds the limits to be specified for this purpose in the Subsidiary Arrangements; or

   b)   If the containment has unexpectedly changed from that specified in the Subsidiary Arrangements to the extent that unauthorised removal of *nuclear material* has become possible.

....

INTERNATIONAL TRANSFERS

General

91.   The Agreement should provide that *nuclear material* subject or required to be subject to safeguards thereunder which is transferred internationally shall, for purposes of the Agreement, be regarded as being the responsibility of the State:

   a)   In the case of import, from the time that such responsibility ceases to lie with the exporting State, and no later than the time at which the *nuclear material* reaches its destination; and

   b)   In the case of export, up to the time at which the recipient State assumes such responsibility, and no later than the time at which the *nuclear material* reaches its destination.

The Agreement should provide that the States concerned shall make suitable arrangements to determine the point at which the transfer of responsibility will take place. No State shall be deemed to have such responsibility for *nuclear material* merely by reason of the fact that the nuclear material is in transit on or over its territory or territorial waters, or that it is being transported under its flag or in its aircraft.

Transfers out of the State

92.   The Agreement should provide that any intended transfer out of the State of safeguarded *nuclear material* in an amount exceeding one *effective kilogram*, or by successive shipments to the same State within a period of three months each of less than one *effective kilogram* but exceeding in total one *effective kilogram*, shall be notified to the Agency after the conclusion of the

contractual arrangements leading to the transfer and normally at least two weeks before the *nuclear material* is to be prepared for shipping. The Agency and the State may agree on different procedures for advance notification. The notification shall specify:

   a) The identification and, if possible, the expected quantity and composition of the *nuclear material* to be transferred, and the *material balance area* from which it will come;
   b) The State for which the *nuclear material* is destined;
   c) The dates on and locations at which the *nuclear material* is to be prepared for shipping;
   d) The approximate dates of dispatch and arrival of the *nuclear material*; and
   e) At what point of the transfer the recipient State will assume responsibility for the *nuclear material*, and the probable date on which this point will be reached.

93.　　The Agreement should further provide that the purpose of this notification shall be to enable the Agency if necessary to identify, and if possible verify the quantity and composition of, *nuclear material* subject to safeguards under the Agreement before it is transferred out of the State and, if the Agency so wishes or the State so requests, to affix seals to the *nuclear material* when it has been prepared for shipping. However, the transfer of the *nuclear material* shall not be delayed in any way by any action taken or contemplated by the Agency pursuant to this notification.

94.　　The Agreement should provide that, if the *nuclear material* will not be subject to Agency safeguards in the recipient State, the exporting State shall make arrangements for the Agency to receive, within three months of the time when the recipient State accepts responsibility for the *nuclear material* from the exporting State, confirmation by the recipient State of the transfer.

. . . .

Special reports

97.　　The Agreement should provide that in the case of international transfers a special report as envisaged in paragraph 68 above shall be made if any unusual incident or circumstances lead the State to believe that there is or may have been loss of *nuclear material*, including the occurrence of significant delay during the transfer.

ANNEX IV

EXTRACTS FROM THE AGREEMENT BETWEEN BELGIUM, DENMARK, THE FEDERAL REPUBLIC OF GERMANY, IRELAND, ITALY, LUXEMBOURG, THE NETHERLANDS, THE EUROPEAN ATOMIC ENERGY COMMUNITY AND THE AGENCY IN CONNECTION WITH THE TREATY ON THE NON-PROLIFERATION OF NUCLEAR WEAPONS

(So-called Verification Agreement, 5th April 1973)

TRANSFER OF NUCLEAR MATERIALS OUT OF THE STATES

Article 12

The Community shall give the Agency notification of transfers of nuclear material subject to safeguards under this Agreement out of the States, in accordance with the provisions of this Agreement. Safeguards under this Agreement shall terminate on nuclear material when the recipient State has assumed responsibility therefor as provided for in this Agreement. The Agency shall maintain records indicating each transfer and, where applicable, the re-application of safeguards to the transferred nuclear material.

....

TRANSFERS INTO OR OUT OF THE STATES

Article 91

General provisions

Nuclear material subject or required to be subject to safeguards under this Agreement which is transferred into or out of the States shall, for purposes of this Agreement, be regarded as being the responsibility of the Community and of the State concerned:

a) In the case of transfers into the States, from the time that such responsibility ceases to lie with the State from which the material is transferred, and no later than the time at which the material reaches its destination; and

b) In the case of transfers out of the States up to the time at which the recipient State has such responsibility, and no later than the time at which the nuclear material reaches its destination.

The point at which the transfer of responsibility will take place shall be determined in accordance with suitable arrangements to be made by the Community and the State concerned, on the one hand, and the State to which or from which the nuclear material is transferred, on the other hand. Neither the Community nor a State shall be deemed to have such responsibility for nuclear material merely by reason of the fact that the nuclear material is in transit on or over a State's territory, or that it is being transported on a ship under a State's flag or in the aircraft of a State.

TRANSFERS OUT OF THE STATES

Article 92

a) The Community shall notify the Agency of any intended transfer out of the States of nuclear material subject to safeguards under this Agreement if the shipment exceeds one effective kilogram, or, for facilities which normally transfer significant quantities to the same State in shipments each not exceeding one effective kilogram, if so specified in the Subsidiary Arrangements.

b) Such notification shall be given to the Agency after the conclusion of the contractual arrangements leading to the transfer and within the time limit specified in the Subsidiary Arrangements.

c) The Agency and the Community may agree on different procedures for advance notification.

d) The notification shall specify:

    i) The identification and, if possible, the expected quantity and the composition of the nuclear material to be transferred, and the material balance area from which it will come;

    ii) The State for which the nuclear material is destined;

    iii) The dates on and locations at which the nuclear material is to be prepared for shipping;

    iv) The approximate dates of dispatch and arrival of the nuclear material; and

    v) At what point of the transfer the recipient State will assume responsibility for the nuclear material for the purpose of this Agreement, and the probable date on which that point will be reached.

Article 93

The notification referred to in Article 92 shall be such as to enable the Agency to make, if necessary, an ad hoc inspection to identify, and if possible verify the quantity and composition of the nuclear material before it is transferred out of the States, except for transfers within the Community and, if the Agency so wishes or the Community so requests, to affix seals to the nuclear material when it has been prepared for shipping. However, the transfer of the nuclear material shall not be delayed in any way by any action taken or contemplated by the Agency pursuant to such a notification.

Article 94

If nuclear material will not be subject to Agency safeguards in the recipient State the Community shall make arrangements for the Agency to receive within three months of the time when the recipient State accepts responsibility for the nuclear material, confirmation by the recipient State of the transfer.

TRANSFERS INTO THE STATES

Article 95

a) The Community shall notify the Agency of any expected transfer into the States of nuclear material required to be subject to safeguards under this Agreement if the shipment exceeds one effective kilogram, or, for facilities to which significant quantities are normally transferred from the same State in shipments each not exceeding one effective kilogram, if so specified in the Subsidiary Arrangments.

b) The Agency shall be notified as much in advance as possible of the expected arrival of the nuclear material, and in any case within the time limits specified in the Subsidiary Arrangements.

c) The Agency and the Community may agree on different procedures for advance notification.

d) The notification shall specify:

    i) The identification and, if possible, the expected quantity and composition of the nuclear material;

    ii) At what point of the transfer the Community and the State concerned will have responsibility for the nuclear material for the purpose of this Agreement, and the probable date on which that point will be reached; and

    iii) The expected date of arrival, the location where, and the date on which, the nuclear material is intended to be unpacked.

Article 96

The notification referred to in Article 95 shall be such as to enable the Agency to make, if necessary, an ad hoc inspection to identify, and if possible verify the quantity and composition of the nuclear material transferred into the States, except for transfers within the Community, at the time the consignment is unpacked. However, unpacking shall not be delayed by any action taken or contemplated by the Agency pursuant to such a notification.

Article 97

Special reports

The Community shall make a special report as envisaged in Article 68 if any unusual incident or circumstances lead the Community to believe that there is or may have been loss of nuclear material, including the occurrence of significant delay, during a transfer into or out of the States.

ANNEX V

EXTRACTS FROM THE COMMISSION OF THE EUROPEAN COMMUNITIES' REGULATION (EURATOM) No. 3227/76 CONCERNING THE APPLICATION OF THE PROVISIONS ON EURATOM SAFEGUARDS

(19th October 1976)

TRANSFERS: IMPORTS/EXPORTS

(Part III)

Article 24

a)   The persons and undertakings referred to in Article 1 which export source or special fissile materials to a non-Member State shall give advance notification to the Commission of every such export. Similarly, advance notification shall be given to the Commission:

— in the case of any export from a Member State party to the Agreement to a Member State not party to the Agreement, and

— in the case of any export from the United Kingdom to a Member State party to the Agreement.

However, advance notification is required only:

   i) where the consignment exceeds one effective kilogramme;

   ii) where the "particular safeguard provisions" referred to in Article 7 so specify, in the case of installations habitually transferring large total quantities of materials to the same State, even though no single consignment exceeds one effective kilogramme.

b)   Such notification shall be given after the conclusion of the contractual arrangements leading to the transfer and in any case in time to reach the Commission eight working days before the material is to be prepared for shipment.

c)   Such notification shall be given in accordance with the form set out in Annex V to this Regulation and shall state, *inter alia*,

— the identification and, if possible, the expected quantity and composition of the material to be transferred, and the material balance area from which it will come,

— the State to which the nuclear material is to be sent,

— the dates on and locations at which the nuclear material will be prepared for shipment,

— the approximate dates of dispatch and arrival of the nuclear material,

— the use which the persons or undertakings concerned had made of the nuclear material.

d)   If so required for reasons of physical protection, special arrangements concerning the form and transmission of such notification may be agreed upon with the Commission.

Article 25

a)   The persons and undertakings referred to in Article 1 which import source or special fissile materials from a non-member State shall give advance notification to the Commission of every such import. Similarly, advance notification shall be given to the Commission:

— in the case of any import into a Member State party to the Agreement from a Member State not party to the Agreement, and

— in the case of any import into the United Kingdom from a Member State party to the Agreement.

However, advance notification is required only:

   i) where the consignment exceeds one effective kilogramme;

   ii) where the "particular safeguard provisions" referred to in Article 7 so specify, in the case of installations to which large total quantities of materials are habitually transferred from the same State, even though no single consignment exceeds one effective kilogramme.

b) Such notification shall be given as far in advance as possible of the expected arrival of the nuclear material and, in any case, on the date of receipt and in time to reach the Commission five working days before the material is unpacked.

c) Such notification shall be given in accordance with the form set out in Annex VI and shall state, *inter alia*:
  — the identification and, if possible, the expected quantity and composition of the material,
  — the expected date of arrival, the location where and the date on which the nuclear material is expected to be unpacked.

d) If so required for reasons of physical protection, special arrangements concerning the form and transmission of such notification may be agreed upon with the Commission.

Article 26

When persons or undertakings not subject to Article 1 decide to export or import nuclear materials referred to in Articles 24 and 25, these persons or undertakings are required to make the notifications foreseen in Articles 24 and 25.

Article 27

A special report as provided for in Article 17 shall be prepared by the persons or undertakings covered by Articles 24 and 25 if, following exceptional circumstances or an incident, they have received information that nuclear materials have been or appear to be lost, particularly when there has been a considerable delay during transfer. In the same circumstances persons or undertakings covered by Article 26 are also required to inform the Commission.

Article 28

Any change of date in the preparation for shipment, in the shipment or in the unpacking of nuclear materials with respect to the dates given in the notifications provided for in Articles 24 and 25, but not a change that gives rise to special reports, shall be communicated without delay, with an indication of the revised dates, if known.

....

SPECIFIC PROVISIONS APPLICABLE IN THE TERRITORIES OF MEMBER STATES WHICH ARE NUCLEAR-WEAPON STATES

Article 35

1. The provisions of this Regulation shall not apply:
    a) to installations or parts of installations which have been assigned to meet defence requirements and which are situated on the territory of a Member State not party to the Agreement; or
    b) to nuclear materials which have been assigned to meet defence requirements by that Member State.

2. For nuclear materials, installations or parts of installations which are liable to be assigned to meet defence requirements and which are situated on the territory of a Member State not party to the Agreement, the extent of the application of this Regulation and the procedures under it shall be defined by the Commission in consultation and in agreement with the Member State concerned, taking into account the provisions of the second paragraph of Article 84 of the Treaty.

3. It is understood in any event that:
    a) the provisions of Articles 1 to 4, 7 and 8 shall apply to installations or parts of installations which at certain times are operated exclusively with nuclear materials liable to be assigned to meet defence requirements but at other times are operated exclusively with civil nuclear materials;
    b) the provisions of Articles 1 to 4, 7 and 8 shall apply, with exceptions for reasons of national security, to installations or parts of installations to which access could be restricted for such reasons but which produce, treat, separate, reprocess or use in any other way simultaneously both civil nuclear materials and nuclear materials assigned or liable to be assigned to meet defence requirements;
    c) the provisions of Articles 6, and 9 to 37 shall apply in relation to all civil nuclear materials situated in installations or parts of installations as referred to in subparagraphs (a) and (b) above.

ANNEX V

COMMISSION OF THE EUROPEAN COMMUNITIES
EURATOM SAFEGUARDS

ADVANCE NOTIFICATION OF EXPORTS OF NUCLEAR MATERIAL

(1) Material balance area

Code : ....

(2) Installation:
     (Shipper)

Installation :
(Receiver)

....

....

(3) Quantities: ....

....

(4) Chemical composition: ....

(5) Enrichment or isotopic composition: ....

(6) Physical form: ....

(7) Number of items: ....

(8) Description of containers and seals:..

....

(9) Shipment identification data: ....

(10) Means of transport: ....

(11) Location where material will be stored or prepared: ....

....

(12) Last date when material can be identified: ....

(13) Approximate dates of dispatch: ....

       Expected dates of arrival: ....

(14) Use: ....

....

(15) International agreement: . ...

       — Commission authorization: ....

       — Intervention of Supply Agency: ....

Name and position of responsible signatory:

....

Date and place of dispatch of notification:

....

Signature: ....

EXPLANATORY NOTES

(1) Code of the reporting material balance area notified to the installation concerned in the particular safeguard provisions.

(2) Name, address and country of the installation shipping and of the installation receiving the nuclear material. In the case of export out of the United Kingdom, the receiver of ultimate destination should also be indicated where applicable.

(3) The total weight of the elements shall be identified in kilogrammes for natural and depleted uranium and for thorium, and in grammes for enriched uranium and plutonium. The weight of fissile isotopes shall be identified, if applicable.

(4) Chemical composition shall be identified.

(5) If applicable, the degree of enrichment or the isotopic composition shall be identified.

(6) Use the description of materials as laid out in Annex II (10) to this Regulation.

(7) The number of items included in the shipment shall be identified.

(8) Description (type) of containers, including features that would permit sealing.

(9) Shipment identification data (e.g. container markings or numbers).

(10) Indicate, as applicable, the means of transport.

(11) Indicate the location within the material balance area where the nuclear material is prepared for shipping and can be identified, and where its quantity and composition can if possible be verified.

(12) Last date when material can be identified and when its quantity and composition can if possible be verified.

(13) Approximate dates of dispatch and of expected arrival at destination.

(14) Indicate the use to which the nuclear material has been assigned.

(15) Give in particular as appropriate :

— the Agreement concluded by the Community with a non-Member State or an international organization under which the material is transferred;

— the Commission authorization under Article 59 of the Treaty;

— the date on which the contract was concluded or considered as concluded by the Supply Agency and any useful references;

— for jobbing contracts (Article 75 of the Treaty) and for contracts for the supply of small quantities of material (Article 74 of the Treaty and Commission Regulation No 17/66/Euratom, as amended by Regulation (Euratom) No 3137/74), date of notification to the Supply Agency and any useful references.

NB: Pursuant to Article 79 of the Treaty, those subject to safeguard requirements shall notify the authorities of the Member State concerned of any communications they make to the Commission pursuant to that Article.

This form, duly completed and signed, must be forwarded to the Commission of the European Communities, Euratom Safeguards Directorate, 'Jean Monnet' Building, Kirchberg, Luxembourg (Grand Duchy of Luxembourg).

COMMISSION OF THE EUROPEAN COMMUNITIES
EURATOM SAFEGUARDS

ADVANCE NOTIFICATION OF IMPORTS OF NUCLEAR MATERIAL

(1) Material balance area

Code: ....

(2) Installation: Installation:
(Receiver)   (Shipper)

....

....

(3) Quantities: ....

....

(4) Chemical composition: ....

(5) Enrichment or isotopic composition: ....

(6) Physical form: ....

(7) Number of items: ....

(8) Description of containers and seals: ....

....

(9) Means of transport: ....

(10) Date of arrival: ....

(11) Location where materials will be unpacked ....

(12) Date(s) when material will be unpacked: ....

(13) — International agreement: ....

— Intervention of Supply Agency: ....

Name and position of responsible signatory:

....

Date and place of dispatch of notification:

....

Signature: ....

EXPLANATORY NOTES

(1) Code of the reporting material balance area notified to the installation concerned in the particular safeguard provisions.

(2) Name, address and country of the installation receiving and of the installation shipping the nuclear material.

(3) The total weight of the elements shall be identified in kilogrammes for natural and depleted uranium and for thorium, and in grammes for enriched uranium and plutonium. The weight of fissile isotopes shall be identified if applicable.

(4) Chemical composition shall be identified.

(5) If applicable, the degree of enrichment or the isotopic composition shall be identified.

(6) Use the description of materials as laid out in Annex II (10) to this Regulation.

(7) The number of items included in the shipment shall be identified.

(8) Description (type) of containers, and if possible, of the seals affixed.

(9) Indicate as applicable, the means of transport.

(10) Expected or actual date of arrival in the reporting material balance area.

(11) Indicate the location within the material balance area where the material will be unpacked and can be identified and where its quantity and composition can be verified.

(12) Date(s) when material will be unpacked.

(13) Give in particular as appropriate:

— the Agreement concluded by the Community with a non-Member State or an international organization under which the material is transferred;

— the date on which the contract was concluded or considered as concluded by the Supply Agency and any useful references;

— for jobbing contracts (Article 75 of the Treaty) and for contracts for the supply of small quantities of material (Article 74 of the Treaty and Commission Regulation No 17/66/Euratom, as amended by Regulation (Euratom) No 3137/74), date of notification to the Supply Agency and any useful references.

NB: Pursuant to Article 79 of the Treaty, those subject to safeguard requirements shall notify the authorities of the Member State concerned of any communications they make to the Commission pursuant to that Article.

This form, duly completed and signed, must be forwarded to the Commission of the European Communities, Euratom Safeguards Directorate, 'Jean Monnet' Building, Kirchberg, Luxembourg (Grand Duchy of Luxembourg).

ANNEX VII

COMMISSION OF THE EUROPEAN COMMUNITIES
EURATOM SAFEGUARDS

(1) DECLARATION OF ORE SHIPMENTS/EXPORTS

(2) Undertaking: ....

(3) Mine: .... (4) Code: ....

Year: ....

Date	Consignee	Quantity contained in kg:		Remarks
		of uranium	of thorium	

Date and place of dispatch of declaration: Name and position of responsible signatory:

....

Signature: ....

74

EXPLANATORY NOTES

(1) The declaration of shipments is to be made at the latest at the end of January of each year for the previous year separately for each consignee. The declaration of export is to be made for each export at the date of shipment.

(2) Name and address of the reporting undertaking.

(3) Name of the mine in respect of which the declaration is made.

(4) Code of the mine as notified to the undertaking by the Commission.

NB: Pursuant to Article 79 of the Treaty, those subject to safeguard requirements shall notify the authorities of the Member State concerned of any communications they make to the Commission pursuant to that Article.

This form, duly completed and signed, must be forwarded to the Commission of the European Communities, Euratom Safeguards Directorate, 'Jean Monnet' Building, Kirch-Luxembourg (Grand Duchy of Luxembourg).

CHAPTER FIVE

REINFORCING THE CONTROL OF NUCLEAR TRADE AND ASSURING SUPPLIES

A. THE NUCLEAR EXPORTERS COMMITTEE — ZANGGER COMMITTEE

I. THE FRAMEWORK FOR NEGOTIATIONS

After the entry into force, on 5th March 1970, of the Treaty on the Non-Proliferation of Nuclear Weapons, a number of States which were either already, or intending to become Party to the Treaty, and actual or potential exporters of nuclear materials and equipment, decided to enter into consultations; their goal was to reach agreement on the procedures and rules to apply to their exports to non-nuclear-weapon States, in order to meet the obligations laid down in Article III.2 of the Treaty[1].

These discussions took place, outside the framework of the IAEA, in a Committee chaired by a Swiss expert, Professor Claude Zangger. Based on the consensus they reached, participating States agreed to implement their decision by means of a letter signed, in each case, by the Resident Representative to the IAEA of the State concerned, addressed to the Director General of the Agency. These letters, dated 22nd August 1974, were sent together with attachments: Memorandum A, Memorandum B and an Annex. The Agency was requested by these States to communicate the letters and documents attached to all Member States (reproduced in document INFCIRC/209 of 3rd September 1974, "Communications received from Members regarding the Export of Nuclear Material and of certain Categories of Equipment and other Material"). Subsequently, other States sent similar letters to the Director General of the Agency (see *annexes I and II* to this chapter).

II. DECLARATIONS OF POLICY

In each of the above-mentioned letters, written in identical terms, the Government of the country in question indicated that it had been called upon to examine procedures in relation to exports:
— of source or special fissionable material,
— of equipment and material especially designed or prepared for the processing, use or production of special fissionable material,

in the light of its commitment under Article III paragraph 2 of the NPT not to provide such items to any non-nuclear-weapon State for peaceful purposes, unless the source or special fissionable material is subject to safeguards under an agreement with the IAEA.

The letters finished by saying that the Governments in question had decided to act in accordance with the provisions of the memoranda attached.

On the same day, two Member States of the European Community (Denmark and the United Kingdom) sent complementary letters specifying that, so far as trade within the European Community was concerned, they would implement paragraphs 5 of both memoranda, relating to retransfers, in the light of their commitments under the Treaty of Rome. Two other Member States of the European Community (the Netherlands and the Federal Republic of Germany) chose rather to send, again on the same day, a single letter, combining the common model with the said reservation about commitments under the Treaty of Rome.

For its part, the United States sent two complementary letters. In the first, also dated 22nd August 1974, the United States Government declared its intention to apply paragraph 6 of Memorandum B by extending the safeguards requirement to equipment and material, exported from the United States, in addition to those specified in paragraph 2 of the Memorandum, i.e. the "Trigger List". In the second, dated 3rd October 1974, the United States Government indicated that deliveries provided for under contracts concluded pursuant to agreements in force with the European Community would be carried out in the light of the expected entry into force of the so-called Verification Agreement concluded on 5th April 1973 between the IAEA, EURATOM and seven EURATOM Member States Party to the NPT. By letter, also dated 3rd October 1974, the USSR stressed the importance it attached to completion of the process of accession to the Treaty by the Member States of the European Community which had signed it, and to the entry into force of the said Safeguards Agreement (INFCIRC/209/Add.2).

III. MEMORANDUM A

Memorandum A, which is attached to the declarations of policy, deals with source and special fissionable material, and in explaining these terms, adopts the definition contained in Article XX of the IAEA Statute[2]. Memorandum A states that "the Government is solely concerned with ensuring .... the application of safeguards in non-nuclear-weapon States not party to the Treaty on the Non-Proliferation of Nuclear Weapons with a view to preventing diversion of the safeguarded nuclear material from peaceful purposes to nuclear weapons or other nuclear explosive devices".

Consequently, each Government undertakes to specify to the recipient State, as a condition of supply, that the source or special fissionable material, or special fissionable material produced in or by the use thereof, shall not be diverted to nuclear weapons or other nuclear explosive devices. It also undertakes to satisfy itself that safeguards to that end, under an agreement with the Agency and in accordance with its safeguards system, will be applied to the source or special fissionable material in question. As regards direct exports, authorisation for export will not be given unless the Government is satisfied that the safeguards agreement with the Agency will apply, at the latest, when the material reaches its destination. It will also require satisfactory assurances that safeguards will apply in the event of material being re-exported (re-transferred) to a non-nuclear-weapon State not party to the NPT.

Paragraph 6 of the Memorandum provides, however, that small quantities of material, delivered within a period of twelve months, shall be disregarded for the purpose of these procedures.

IV. MEMORANDUM B

Memorandum B deals with equipment or material especially designed or prepared for the processing, use or production of special fissionable material, and includes:

i) a "Trigger List" of such equipment, with clarifications of some of the items contained in an Annex. This List includes first of all reactors and equipment therefor, but non-nuclear materials are also included, i.e. deuterium and heavy water, graphite and also reprocessing plants, plants for the fabrication of fuel elements and equipment especially designed or prepared for the separation of isotopes of uranium;

ii) a statement of the main principles for granting export and retransfer authorisations similar to those mentioned above under Memorandum A.

Paragraph 6 of Memorandum B stipulates that "the Government" reserves to itself complete discretion as to the interpretation and implementation of its commitment to subject the exports in question to the application of safeguards, and as to the right to require that safeguards be applied to items other than those specified in the Trigger List.

V. RESUMPTION OF ZANGGER COMMITTEE ACTIVITIES AFTER 1978

After a period during which co-ordination of nuclear export policies was effected within the framework of the London Club, the countries participating in the Zangger Committee work started consultations again.

Memorandum B was clarified and completed, by identical letters dated 1st September 1978 addressed to the Director General of the IAEA and distributed by the Agency to all its Member States (INFCIRC/209/Mod. 1, December 1978 — see annex III to this chapter), with regard to zirconium tubes, plants for the production of heavy water, deuterium and deuterium compounds and certain items of equipment for isotope separation plants.

Letters were again sent to the Director General, as from January 1984, developing and clarifying the Trigger List with regard to assemblies and components especially designed or prepared for use in isotope separation facilities using the gas centrifuge enrichment process (INFCIRC/209/Mod. 2 — February 1984 — see annex IV to this chapter).

Lastly, on 1st July 1985, the Annex to a new series of letters clarified provisions concerning equipment performing essential tasks in reprocessing plants (INFCIRC/209/Mod. 3 — see annex V to this chapter).

It should be noted that in the letters of January 1984 and 1st July 1985, each Government reserved to itself discretion as to the interpretation and implementation of the procedures provided for in the documents and the right to control, if it wished, the export of items, used either in the gas centrifuge enrichment process or in reprocessing plants, other than those specified in the attached documents.

Discussions on other types of installations are currently proceeding in the Zangger Committee, which includes twenty-one countries to date.

B. NUCLEAR SUPPLIERS GROUP — LONDON CLUB

I. THE FRAMEWORK FOR NEGOTIATIONS

On the initiative of the United States, seven countries met in London early in 1975 to make up the Nuclear Suppliers' Group more commonly known as the "London Club". The purpose of the consultations within the expert group is to try to harmonise export policies from the safeguards and control angle for transfers of "nuclear items", outside the framework of the IAEA and the NPT; this also applies to technology transfers, control of retransfers and physical protection. The Indian explosion in 1974 and more recently, the interest accorded to physical protection had brought to light certain gaps in the NPT regarding this latter question. Having succeeded, in 1976, in negotiating an outline agreement, these countries then invited eight other States to participate in their consultations (see *annex VI* to this chapter). A compromise was reached in September 1977 (certain participating countries would have preferred full scope safeguards), following which the States involved, in the absence of a formal agreement, adopted the procedure of parallel unilateral commitments previously used by the Zangger Committee. To this end, on 11th January 1978, each of the States involved sent a letter to the Director General of the IAEA to inform him of its decision to act in accordance with the principles contained in the document attached to the letter with regard to the export of nuclear material, equipment or technology, and asking him to inform all Member States of this fact (INFCIRC/254, February 1978, "Guidelines for the Export of Nuclear Material, Equipment or Technology" — see *annex VII* to this chapter). Subsequently, a number of States which had not taken part in the work of the London Club informed the Director General of the IAEA that they approved of this initiative and had also decided to act in accordance with the above-mentioned principles. More recently, certain countries have published measures supplementing the "London Club" Guidelines[3].

II. DECLARATIONS OF POLICY AND RESERVATIONS

It should be noted that the decision to act in accordance with the said principles is couched in identical or similar terms in the above-mentioned letters. It was agreed that any changes to the Guidelines require unanimous consent, especially those which might arise from a reconsideration of common safeguards requirements. Either in their principal letter or an additional one, certain States, however, expressed reservations or made known their intention to impose additional requirements.

In some cases, full implementation of the London Club Guidelines meant that countries had to change their domestic legislation, and they therefore made an appropriate reservation.

With the exception of France, the Member States of the European Community specified that they would apply the Guidelines to intra-Community trade in the light of commitments undertaken by virtue of the EURATOM Treaty. Furthermore, the USSR and certain socialist countries expressed their intention to make the supply to a non-nuclear-weapon State of the items mentioned in the Trigger List conditional upon the application of IAEA safeguards to all nuclear activities in the State in question. For its part, Canada reserved the right to act in accordance with "other principles considered pertinent by it". Similarly, Australia reserved the right to take certain other requirements into account. The United States adopted a different approach from that taken following the initial work of the Zangger Committee and did not introduce any reservation in its letters, but in fact subsequently introduced in its legislation (NNPA) stricter requirements than those in the Guidelines.

At a more general level, it may be noted that the Guidelines represent consensus at the level of the lowest common denominator without preventing Governments from applying a wide interpretation. Thus, the Annex entitled "Clarifications of Items on the Trigger List", which applies in theory to the export of the whole set of major items making up reactors, pressure vessels, and fuel processing or fabrication plants, allows Governments to apply the procedures of the Guidelines to separate deliveries of items of such reactors, vessels or plants.

III. DOCUMENTS ATTACHED TO THE DECLARATIONS OF POLICY

There are three documents attached to the letters sent by the above-mentioned States:

— an *Appendix* entitled "Guidelines for Nuclear Transfers";
— *Annex A* entitled "Trigger List referred to in Guidelines", to which is attached a separate Annex entitled "Clarifications of Items on the Trigger List";
— *Annex B* entitled "Criteria for Levels of Physical Protection", to which is attached a Table categorising nuclear material.

The purpose of the *Guidelines* is to lay down the fundamental principles relating to the safeguards and controls which must be applied to any export of nuclear items to non-nuclear-weapon States. They are based on the Trigger List drawn up by the supplier States which took part in the work of the London Club and on common criteria, agreed between these States, relating to technology transfers. In addition, they define application procedures for certain special cases and provide for "supporting activities".

It is recalled that any changes to these Guidelines, especially those which might result from a reconsideration of common safeguards requirements, require unanimous consent. However, exceptions made to safeguards duration and coverage provisions only require prior consultation with the parties to this understanding.

1. The "fundamental principles" of general application

In the first place, the Guidelines provide that suppliers should authorise transfer of items identified in the Trigger List only upon formal governmental assurances from recipients explicitly excluding uses which would result in any nuclear explosive device.

Secondly, the Guidelines reflect the concern, relatively recent at that time, about physical protection and lay down the principle that all nuclear materials and facilities identified in the Trigger List should be placed under effective physical protection to prevent unauthorised use and "handling". These provisions of the Guidelines, together with the criteria for levels of physical protection constituting Annex B, will be examined in chapter seven of this Study.

Thirdly, the Guidelines lay down the principle that suppliers should transfer Trigger List items only when these are covered by IAEA safeguards, with duration and coverage provisions in accordance with the Recommendations already contained in a Memorandum from the Director General of the IAEA to the Board of Governors (GOV/1621 of 20th August 1973) aiming to encourage more standardised safeguards agreements as concerns their duration and termination (see *annex VIII* to this chapter).

In order to harmonize the duration and termination clauses of safeguards agreed within the framework of the pre-NPT IAEA system (INFCIRC/66/Rev.2), the Memorandum provides:

— that the duration of the safeguards agreement should be related to the period of actual use of the supplied items in the recipient State; and
— that the provisions for terminating the agreement should be formulated in such a way that the rights and obligations of the Parties continue to apply in connection with supplied nuclear material and with special fissionable material obtained, produced, processed or used in or in connection with supplied nuclear material, equipment, facilities or non-nuclear material (or obtained through such items), until such time as the Agency has terminated the application of safeguards thereto, in accordance

with its Safeguards System (return to supplier country or transfer under safeguards to a third country; no longer usable for any nuclear activity).

Exceptions should be allowed only after consultations between the Parties.

In conclusion, provision is made to apply the whole safeguards system of the Agency, but only to materials and equipment supplied. It is only by means of a joint reconsideration of the Guidelines that an integral safeguards system could be introduced as a common requirement.

2. Clarifications regarding the conditions for applying the principles (problem of "sensitive" exports)

The original feature of the London Club Guidelines is that they have created a category of "sensitive" materials, equipment and technology whose export is restricted and subject to additional, particularly restrictive conditions. This is a compromise between countries which, like the United States, wished to prohibit such exports, and other countries in favour of recognising their validity in accordance with Article IV of the NPT. From that angle the interest of the Guidelines is that they have clarified the conditions for applying the three above-mentioned principles (no uses which would result in any nuclear explosive device; application of physical protection measures and safeguards) to the most sensitive situations, given the general interests which these principles are intended to safeguard. These situations are the following:

i) Safeguards brought into question by the transfer of certain technology (paragraph 6 of the Appendix)

The technology transfers involved concern facilities for reprocessing, enrichment, or heavy-water production, utilising technology directly transferred by the supplier or derived from transferred facilities or "major critical components thereof" (see chapter nine on definitions in this Study).
Such transfers of technology, facilities or "major critical components thereof" should only be effected if assurance has been given that:
— IAEA safeguards will be applied to any facility of the same type, i.e. based on the same or similar physical or chemical processes, constructed during an agreed period in the recipient country (by duplication or derivation); and
— there will at all times be in effect an agreement permitting the IAEA to apply Agency safeguards to such facilities using transferred technology, as identified by the recipient or the supplier, in which latter case after consultation with the recipient.

ii) Special controls on sensitive exports (paragraph 7)

This paragraph concerns the transfer of sensitive materials, facilities, technology and weapons-usable materials. In the absence of a precise definition of these terms it can, however, be deduced from paragraph 7 that they include, in the first place, facilities, equipment or technology for enrichment, or reprocessing. As regards materials which, because of their nature or degree of enrichment, have a military quality, they too are not explicitly defined but the context (in particular paragraph 8 of the Appendix and the categorisation of materials for physical protection purposes) indicates sufficiently that this concerns unirradiated plutonium (except that with a high Pu 238 content) and uranium enriched to 20 per cent or more. Although paragraph 7 does not expressly refer to heavy-water production facilities it would seem that such facilities should also be considered as sensitive exports for application of controls as provided by this paragraph. "Sensitive" technologies are those referred to in paragraph 6 of the Guidelines and defined in Annex A, Part B, paragraph 1 thereof (Trigger List).

As for the transfers relating to enrichment and reprocessing, the basic principle is that suppliers should limit such transfers and that if, nevertheless, they are to be effected, suppliers should encourage recipients to accept, as an alternative to the construction of plants on a strictly national basis, supplier involvement and/or other appropriate multinational participation in resulting facilities. Suppliers should also promote projects for multinational fuel cycle centres (see section E, below).

iii) Special controls on the export of enrichment facilities, equipment and technology (paragraph 8)

Under the Guidelines, the recipient nation must agree that neither the transferred facility, nor any facility based on the technology involved, will be designed or operated for the production of uranium enriched to more than 20 per cent enriched uranium without the consent of the supplier nation, of which the IAEA should be advised.

iv) Controls on supplied or derived weapons-usable material (paragraph 9)

Suppliers recognise the importance of including in agreements on supply of nuclear materials or of facilities which produce weapons-usable material, provisions "calling for" mutual agreement between the supplier and recipient on arrangements for reprocessing, storage, alteration, use, transfer or retransfer of any weapons-usable material involved. This complex phrasing in effect concerns "prior consent" to reprocessing which certain countries wished to impose when supplying reactors or their fuels. However, since no consensus was reached in this respect, the last sentence of paragraph 9 provides simply that suppliers "should endeavour to include such provisions whenever appropriate and practicable".

v) Controls on retransfer (paragraph 10)

Suppliers should transfer Trigger List items, including technology in respect of reprocessing, enrichment, or heavy-water production facilities, only upon assurance that in the case of:

— retransfer of such items; or
— transfer of Trigger List items derived from facilities originally transferred by the supplier, or with the help of equipment or technology originally transferred by the supplier,

the recipient will have provided the same assurances as those required by the supplier for the original transfer.

In addition, the supplier's consent should be required for:

— any retransfer of the facilities, "major critical components" or technology relating to reprocessing, enrichment, or heavy-water production facilities;
— any transfer of facilities or "major critical components" derived from those items;
— any retransfer of heavy water or weapons-usable material.

3. Supporting activities

The primary objective of suppliers is to promote international co-operation concerning exchanges of "physical security" information, a term which seems, given the context, to mean a facet of physical protection, indicating measures intended to make it more difficult to use nuclear materials for explosive purposes; such exchanges cover both the protection of nuclear materials in transit, and recovery of stolen nuclear materials and equipment (see chapter eight of this Study).

In addition, suppliers should make "special efforts" in support of effective implementation of IAEA safeguards, and should also support the Agency's efforts to increase the technical effectiveness of safeguards and to assist Member States in the improvement of their national systems of accounting and control of nuclear materials, systems on which safeguards are in part based.

Furthermore, each supplier State is obliged to encourage the "designers" and makers of sensitive equipment to construct it in such a way as to facilitate the application of safeguards.

When considering transfers, suppliers are also asked to exercise prudence, having regard to all the circumstances of each case, and should in particular assess whether there is any risk that technology transfers or retransfers, other than sensitive technology (reprocessing, enrichment, or heavy-water production), might result in the production of unsafeguarded nuclear materials. Lastly, suppliers should consult through regular channels, either amongst themselves on questions relating to the implementation of the Guidelines, or with the

Governments concerned on specific cases which, due to a risk of conflict or international disorders, are particularly sensitive.

In the event of a breach of supplier/recipient agreements relating to the implementation of the Guidelines — particularly in the case of an explosion of a nuclear device, or illegal termination or violation of IAEA safeguards by a recipient — suppliers should consult promptly through diplomatic channels in order to determine and assess the reality and extent of the alleged breach, and agree, in the spirit of Article XII of the IAEA Statute concerning safeguards, on an appropriate response, which could include termination of nuclear transfers to the recipient in question.

4. Trigger List and clarifications of items on the List

Annex A to the Guidelines constitutes the "Trigger List" referred to therein. An Annex to Annex A in turn clarifies certain items on the Trigger List.

The Trigger List itself is divided into two parts: *Part A* deals with material and equipment. Source or special fissionable material is defined as in Article XX of the IAEA Statute. However, "small quantities" exported to a given country within a period of twelve months are exempted from application of the Guidelines.

The list of equipment and non-nuclear materials for reactors, together with the above-mentioned "small quantities", is copied from that in Memorandum B of the Zangger Committee of 22nd August 1974 (see section A above); like this Memorandum, the list also includes plants for the fabrication and reprocessing of fuel elements as well as "equipment ... especially designed or prepared for the separation of isotopes of uranium". also included in the list are "plants for the production of heavy water, deuterium and deuterium compounds and equipment especially designed or prepared therefor". This was no doubt added as a result of the explosion, between the dates when the two documents were drawn up, of a nuclear device in India.

Part B of the Trigger List (Annex A) lays down the common criteria for the transfer of sensitive technology. It will be recalled that paragraph 6 of the Guidelines, in making the transfer of technology relating to facilities for reprocessing, enrichment or heavy-water production subject to conditions concerning the prohibition on obtaining nuclear explosive devices, on physical protection measures and on the application of safeguards, emphasizes the need to extend these obligations to "major critical components", to technology resulting from transferred facilities and to facilities "of the same type", constructed in the recipient country during an agreed period.

The purpose of Part B is to define the concepts underlying the requirements laid down by the above-mentioned paragraph 6 of the Guidelines. To this end, Part B specifies the meaning to be given to the word "technology" and the terms "major critical components" and facilities "of the same type", it being made clear that these definitions should be interpreted differently from those applicable to Part A (see chapter nine on definitions in this Study). In addition, criteria are laid down which give a fairly wide interpretation of the scope of paragraph 6 in two situations which might have given rise to doubt:

— firstly, by specifying that a transfer in the aggregate of a significant fraction of the items essential to the operation of a facility referred to in paragraph 6, together with the know-how for construction and operation of that facility shall be deemed to be a transfer within the meaning of paragraph 6, even though no major critical component is included in the transfer; and secondly,

— by specifying that a similar approach would be applied to define facilities "of the same type" and "major critical components" in cases where methods different from those based on physical or chemical processes already defined as "of the same type" for the design, construction and operation of reprocessing, enrichment or heavy-water production facilities.

Furthermore, Part B lays down a minimum period of twenty years, from the date of the commissioning of the facility in question, during which IAEA safeguards must be applied to facilities of the same type as the transferred facility, and built in the recipient country.

The Annex to Annex A contains clarifications of items on the Trigger List. The purpose of this Annex is essentially to identify the parts which make up all the major components of a reactor, pressure vessel, reprocessing plant, etc. As already indicated, the Guidelines apply as a rule to the whole set of these items, with Governments reserving to themselves the right to apply the procedures to an item delivered separately.

5. Annex B

Annex B, which constitutes the last part of the Guidelines, lays down the criteria for levels of physical protection. This topic is dealt with in chapter eight, below.

C. LEGAL NATURE OF THE COMMITMENTS UNDERTAKEN PURSUANT TO THE WORK OF THE ZANGGER COMMITTEE AND THE LONDON CLUB

The legal scope of these commitments, especially the London Club Guidelines, has been the subject of conjecture. Depending on the view taken, the Guidelines either do or do not give rise to obligations under international law:

— in the first place, between the States having undertaken these commitments;

— in the second place, between these States and those IAEA Member States which did not do so.

Furthermore, if it is felt that the Guidelines do give rise to obligations, the precise nature and extent of such obligations should be clarified.

The same problem would seem to arise with regard to the commitments undertaken pursuant to the work of the Zangger Committee. The use of the word "commitment" has been adopted here in the absence of a more neutral term, and should not be taken as prejudging the nature and legal effects of the undertakings involved.

In the abundant literature to which the London Club Guidelines have given rise, certain writers have described them as a "gentleman's agreement", or likened them to what has been termed "international soft law".

"Gentleman's agreement" should of course not be taken in this case to mean a personal arrangement by negotiators which is not legally binding on the States represented, as was its original meaning. For, the term has evolved to include an arrangement between States which does create obligations although these, if not performed, do not involve the international liability of the State in question, inasmuch as no legal proceedings can be brought before the International Court of Justice. Neither is a "gentleman's agreement" restricted, in modern times, to declarations of general policy. Certain such agreements interpret or fill in gaps in treaties or the statutes of international organisations; others define the regulatory framework in which relations between States should be established.

There are sound arguments in doctrine for defining the London Club and Zangger Committee Guidelines in this way.

Despite the confidential nature of discussions, especially those of the London Club, there is reason to think that, as early as their 1975 meeting, the "founding" States agreed not to conclude a formal international treaty, which meant that neither the London Club Guidelines nor the Zangger Committee Memoranda needed to be registered with the United Nations Secretariat (Article 102 of the Charter).

It nevertheless remains true that concerted action was taken — a meeting of minds after prior negotiation — as shown not only by the identical or similar nature of the letters addressed to the Director General of the IAEA, but also by the regulatory nature of the Annexes and the decision not to amend the London Club Guidelines except by unanimous consent, and to continue to hold consultations on certain questions of implementation. The Zangger Committee practice shows even more strongly a desire to complete the initial agreement, namely the NPT.

Lastly, if it is accepted that "gentlemen's agreements" are governed by international law, but not the part of that law dealing with international treaties, they must be performed in good faith. Thus, a State which has undertaken a commitment within the framework of a gentleman's agreement is not entitled arbitrarily or secretly to depart from or change it, and this is true both with regard to the other States committed under the agreement as well as, given the publication request to the IAEA, to the other Member States of the Agency. However, it is accepted that a participating State is free to decide to no longer join the consultations or undertake new commitments.

That the commitments published by the IAEA under the reference INFCIRC/209 or 254 are not purely potestative also follows from the other possible interpretation which is supported by various countries: that of unilateral undertakings[4]. The International Court of Justice (ICJ) today recognises that a unilateral undertaking[5] by a State, expressed publicly and with the intention of committing itself has, even if outside the framework of international negotiations, a binding effect. The State concerned is then obliged, by virtue of the principle of good faith, to follow a line of conduct in accordance with this declaration. It follows that while a declaration by a State limiting its freedom of future action must certainly be interpreted restrictively, it does not carry with it an arbitrary power of review. According to some, a State could terminate its undertakings by means of a unilateral action in the same manner as it would relieve itself from obligations derived from a treaty. Thus a fundamental change of circumstances, which may be likened to the *rebus sic stantibus* of the law of treaties could justify a repeal or a review of the obligations entered into.

Following this line of thought, a distinction should perhaps be made between two situations:

— that of those States which participated in the consultations undertaken within the framework of the Zangger Committee or London Club; in this case, the relations between these States would be considered as being regulated by a gentlemen's agreement, whereas those between any such State and States not having taken part in these consultations would be governed by the rules on unilateral undertakings;

— that of the many other States which, subsequently, asked the IAEA to record and publish their decision to act in compliance with the principles laid down in the Zangger Committee or London Club documents. Such a decision would be in the nature of a unilateral undertaking with regard both to countries having participated in these consultations as well as third States. An alternative interpretation would be to consider the commitment undertaken by these States with regard to countries having participated in the consultations as an adhesion to the gentlemen's agreement, subject, wherever necessary, to the relevant provisions of the law of treaties relating to adhesion. Given the present state of the law,

however, this latter interpretation appears somewhat unorthodox.

In conclusion, it may be noted that there is an essential difference between the Zangger Committee and the London Club. The declarations of the former give a common interpretation of an Article of the NPT which involves a specific undertaking by the Contracting Parties. On the other hand, the London Club Guidelines do not refer to a formal provision and simply express the decision of participating countries to act in accordance with principles defined in annex. Therefore, the two initiatives cannot be placed on the same level.

D.   OTHER ELEMENTS IN THE REGULATION OF NUCLEAR TRADE[6]

I.   DECLARATION OF COMMON POLICY BY THE EUROPEAN COMMUNITY

On 20th November 1984 and in the context of European political co-operation (the EEC Council), the Foreign Ministers of the ten Member States of the European Community adopted a declaration of common policy (*annex IX* to this chapter)[7]. After noting, *inter alia*, that all the Member States of the European Community had adopted, as unilateral undertakings, the London Club Guidelines published in document INFCIRC/254, these States declared that "the principles contained in the Guidelines constitute a common, fundamental set of rules for all the Member States in relation to their nuclear exports". They also stated that transfers of nuclear material, equipment and technology could be made without restriction between the Member States, on condition that the provisions of the Treaty of Rome and the competence of the Member States were respected, subject to the following additional arrangements:

— until such time as they are used, separated plutonium and uranium enriched to more than 20 per cent will be stored by the Member States at the place of separation or enrichment, or at the places of fabrication of fuels containing such materials, or in a store established and administered by a Member State, or in a place to be determined by common agreement between the Member States concerned;

— such plutonium and uranium will be transferred by the Member States upon receipt of a certificate from the consignee, drafted in accordance with an attached model form specifying the final destination, the quantities, the approximate date of delivery, the timetable for utilisation, the form in which delivery is to take place and the allocation of the material to one or other of the uses specified in the Declaration: fuel supply for a reactor, fabrication of fuel elements, research and development, or utilisation in any other installation connected with an energy or research and development programme; these requirements also apply to any intermediate storage;

— the Government of the Member State to which the consignee belongs must confirm the accuracy of the information contained in the said certificate;

— such plutonium and uranium will not be retransferred to a third State without mutual agreement between the Member State that has separated or enriched them and the Member State desiring to effect the retransfer, "without prejudice to any other rights of prior consent that may exist".

Exceptions have, however, been made in the case of plutonium having an isotopic concentration of plutonium 238 above 80 per cent and of special fissile materials used in quantities of the order of 1 gramme as a component of measuring instruments.

On this basis, and after also stating that they would apply physical protection measures at least equal to those recommended by the London Club, the Member States of the Community declared that transfers between them would be carried out in a manner compatible with the requirements of both non-proliferation and the free movement of goods.

II.   NOTIFICATIONS TO THE IAEA OF EXPORTS AND IMPORTS OF NUCLEAR MATERIAL BY NUCLEAR-WEAPON STATES

With a view to helping the Agency with its safeguards activities, the Resident Representatives of the United States, the United Kingdom and the Soviet Union informed the Director General of the IAEA, on 10th July 1974, of the decision of their respective Governments henceforth to communicate to the Agency information concerning their exports and imports of nuclear materials (INFCIRC/207 — 26th July 1974). By letter of 7th February 1984, the Governor representing France on the IAEA Board of Governors informed the Director General that the French Government would do the same

(INFCIRC/207/Add. 1 — May 1984) (see *annex X* to this chapter).

III. FINAL DECLARATION OF THE CONFERENCE OF MINISTERS OF FOREIGN AFFAIRS OF THE NON-ALIGNED COUNTRIES, SEPTEMBER 1985

The conditions imposed on nuclear trade resulting from the discussions among exporting countries within the Zangger Committee and the London Club gave rise to strong criticism from developing countries, and especially from certain "newly industrialised countries" with ambitious nuclear programmes, such as India, Brazil and Argentina. These criticisms were aired in the form of certain principles debated in the IAEA Committee on Assurances of Supply (CAS) or on the occasion of NPT Review Conferences.

Similar criticisms were also voiced at the Conference of Ministers of Foreign Affairs of the Non-Aligned Countries held at Luanda from 4th to 7th September 1985. Expressing their "concern at the obstacles which the developed countries place in the way of transfer of technology related to the peaceful uses of nuclear energy by fixing conditions which are incompatible with the sovereignty of developing countries", the Ministers "affirmed the inalienable right of all States to apply and develop their programmes for peaceful uses of nuclear energy for economic and social development, in conformity with their priorities, interests and needs". They added that it was necessary to comply with the principles of non-discrimination and free access to nuclear technology; non-proliferation should not be used as an excuse to prevent States from exercising their rights fully (INFCIRC/332, March 1986, see *annex XI* to this chapter).

In conclusion, note may also be taken of a communication from South Africa concerning its policy with regard to nuclear exports[8].

E. INTERNATIONAL NUCLEAR FUEL CYCLE EVALUATION

I. OBJECTIVES

Various events, such as the explosion in India (1974) of a nuclear device as well as competition between countries exporting nuclear facilities, led the United States, in the 1970s, to undertake a comprehensive review of its policy with regard to the non-proliferation of nuclear weapons. One of the aspects of this reassessment led to the theory that by resorting to certain types of fuel cycle (thorium, for example), giving up certain types of reactor (fast breeder reactors), or making enrichment and reprocessing plants subject to international intervention and inspection, meeting the needs of nuclear energy could be reconciled with reducing the risk of proliferation[9]. To this end, the United States undertook a review of its legislation (Nuclear Non-Proliferation Act of 1978 — in particular, its Section 104 concerning an International Fuel Authority Project) and of its domestic policy (ceasing the reprocessing of spent (irradiated) fuel, for example). It also endeavoured to convert other countries to its point of view. Indeed, in April 1977, the President of the United States had proposed negotiations on fuel cycle issues. Following this proposal, a Conference was held in October of the same year under the title of International Nuclear Fuel Cycle Evaluation — INFCE.

However, as certain participating countries refused, a priori, to follow the policies implemented in the United States (for example, giving up the use of plutonium), it was agreed, before the opening of this Conference, that INFCE would not constitute formal negotiations and that, consequently, Participating States would not be bound by its conclusions. In addition, it was recognised that the decision by a State to manufacture nuclear weapons was essentially political in nature, motivated by national security considerations, and it was also agreed that INFCE would not discuss such questions. Therefore, INFCE was to be an objective technical and analytical study respecting the choice of fuel cycle operated by each country subject to the application of a mutually acceptable security control.

Within these limits, agreed in advance, INFCE involved a considerable amount of work, making use of more than five hundred experts representing sixty-six countries and five international organisations. This undertaking, directed by a Technical Co-ordinating Committee, was carried out in eight Working Groups. It was given concrete form by the publication, under the auspices of the IAEA in 1980, of reports by these Groups and a summary and overview report by the Technical Co-ordinating Committee after its approval in a Plenary Conference held in February 1980 in Vienna[10].

II. THE FINAL REPORT

The Conference recognised that one of the most noteworthy aspects of the INFCE study was that it had been carried out in a climate of mutual understanding and co-operation. The hope was expressed that this climate would continue to prevail in respect of the international co-operation required to achieve the results which were felt to be attainable with regard to:

— increasing the role of nuclear energy to satisfy world energy requirements;
— adopting effective measures to satisfy the specific needs of developing countries in respect of the peaceful use of nuclear energy;
— adopting effective measures to minimise the danger of the proliferation of nuclear weapons without, however, endangering energy supplies or the development of nuclear energy for peaceful purposes[11].

In substance, the INFCE study highlighted the interdependence of assuring supplies on the one hand and non-proliferation on the other. When regular supply is properly assured, consumer countries have less incentive to build enrichment and reprocessing facilities. Further, if such facilities, or repositories for plutonium, spent fuel or nuclear waste, were built and operated on a multinational or international basis, this would reduce the number of such facilities required, facilitating the application of safeguards and bringing the economic benefits that come with size. Nevertheless, it was recognised that, given the complexity of the siting problems to be solved with the country hosting such facilities, such projects would require further consideration.

III. INFCE STUDY FOLLOW-UP

Several expert groups were constituted by the Director General of the IAEA in order to examine in greater detail the various issues addressed during the INFCE study.

1. Expert Group on International Spent Fuel Management (ISFM)

This Group on ISFM which was formed in 1979, consisted of experts from twenty-four Member States and representatives of the EEC and NEA. Its report was submitted on 6th July 1982[12].

The Group was charged with examining possibilities for international co-operation in the field of spent fuel management and with assisting the IAEA in defining what role the Agency might play in solving the problems created by the growing accumulation of spent fuel. Two Subgroups were formed: Subgroup A dealt with technical and economic considerations, while Subgroup B was asked to deal with institutional, legal and procedural considerations.

The Group limited its work to the phase of the fuel cycle between the unloading of a reactor and the reprocessing or conditioning of the spent fuel, i.e. to the intermediate storage phase, during which the fuel retains its original form.

As concerns more especially the institutional and legal aspects of the problem, the Group recognised the advantages of two-tier agreements for the storage of spent fuel, the first taking the form of an inter-governmental "umbrella" agreement regulating policy issues (non-proliferation, for example), the second tier being a commercial or industrial agreement covering operational aspects.

The Group studied the question of incentives to conclude such agreements. Having observed that the bilateral framework for storage gave rise to a host country/customer country dependence, it suggested that this could be reduced by means of a financial participation by the customer in the host storage facility, especially in the event of extension of such a facility. More generally, a preference was shown for multinational agreements. It was observed that agreements are also strengthened by an assurance of the retrievability of fuel after a predetermined period. Lastly, it was recognised that a potential host country would have a better chance of being selected if, along with storage, it offered other services such as reprocessing, repackaging and waste conditioning.

2. Expert Group on International Plutonium Storage (IPS)

Article XII.A.5 of the IAEA Statute provides, with regard to safeguards, that any excess special fissionable materials recovered over and above what is needed for research or in reactors must be deposited with the Agency. With a view to effective implementation of this Article, the Director General of the Vienna Agency asked the Secretariat to carry out a study in 1976. This study was submitted to Member States, along with a covering letter dated 13th July 1978 inviting them to designate experts to participate in a meeting to formulate proposals for the drawing up of plans for the international management and storage of plutonium — IPS.

From the 1978 meeting till October 1982, experts from thirty-seven countries studied the technical and practical aspects of this problem. A technical report was submitted to the IAEA Board of Governors in November 1982, after which consideration of the question was to continue by way of informal consultation between Members of the Board[13].

In fact, the experts did not succeed in reaching consensus on the various possible approaches because of different conceptions of the relations to be established between States and the Agency. Agreement was, however, reached on the objective to be attained: implementation of Article XII.A.5 should promote the development of a fuel cycle using plutonium and not interfere with national energy programmes, and at the same time facilitate, without discrimination between States, the application of Agency safeguards.

In this perspective, IPS would be considered as an element of the existing Agency safeguards system, and not as a separate system; consequently, the reports and inspection procedures provided for under the present system should be used to the fullest extent possible.

In order both to limit the transport of plutonium as far as possible — this being an expensive operation, also requiring physical protection precautions — and because the construction of plutonium stores specifically for the IAEA would be unnecessarily costly, stores should be built at places where the plutonium is already located [for example, plants for the reprocessing or fabrication of MOX fuel (U/Pu mixed oxide)]. This would mean that the respective rights and obligations of the Agency and the State on whose territory the store is situated would, as concerns the custody and supervision of plutonium deposited by the Agency, have to be specified in a bilateral agreement. The possibility of multilateral agreements would, nevertheless, deserve more detailed consideration.

F. COMMITTEE ON ASSURANCES OF SUPPLY

I. MANDATE OF THE COMMITTEE

Following the work of INFCE, the IAEA Board of Governors, on 20th June 1980, established a Committee on Assurances of Supply (CAS), open to all Member States. This body was instructed to consider and advise the Board of Governors on:

— ways by which supplies of nuclear material, equipment and technology and fuel cycle services can be assured on a more predictable and long-term basis in accordance with mutually acceptable considerations of non-proliferation; and
— the Agency's role and responsibilities in relation thereto.

Since September 1980, the Committee has met on numerous occasions, with participants including representatives from some fifty Member States as well as from the NEA and CEC. Priority has been given to studying:

— "principles of international co-operation in the field of nuclear energy" in accordance with the above-mentioned mandate from the Board of Governors;
— a project for emergency and back-up mechanisms in the event of an interruption or impending delay in supply;
— mechanisms for reviewing intergovernmental nuclear co-operation agreements.

It has also examined the practical, technical and administrative problems related to international shipments of nuclear materials and equipment. A summary is given below of the stage which work had reached at the date of publication of this study; it should be noted at the outset that the principles formulated by CAS do not have the status of official documents and are not published, since all participants have not withdrawn their reservations as to the contents of these principles.

At the close of last meeting of CAS in May 1987 (twenty-first session), the members of the Committee noted the impossibility of reaching agreement on the principles of international co-operation in the nuclear field, the UNPICPUNE Conference (see chapter one of this Study) also having been unable to achieve a consensus in this respect. It was therefore agreed to suspend the work of the Committee for the time being, while pursuing at the same time consultations to find a way to reach an agreement. The following paragraphs provide an indication on the status of work of the Committee.

II. PRINCIPLES OF INTERNATIONAL CO-OPERATION IN THE FIELD OF NUCLEAR ENERGY

After an introductory statement on the meaning given by the various members of the Committee to the term "non-proliferation", the draft is divided into two parts. The first deals with the principles on which supplies of nuclear material, equipment and technology and fuel cycle services can be assured on a more predictable and long-term basis.

The first of these principles is that all States have the right to use nuclear energy for peaceful purposes in order to promote their economic and social development. This implies that each State has the right to choose a fuel cycle in conformity with its national interests and needs, and to determine the role it intends to play in international nuclear trade.

Consequently, nuclear supplies and international nuclear trade in general should be assured on a long-term, international basis that is widely acceptable and in a non-discriminatory, predictable, consistent and ultimately uniform manner. Effective measures should be taken to meet the specific needs of States whose nuclear programmes are still in an early stage of development.

States should fulfil their obligations under agreements in the nuclear field and any modification of such agreements should be made only by mutual consent of the parties, through previously agreed revision mechanisms. Similarly, prior consent clauses should be implemented as agreed upon by the parties and in an established and predictable manner. The retransfer of nuclear materials, materials and equipment should be possible when these are subject, in the recipient State, to the same commitments as those accepted by the retransferring State in relation to the original transfer.

The availability of international mechanisms to ensure assistance in cases of supply emergencies should be an important element of backing up assurances of supplies.

The second series of principles deals with the links between assurances of supply and mutually acceptable considerations of non-proliferation. The interdependence of assurances of supply and assurances of non-proliferation is recognised as essential for international co-operation in the peaceful utilisation of nuclear energy.

It is recognised that fulfilment of a common basic objective of non-proliferation would enhance the interests of all States in stable international nuclear trade and co-operation. The implementation of non-proliferation assurances should be compatible with the right of every State to use nuclear energy for peaceful purposes.

Such non-proliferation assurances for securing nuclear supply should be based on a national commitment that is internationally binding, together with verification of compliance with this commitment by the application of Agency safeguards. The exact wording of this principle (the fourth in the draft) remains, however, under consideration, and no conclusion was reached on this fourth draft principle or on an Introduction to the Principles during the Committee's recent work.

The Committee has also studied the role and responsibilities of the IAEA. It agreed that the Agency was the primary multilateral forum for the development of international co-operation in the field of nuclear energy, and that its safeguards had a basic role in facilitating such co-operation. The Agency keeps the nuclear supply situation under permanent review, and States may inform it of abnormal situations of concern to them regarding supply assurances. At the request of a State, the Agency may provide summary statements on the results of the implementation of safeguards in the requesting State, which should constitute an important element of confirmation of that State's compliance with its non-proliferation commitments.

III. EMERGENCY AND BACK-UP MECHANISMS FOR MEMBER STATES

The purpose of the proposed mechanisms is to ensure "substitute supplies" for Member States in the event of an "emergency situation". Detailed discussions took place concerning the meaning of these two expressions.

The Committee defines "emergency situation" as "an interruption or delay, or an impending interruption or delay, in providing contracted supplies, not caused by a breach of agreement by the recipient, which has led, or is likely to lead, to a disruption or delay in the construction or operation of a nuclear facility in a State which is in compliance with its non-proliferation and safeguards commitments".

"Substitute supplies" are defined by the Committee as "supplies offered by participating States to be available upon request by Member States facing an emergency situation as defined above".

The Committee agreed to recommend that a system be established within the IAEA, in order to:
— receive, register and keep records on supplies made available for a back-up mechanism and on the conditions for drawing on such supplies;
— make this information available to Member States wishing to participate in the system.

Depending on the circumstances, the Agency's role would consist merely of providing information to the States concerned which would conclude a bilateral agreement between themselves, or would go as far as having the Agency act as an intermediary in concluding such an agreement, as provided for under Article III.A.1 of its Statute or, lastly, providing supplies in compliance with the provisions of Articles IX and XI of the Statute.

The Committee also recommended that the main characteristics of the emergency and back-up system should include automatic access to substitute supplies by all Member States facing an emergency situation, and non-interference with the commercial market.

In February 1985, the IAEA Board of Governors took note of the Secretariat's report on the financial and organisational implications of the system, but decided to engage in informal consultations before reaching any decision.

IV. CO-OPERATION AGREEMENT REVISION MECHANISMS

The Committee also succeeded in reaching conclusions on the procedures for revising intergovernmental nuclear co-operation agreements. However, certain of the States participating in this work reserved the right to return to this topic at a later date.

The four conclusions reached by the Committee were:
— a nuclear co-operation agreement could be amended only by mutual agreement of all the States parties, which should negotiate in good faith. Subject to principles of international law, supplies under nuclear co-operation agreements should continue during consultations and negotiations on amendments;
— nuclear co-operation agreements should contain provisions for consultation among the parties; such consultations might have the advantage of avoiding the need for formal amendments;

— although questions concerning arbitration, legal recourse and procedures in the event of non-compliance lay outside the scope of the consideration of revision mechanisms, the Committee was nevertheless of the opinion that the proposal of an amendment should not affect the rights of recourse to legal remedies and that failure to agree on a proposed amendment should not in itself create new rights as regards continuation and implementation of the existing agreement;

— the Agency could, to the extent requested by the States parties, facilitate consultations, and serve as an independent source of information for any State requiring such information. On this topic, the Secretariat observed that the Agency might provide documents in addition to those already supplied, such as a statement that no proceedings foreseen by Article XII.A.7 of the Statute had ever been initiated, on condition that the State requesting such information communicate it to the supplier State.

V. EXISTING PRACTICAL, TECHNICAL AND ADMINISTRATIVE PROBLEMS IN INTERNATIONAL SHIPMENTS OF NUCLEAR MATERIAL AND EQUIPMENT

At the request of the Committee, a report was prepared by the Secretariat and a Group of Experts on these questions which had caused longer lead times and higher costs for supplies than would appear necessary. As far as national, and sometimes local regulations and administrative practices were concerned, there was a tendency to diverge from the provisions accepted at international level, such as the IAEA Regulations on the Safe Transport of Radioactive Materials; this indicated that there was a need for further efforts to achieve harmonization and consistent application of several co-operation agreements to a given batch of nuclear materials ("multiple labelling"), and for increased co-operation between the national authorities concerned.

G. CO-ORDINATING COMMITTEE ON EXPORT CONTROLS (COCOM)

The Co-ordinating Committee on Export Controls (COCOM), created in 1950, is not strictly speaking an international organisation. It was not set up under a treaty and has no statute. It is, in effect, a forum for direct and flexible consultations between allied countries. COCOM, located in Paris with a small secretariat, operates on a confidential basis, according to an informal agreement between Members. The Committee formulates resolutions which do not constitute international obligations. This does not mean, however, that these resolutions have no practical implications for exporters since in fact, the Member countries apply them in substance in their regulations on foreign trade.

COCOM allows participating governments to discuss trade matters of strategic importance and to reach agreements reinforcing controls on exports to certain proscribed countries. COCOM holds regular meetings of accredited national representatives as well as meetings of a senior Advisory Group; it also organises frequent ad hoc meetings of national delegates to deal with political or technical questions.

COCOM owes its origin to the embargo on sales to Eastern European countries of goods or materials considered "strategic", a policy laid down by the United States' Economic Act of 1948 and the Mutual Defence Assistance Control Act of 1951 (Battle Act). These Acts provided that the United States would grant military, economic or financial assistance only to a foreign country which placed an embargo on such exports. These considerations led, from the end of 1948, to negotiations and to the meeting of an Advisory Group in January

1950, and to the creation of COCOM in that same year. Although there is no legal link with NATO, the Committee and the Advisory Group include representatives of Member countries of that Organisation, with the exception of Iceland, and with participation by Japan since 1952. Those countries are: Belgium, Canada, Denmark, France, the Federal Republic of Germany, Greece, Italy, Luxembourg, the Netherlands, Norway, Portugal, Spain (1985), the United Kingdom, the United States of America, and Turkey. Another group of countries (Austria, Finland, Sweden, Switzerland, Yugoslavia, and the Territory of Hong Kong) are not part of the Committee but "co-operate" in some of its undertakings.

COCOM's objective therefore, is to control the export of goods likely to strengthen significantly the military power of Socialist bloc countries, whether directly or through the development of advanced technology which could be used for military purposes. To this effect COCOM:

— establishes and periodically reviews, according to the state of East-West relations and to the respective technical and industrial developments, lists of "goods" to which participating countries should apply embargo or control measures. Three lists have been thus established (war materials or items specific to the design or production of such materials, *materials relating to nuclear energy*[14], and industrial materials or goods);

— rules on requests by Member countries for exemptions from the embargo[15].

When establishing or reviewing the lists as well as deciding on requests for derogations, COCOM and the senior Advisory Group act by unanimity. However, participating countries unilaterally may establish more exhaustive or stricter embargo lists or, on the contrary, not conform entirely to COCOM recommendations if they consider that their national interest may be seriously affected. Member countries may also co-operate within COCOM for implementation of controls relating to their export of goods on the lists[16].

NOTES AND REFERENCES

1. Article III.2 of the NPT provides: "each State Party to the Treaty undertakes not to provide: *a)* source or special fissionable material, or *b)* equipment or material especially designed or prepared for the processing, use or production of special fissionable material, to any non-nuclear-weapon State for peaceful purposes, unless the source or special fissionable material shall be subject to the safeguards required by this article".

2. See, on this topic, chapter nine of this Study.

3. To supplement the London Club Guidelines, France, the United States, the United Kingdom, Italy, the Federal Republic of Germany, Canada and Japan took part in discussions, starting in 1982, to harmonize their respective policies for controlling the export of equipment and technology which could help develop ballistic vehicles capable of carrying nuclear warheads. These consultations gave rise to the drafting of Guidelines and a Technical Annex and to simultaneous declarations, on 16th April 1987, by the above-mentioned States, on the policy they intended to follow with regard to such exports. The hope was expressed that other potential suppliers of such equipment and technology would adopt similar provisions. While the objective pursued by these policies is identical to that of the provisions studied in this chapter, the activities subject to the new controls fall into the sphere of space rather than into the nuclear field. It cannot, however, be excluded that certain materials, equipment and technologies will be covered both by the 1987 Guidelines and by the provisions examined in this Study.

4. The message of the Swiss Government to Parliament may be mentioned to back up this interpretation — the message concerned the conclusion of Agreements with Australia and the People's Republic of China (20th May 1987) and said the following: The Zangger Committee and the London Club documents are not part of international public law, they are arrangements applied by Member States autonomously, in accordance with their respective domestic laws. Together with the NPT, they constitute today the international non-proliferation regime, agreed multilaterally. This regime is not determined once and for all but is adapted, where necessary, to evolving technical and political conditions. The message adds that as the Directives have no legal scope *per se*, they must be transposed into national legislation for purposes of implementation.

5. ICJ, 20th December 1974 — French Nuclear Tests case — ICJ Reports 1974, p. 457.

6. As seen from the angle of the questions dealt with in this chapter, the results of the successive Review Conferences of the NPT and the United Nations Conference for the Promotion of International Co-operation in the Peaceful Uses of Nuclear Energy (UNPICPUNE) have been briefly reviewed in chapter one of this Study.

7. On 22nd March 1985, this Declaration was forwarded by Italy to the Director General of the IAEA for communication to the Member States of the Agency (INFCIRC/ 322, April 1985). By letter of 17th June 1986 addressed to the Director General of the IAEA, Portugal indicated that it also adopted this Declaration of Common Policy.

8. By letter of 31st January 1984, the Resident Representative of South Africa transmitted to the Director General of the IAEA a declaration by the Chairman of the Atomic Energy Corporation of South Africa Limited, in the form of a press communiqué. At the request of the Resident Representative of South Africa, this declaration was subsequently communicated to Member States (INFCIRC/ 314, February 1984). South Africa gave the assurance that it would conduct its nuclear affairs in compliance with the spirit, principles and objectives of the NPT and the London Club Guidelines. More exactly:

    a) South Africa would not sell uranium to non-nuclear-weapon States without IAEA or EURATOM safeguards;

    b) South Africa would not supply sensitive technology to another country without Agency or EURATOM safeguards;

    c) South Africa would not sell enriched uranium or nuclear equipment without Agency or EURATOM safeguards.

    The recipient country would have to guarantee that the technology, materials or equipment would not be used for nuclear explosive devices, but only for peaceful purposes.

9. After it had been recognised, in the final declaration of the 1975 NPT Review Conference, that it would be desirable, both from the economic and the safeguards and physical protection viewpoints, to set up regional or multinational fuel cycle centres, the IAEA undertook a study, completed in July 1976, on the institutional and legal aspects of setting up and operating such centres.

    (IAEA-RFCC/2 — Regional Nuclear Fuel Cycle Centre Study: Institutional-Legal Framework Aspects — Report regarding preliminary exploratory phase; July 1976).

10. IAEA-INFCE — Summary Volume — STI/PUB/534, IAEA, Vienna, 1980.

11. The following conclusions of the INFCE Report are of relevance to the present study: as concerns the availability of mineral resources and fuel cycle services, the mining industry should not, given the forecasts for nuclear energy demand and assuming that the necessary prospecting and investment activities are carried out, encounter insurmountable difficulties in satisfying demand till 2000 or, on certain conditions, till 2025. However, it was said in this report that the breakdown of supply and demand among the different regions of the world gives rise to concern from the point of view both of consumers (with regard to assurances of supply) and producers (as concerns outlets). This may lead to different strategies with regard to the choice of reactor-type and the degree of use of uranium resources, choices in which political and commercial factors are as important as technical ones. As concerns fuel cycle services, actual and forecast enrichment capacity should satisfy needs till 1995.

In accordance with its mandate, INFCE studied the proliferation problem from the technical aspect only and, after in-depth examination, concluded that the construction of a fuel cycle facility was not the easiest way to acquire the materials necessary to manufacture weapons. Nevertheless, the possibility of materials being diverted for this purpose could not be totally excluded and, indirectly, the knowledge and know-how resulting from the implementation of a nuclear programme could prove useful in carrying out a military programme (the most sensitive phases naturally being isotope separation technology, and the possession of a sufficient quantity of plutonium 239 and even of "commercial" plutonium made up of a mixture of isotopes).

With regard to ways of assuring the long-term supply of technology, fuel and services in order to meet national needs while complying with the non-proliferation objective, Working Group 3 examined both the commercial and governmental aspects of this question (Assurances of long-term supply of technology, fuel and heavy water and services in the interest of national needs consistent with non-proliferation — Report of INFCE Working Group 3 — IAEA, Vienna, 1980, STI/PUB/534, and INFCE Summary Volume, *op. cit.*, p. 121 *et seq.*).

With regard to the commercial aspects, the Group recognised that contractual mechanisms had, especially since a balance between supply and demand had been established, made it possible to solve satisfactorily the problem of the assurance of short-term supplies (stocks, exchanges, loans). Improved information and the existence of a healthy spot transactions market were nevertheless desirable.

It was noted that, for the purposes of assurances of long-term supply, contractual provisions should be made more flexible in order to allow the necessary adaptations to be made as well as an equitable sharing between supplier and consumer of the costs and risks of market fluctuations. It also appeared that the balance between supply and demand in the future would be facilitated if consumers participated in or made a financial contribution towards the development of new uranium resources and the construction of separation and reprocessing facilities.

The Group explored two possible formulas for emergency mechanisms to deal with any market failure. The first, called the "Uranium Emergency Safety Network", would perfect the solutions introduced by various users by giving them a more institutional form. It would involve participants pooling part of their stocks. Withdrawals could be made from this pool in the event of an interruption of supply, under conditions to be specified. The second, called an "international nuclear fuel bank", was based on a United States proposal. A bank, constituted by supplier States and consumers, would possess a stock of natural or slightly enriched uranium from which consumer States whose supply had been interrupted as a result of contractual problems could make withdrawals. The Group also felt that emergency transfers should be rapid and automatic, in accordance with rules agreed in advance with supplier States.

Many of the supply difficulties encountered result from government intervention, motivated by non-proliferation or domestic policy considerations. Such interventions take the form of the control of nuclear exports and imports, and prior consent procedures. To reduce the resulting uncertainty for consumer countries arising from the unilateral exercise — especially in the case of changes of internal policy — of these rights by supplier States, it was generally agreed that concrete criteria for using these powers should be specified in advance, and they should be made less discretionary in nature by means of international, bilateral or multilateral agreements.

12. Final Report of the Expert Group on International Spent Fuel Management, IAEA — ISFM/EG/26.Rev.1, July 1982.

13. Expert Group on International Plutonium Storage — Report to the Director General, 1st November 1982, IAEA-IPS/EG/140(Rev.2).

14. From 1952 to 1957, following the Korean War, a Committee established along the same principles — CHINCOM — had as an objective the elaboration of embargo measures more restrictive than those of COCOM with regard to the People's Republic of China and North Korea. These special controls were abolished in 1957 and the powers of CHINCOM was merged with those of COCOM. Since 1985, there is a special procedure for the "China" list.

15. Although the lists define the goods which, in principle, are prohibited from export to proscribed countries, derogations can be obtained pursuant to such procedures as "general exceptions" or "administrative facility".

16. The Atomic Energy List (List Annex D — International Atomic Energy List, 1984), covers fissile material and materials and diverse equipment used in the nuclear industry. It is not reproduced as such in this Study because it is classified confidential. It is nevertheless well-known that some Member countries reproduce the content of the list in their national export legislation, from the point of view of controlling final destination. It should be noted in this respect that the controls and restrictions resulting from these national regulations do not necessarily apply only to those Socialist countries initially proscribed by COCOM resolutions but may also be extended to other "sensitive" countries.

ANNEX I

STATES HAVING UNDERTAKEN COMMITMENTS FOLLOWING CONSULTATIONS IN THE ZANGGER COMMITTEE

	INFCIRC/209 6.9.1974	INFCIRC/209 Mod 1 December 1978	INFCIRC/209 Mod 2 February 1984	INFCIRC/209 Mod 3 August 1985
* Australia	x	x	x	x
Austria	x (2.2.1976)	x		
* Canada	x	x	x	x
* Czechoslovakia	x (5.5.1975)	x	x	x
* Denmark	x +letter II	x	x	x
* Finland	x	x	x	x
* German Democratic Republic	x (16.9.1974)	x	x	x
* Germany, Federal Republic of	x (letter IV)	x	x	x
* Greece	x (19.9.1984)			
* Hungary	x (25.11.1974)	x (28.1.1986)	x (28.1.86)	x (28.1.1986)
* Ireland	x (26.5.1975) (letters I and IV)	x	x	x
* Japan	x (15.4.1977)	x	x	x
* Luxembourg	x (10.9.1975)	x		
Norway	x	x	x	x
* Netherlands	x (letter IV)	x	x	x
* Poland	x (16.9.1974)	x	x	x
* Sweden	x (10.4.1975)	x	x	x
* United Kingdom	x (letter II)	x	x	x
* United States of America	x +letter III of 22.8.1974 +letter A of 3.10.1974	x	x	x
* Union of Soviet Socialist Republics	x +letter B of 3.10.1974	x	x	x

* This country has also undertaken commitments agreed by the London Club.

ANNEX II-A

COMMUNICATIONS RECEIVED FROM MEMBERS REGARDING THE EXPORT OF NUCLEAR MATERIAL AND OF CERTAIN CATEGORIES OF EQUIPMENT AND OTHER MATERIAL

(INFCIRC/209 — 3rd September 1974)

1.   On 22nd August 1974 the Director General received letters, all dated that day, from the Resident Representatives to the Agency of Australia, Denmark, Canada, Finland, Norway, the Union of Soviet Socialist Republics, the United Kingdom of Great Britain and Northern Ireland and the United States of America, relating to the commitments of these eight Members under Article III, paragraph 2 of the Treaty on the Non-Proliferation of Nuclear Weapons[1]. In the light of the wish expressed at the end of each of those letters, their text is reproduced below as Letter I.

2.   On the same day, the Resident Representatives of Denmark and of the United Kingdom also addressed complementary letters to the Director General, the text of which is reproduced below as Letter II. On that day also the Resident Representative of the United States sent a complementary letter, the text of which is reproduced as Letter III.

3.   Also on 22nd August, the Resident Representatives of the Federal Republic of Germany and of the Netherlands each addressed to the Director General a letter analogous to the above-mentioned Letters I and II, the text of which is reproduced below as Letter IV.

4.   The attachments to the Letters I and IV, which consist in both cases of the same memoranda, are reproduced in the Appendix.

*
**

Letter I

I have the honour to inform you that the Government of .... has had under consideration procedures in relation to exports of *(a)* source or special fissionable material, and *(b)* equipment and material especially designed or prepared for the processing, use or production of special fissionable material, in the light of its commitment under Article III paragraph 2 of the Treaty on the Non-Proliferation of Nuclear Weapons not to provide such items to any non-nuclear-weapon State for peaceful purposes, unless the source or special fissionable material is subject to safeguards under an agreement with the International Atomic Energy Agency.

The Government of .... has decided to act in this context in accordance with the attached memoranda.

I shall be grateful if you will bring this information to the attention of all Members of the Agency.

Letter II

I have the honour to refer to my letter of today's date, and to inform you that, so far as trade within the European Community is concerned, the Government of .... will, where necessary, implement paragraphs 5 of the Memoranda enclosed with that letter in the light of its commitments under the Treaties of Rome.

Letter III

With reference to my letter of this date, concerning procedures of the Government of the United States of America in relation to exports of source and special fissionable material and of equipment and material especially designed or prepared for the processing, use or production of special fissionable material, I shall provide you shortly with additional information concerning the implementation by my Government of such procedures.

---

[1]   Reproduced in document INFCIRC/140.

94

I would like to call attention to paragraph 6 of Memorandum B, enclosed with my letter, and to note that, in accordance with existing procedures of my Government, safeguards are required in relation to items of equipment and material exported from the United States of America, in addition to those specified in paragraph 2 of that Memorandum.

I shall be grateful if you will bring this information to the attention of all Members of the Agency.

Letter IV

I have the honour to inform you that the Government of .... has had under consideration procedures in relation to exports to any non-nuclear-weapon State for peaceful purposes of *(a)* source or special fissionable material, and *(b)* certain categories of equipment and material especially designed or prepared for the processing, use or production of special fissionable material.

The Government of .... has decided to act in this context in accordance with the attached Memoranda. So far as trade within the European Community is concerned, the Government of .... will, where necessary, implement paragraphs 5 of the Memoranda in the light of its commitments under the Treaties of Rome.

I shall be grateful if you will bring this information to the attention of all Members of the Agency.

MEMORANDUM A

INTRODUCTION

1.   The Government has had under consideration procedures in relation to exports of nuclear materials in the light of its commitment not to provide source or special fissionable material to any non-nuclear-weapon State for peaceful purposes unless the source or special fissionable material is subject to safeguards under an agreement with the International Atomic Energy Agency.

DEFINITION OF SOURCE AND SPECIAL FISSIONABLE MATERIAL

2.   The definition of source and special fissionable material adopted by the Government shall be that contained in Article XX of the Agency's Statute[1].

THE APPLICATION OF SAFEGUARDS

3.   The Government is solely concerned with ensuring, where relevant, the application of safeguards in non-nuclear-weapon States not party to the Treaty on the Non-Proliferation of Nuclear Weapons (NPT) with a view to preventing diversion of the safeguarded nuclear material from peaceful purposes to nuclear weapons or other nuclear explosive devices. If the Government wishes to supply source or special fissionable material for peaceful purposes to such a State, it will:

   a)   Specify to the recipient State, as a condition of supply, that the source or special fissionable material, or special fissionable material produced in or by the use thereof, shall not be diverted to nuclear weapons or other nuclear explosive devices; and

   b)   Satisfy itself that safeguards to that end, under an agreement with the Agency and in accordance with its safeguards system, will be applied to the source or special fissionable material in question.

DIRECT EXPORTS

4.   In the case of direct exports of source or special fissionable material to non-nuclear-weapon States not party to NPT, the Government will satisfy itself, before authorising the export of the material in question, that such material will be subject to a safeguards agreement with the Agency, as soon as the recipient State takes over responsibility for the material, but no later than the time the material reaches its destination.

RETRANSFERS

5.   The Government, when exporting source or special fissionable material to a nuclear-weapon State not party to NPT, will require satisfactory assurances that the material will not be re-exported to a non-nuclear-weapon State not party to NPT unless arrangements corresponding to those referred to above are made for the acceptance of safeguards by the State receiving such re-export.

MISCELLANEOUS

6.   Exports of the items specified in sub-paragraph *(a)* below, and exports of source or special fissionable material to a given recipient country, within a period of 12 months, below the limits specified in sub-paragraph *(b)* below, shall be disregarded for the purpose of the procedures described below:

   a)   Plutonium with an isotopic concentration of plutonium 238 exceeding 80 per cent;

   Special fissionable material when used in gram quantities or less as a sensing component in instruments; and

   Source material which the Government is satisfied is to be used only in non-nuclear activities, such as the production of alloys or ceramics;

   b)   Special fissionable material   50 effective grams;
   Natural uranium   500 kilograms;
   Depleted uranium   1 000 kilograms; and
   Thorium   1 000 kilograms.

[1]   See also paragraph 6 below.

MEMORANDUM B

INTRODUCTION

1.   The Government has had under consideration procedures in relation to exports of certain categories of equipment and material, in the light of its commitment not to provide equipment or material especially designed or prepared for the processing, use or production of special fissionable material to any non-nuclear-weapon State for peaceful purposes, unless the source or special fissionable material produced, processed or used in the equipment or material in question is subject to safeguards under an agreement with the International Atomic Energy Agency.

THE DESIGNATION OF EQUIPMENT OR MATERIAL ESPECIALLY DESIGNED OR PREPARED FOR THE PROCESSING, USE OR PRODUCTION OF SPECIAL FISSIONABLE MATERIAL

2.   The designation of items of equipment or material especially designed or prepared for the processing, use or production of special fissionable material (hereinafter referred to as the "Trigger List") adopted by the Government is as follows (quantities below the indicated levels being regarded as insignificant for practical purposes):

2.1.   *Reactors and equipment therefor:*

    2.1.1.   Nuclear reactors capable of operation so as to maintain a controlled self-sustaining fission chain reaction, excluding zero energy reactors, the latter being defined as reactors with a designed maximum rate of production of plutonium not exceeding 100 grams per year.

    2.1.2.   Reactor pressure vessels:

    Metal vessels, as complete units or as major shop-fabricated parts therefor, which are especially designed or prepared to contain the core of a nuclear reactor as defined in paragraph 2.1.1 above and are capable of withstanding the operating pressure of the primary coolant.

    2.1.3.   Reactor fuel charging and discharging machines:

    Manipulative equipment especially designed or prepared for inserting or removing fuel in a nuclear reactor as defined in paragraph 2.1.1 above capable of on-load operation or employing technically sophisticated positioning or alignment features to allow complex off-load fuelling operations such as those in which direct viewing of or access to the fuel is not normally available.

    2.1.4.   Reactor control rods:

    Rods especially designed or prepared for the control of the reaction rate in a nuclear reactor as defined in paragraph 2.1.1 above.

    2.1.5.   Reactor pressure tubes:

    Tubes which are especially designed or prepared to contain fuel elements and the primary coolant in a reactor as defined in paragraph 2.1.1 above at an operating pressure in excess of 50 atmospheres.

    2.1.6.   Zirconium tubes:

    Zirconium metal and alloys in the form of tubes or assemblies of tubes, and in quantities exceeding 500 kg, especially designed or prepared for use in a reactor as defined in paragraph 2.1.1 above, and in which the relationship of hafnium to zirconium is less than 1:500 parts by weight.

    2.1.7.   Primary coolant pumps:

    Pumps especially designed or prepared for circulating liquid metal as primary coolant for nuclear reactors as defined in paragraph 2.1.1 above.

2.2.   *Non-nuclear materials for reactors:*

    2.2.1.   Deuterium and heavy water:

    Deuterium and any deuterium compound in which the ratio of deuterium to hydrogen exceeds 1:5000 for use in a nuclear reactor as defined in paragraph 2.1.1 above in quantities exceeding 200 kg of deuterium atoms for any one recipient country in any period of 12 months.

    2.2.2.   Nuclear grade graphite:

    Graphite having a purity level better than 5 parts per million boron equivalent and with a density greater than 1.50 grams per cubic centimetre in quantities exceeding 30 metric tons for any one recipient country in any period of 12 months.

    2.3.1.   Plants for the reprocessing of irradiated fuel elements, and equipment especially designed or prepared therefor.

2.4.1. Plants for the fabrication of fuel elements.

2.5.1. Equipment, other than analytical instruments, especially designed or prepared for the separation of isotopes of uranium.

Clarifications of certain of the items on the above list are annexed.

THE APPLICATION OF SAFEGUARDS

3.   The Government is solely concerned with ensuring, where relevant, the application of safeguards in non-nuclear-weapon States not party to the Treaty on the Non-Proliferation of Nuclear Weapons (NPT) with a view to preventing diversion of the safeguarded nuclear material from peaceful purposes to nuclear weapons or other nuclear explosive devices. If the Government wishes to supply Trigger List items for peaceful purposes to such a State, it will:

    a) Specify to the recipient State, as a condition of supply, that the source or special fissionable material produced, processed or used in the facility for which the item is supplied shall not be diverted to nuclear weapons or other nuclear explosive devices; and

    b) Satisfy itself that safeguards to that end, under an agreement with the Agency and in accordance with its safeguards system, will be applied to the source or special fissionable material in question.

DIRECT EXPORTS

4.   In the case of direct exports to non-nuclear-weapon States not party to NPT, the Government will satisfy itself, before authorising the export of the equipment or material in question, that such equipment or material will fall under a safeguards agreement with the Agency.

RETRANSFERS

5.   The Government, when exporting Trigger List items, will require satisfactory assurances that the items will not be re-exported to a non-nuclear-weapon State not party to NPT unless arrangements corresponding to those referred to above are made for the acceptance of safeguards by the State receiving such re-export.

MISCELLANEOUS

6.   The Government reserves to itself discretion as to interpretation and implementation of its commitment referred to in paragraph 1 above and the right to require, if it wishes, safeguards as above in relation to items it exports in addition to those items specified in paragraph 2 above.

Annex

CLARIFICATIONS OF ITEMS ON THE TRIGGER LIST

A.   *Complete nuclear reactors*
(Item 2.1.1 of the Trigger List)

1.   A "nuclear reactor" basically includes the items within or attached directly to the reactor vessel, the equipment which controls the level of power in the core, and the components which normally contain or come in direct contact with or control the primary coolant of the reactor core.

2.   The export of the whole set of major items within this boundary will take place only in accordance with the procedures of the memorandum. Those individual items within this functionally defined boundary which will be exported only in accordance with the procedures of the memorandum are listed in paragraphs 2.1.1. to 2.1.5. Pursuant to paragraph 6 of the memorandum, the Government reserves to itself the right to apply the procedures of the memorandum to other items within the functionally defined boundary.

3.   It is not intended to exclude reactors which could reasonably be capable of modification to produce significantly more than 100 grams of plutonium per year. Reactors designed for sustained operation at significant power levels, regardless of their capacity for plutonium production, are not considered as "zero energy reactors".

B. *Pressure vessels*

(Item 2.1.2 of the Trigger List)

4.   A top plate for a reactor pressure vessel is covered by item 2.1.2 as a major shop-fabricated part of a pressure vessel.

5.   Reactor internals (e.g. support columns and plates for the core and other vessel internals, control rod guide tubes, thermal shields, baffles, core grid plates, diffuser plates, etc.) are normally supplied by the reactor supplier. In some cases, certain internal support components are included in the fabrication of the pressure vessel. These items are sufficiently critical to the safety and reliability of the operation of the reactor (and, therefore, to the guarantees and liability of the reactor supplier), so that their supply, outside the basic supply arrangement for the reactor itself, would not be common practice. Therefore, although the separate supply of these unique, especially designed and prepared, critical, large and expensive items would not necessarily be considered as falling outside the area of concern, such a mode of supply is considered unlikely.

C. *Reactor control rods*

(Item 2.1.4 of the Trigger List)

6.   This item includes, in addition to the neutron absorbing part, the support or suspension structures therefor if supplied separately.

D. *Fuel reprocessing plants*

(Item 2.3.1 of the Trigger List)

7.   A "plant for the reprocessing of irradiated fuel elements" includes the equipment and components which normally come in direct contact with and directly control the irradiated fuel and the major nuclear material and fission product processing streams. The export of the whole set of major items within this boundary will take place only in accordance with the procedures of the memorandum. In the present state of technology only two items of equipment are considered to fall within the meaning of the phrase "and equipment especially designed or prepared therefor". These items are:

  a) Irradiated fuel element chopping machines: remotely operated equipment especially designed or prepared for use in a reprocessing plant as identified above and intended to cut, chop or shear irradiated nuclear fuel assemblies, bundles or rods; and

  b) Critically safe tanks (e.g. small diameter, annular or slab tanks) especially designed or prepared for use in a reprocessing plant as identified above, intended for dissolution of irradiated nuclear fuel and which are capable of withstanding hot, highly corrosive liquid, and which can be remotely loaded and maintained.

8.   Pursuant to paragraph 6 of the memorandum, the Government reserves to itself the right to apply the procedures of the memorandum to other items within the functionally defined boundary.

E. *Fuel fabrication plants*

(Item 2.4.1 of the Trigger List)

9.   A "plant for the fabrication of fuel elements" includes the equipment:

  a) Which normally comes in direct contact with, or directly processes, or controls, the production flow of nuclear material, or

  b) Which seals the nuclear material within the cladding.

10.   The export of the whole set of items for the foregoing operations will take place only in accordance with the procedures of the memorandum. The Government will also give consideration to application of the procedures of the memorandum to individual items intended for any of the foregoing operations, as well as for other fuel fabrication operations, such as checking the integrity of the cladding or the seal, and the finish treatment to the solid fuel*.

F. *Isotope separation plant equipment*

(Item 2.5.1 of the Trigger List)

11.   "Equipment, other than analytical instruments, especially designed or prepared for the separation of isotopes of uranium" includes each of the major items of equipment especially designed or prepared for the separation process.

---

*   In accordance with a corrigendum published in INFCIRC/209/Corr.1 the words "solid fuel" were replaced by "sealed fuel".

FURTHER COMMUNICATIONS RECEIVED FROM MEMBERS REGARDING THE EXPORT OF NUCLEAR MATERIAL AND OF CERTAIN CATEGORIES OF EQUIPMENT AND OTHER MATERIAL

(INFCIRC/209/Add.1 — 3rd October 1974)

1.   On 16th September 1974 the Director General received a letter of that date from the Resident Representative of Poland to the Agency in the same terms as the letter reproduced as Letter I in document INFCIRC/209, which deals with communications received from Members regarding the export of nuclear material and of certain categories of equipment and other material. The Resident Representative added that the date from which his Government had decided to act in accordance with the memoranda appended to document INFCIRC/209 was 3rd September.

2.   On 19th September the Director General received a letter of that date in the same terms from the Resident Representative of the German Democratic Republic.

ANNEX II-C

FURTHER COMMUNICATIONS RECEIVED FROM MEMBERS REGARDING THE EXPORT OF NUCLEAR MATERIAL AND OF CERTAIN CATEGORIES OF EQUIPMENT AND OTHER MATERIAL

(INFCIRC/209/Add.2 — 24th October 1974)

1.   On 7th October 1974 the Director General received a letter dated 3rd October 1974 from the Resident Representative of the United States of America to the Agency referring to his two letters of 22nd August regarding the export of nuclear material and of certain categories of equipment and other material[1]. In the light of the wish expressed at the end of that letter its text is reproduced below as Letter A.

2.   On the same day, the Director General also received a letter from the Resident Representative of the Union of Soviet Socialist Republics, dated 3rd October 1974, dealing with the same subject. The text of that letter is reproduced below as Letter B.

*
**

Letter A

"I have the honour to refer to my letters of 22nd August 1974, concerning procedures of my Government in relation to exports of source and special fissionable material and of equipment and material especially designed or prepared for the processing, use or production of special fissionable material.

"Deliveries to the European Atomic Energy Community and to its Members of source and special fissionable material and of equipment and material especially designed or prepared for the processing, use or production of special fissionable material, under contracts made pursuant to existing agreements between the United States of America and the European Atomic Energy Community will continue to be made, in the light of our expectation that the agreement between the International Atomic Energy Agency, the European Atomic Energy Community and certain of its Member States[2], signed on 5th April 1973, will enter into force in the very near future.

"With respect to paragraph 3 of each of the memoranda transmitted on 22nd August[3], I wish to note that the representative of the Government of the United States has placed on the record of meetings of the Board of Governors of the Agency held on 1st March 1972 and on 12th June 1974 the understanding inherent in all of the bilateral agreements for co-operation to which the Government of the United States is a party, that the use of any material or equipment supplied by the United States under such agreements for any nuclear explosive device is precluded; and the understanding inherent in the safeguards agreements related to such co-operation agreements, that the Agency would verify, inter alia, that the safeguarded material was not used for any nuclear explosive device. It was further noted by the United States representative that the continued co-operation of the United States with other countries in the nuclear field is dependent on the assurance that these understandings will continue to be respected in the future.

"I shall be grateful if you will bring this information to the attention of all Members of the Agency."

Letter B

"The Soviet Union is scrupulously fulfilling its obligations under the Treaty on the Non-Proliferation of Nuclear Weapons (NPT), Article I of which, as we know, provides inter alia that nuclear-weapon States party to the Treaty must 'not in any way .... assist, encourage or induce any non-nuclear-weapon State to manufacture or otherwise acquire nuclear weapons or other nuclear explosive devices'. This obligation applies in full to the supply to any non-nuclear-weapon State of the equipment and materials mentioned in Article III.2 of the Treaty, which may not be used in those countries for the manufacture of nuclear weapons or other nuclear explosive devices.

"In connection with the entry into force of the safeguards machinery referred to in Article III.2 of NPT, we deem it necessary to emphasize once more the importance of the speediest possible completion of the process of accession to the Treaty by the countries members of the European Atomic Energy Community which have signed it, and of the entry into force of the appropriate Safeguards Agreement with the Agency."

[1]   Reproduced in document INFCIRC/209 as Letters I and III.
[2]   Reproduced in document INFCIRC/193.
[3]   Reproduced in document INFCIRC/209, Appendix, as Memoranda A and B.

ANNEX III

COMMUNICATIONS RECEIVED FROM MEMBERS REGARDING THE EXPORT OF NUCLEAR MATERIAL AND OF CERTAIN CATEGORIES OF EQUIPMENT AND OTHER MATERIAL

(INFCIRC/209/Mod.1 — December 1978)

1.   During the months of September and October, the Director General received identical letters, all dated 1st September 1978, from the Resident Representatives of Australia, Canada, Denmark, Finland, the Federal Republic of Germany, the Netherlands, Norway, the Union of Soviet Socialist Republics, the United Kingdom of Great Britain and Northern Ireland and the United States of America, relating to the commitments of these Members under Article III, paragraph 2, of the Treaty on the Non-Proliferation of Nuclear Weapons.

2.   In the light of the wish expressed at the end of each of the letters referred to above, the text thereof is reproduced below.

*
* *

1st September 1978

I have the honour to refer to Memorandum B enclosed with the letter of 22nd August 1974 in which the Government of .... informed you that it had decided to act in accordance with certain procedures in relation to exports of nuclear material and of certain categories of equipment and other material, and which you circulated to all Members of the Agency as document INFCIRC/209. The Government of .... has instructed me to inform you that it is now its intention to act in accordance with that Memorandum amended as follows:

a)   Paragraph 2.1.6 to read "Zirconium tubes: Zirconium metal and alloys in the form of tubes or assemblies of tubes, and in quantities exceeding 500 kg per year, especially designed or prepared for use in a reactor as defined in paragraph 2.1.1 above and in which the relationship of hafnium to zirconium is less than 1:500 parts by weight".

b)   A new paragraph following paragraph 2.5.1 to read "2.6.1. Plants for production of heavy water, deuterium and deuterium compounds and equipment especially designed or prepared therefor".

c)   A new sentence following the first sentence of paragraph F to the Annex to read "Such items include:
gaseous diffusion barriers
gaseous diffuser housings
gas centrifuge assemblies, corrosion resistant to UF_6
jet nozzle separation units
vortex separation units
large UF_6 corrosion-resistant axial or centrifugal compressors
special compressor seals for such compressors".

I should be grateful if you would bring this information to the attention of all Members of the Agency.

COMMUNICATIONS RECEIVED FROM MEMBERS REGARDING THE EXPORT OF NUCLEAR MATERIAL AND OF CERTAIN CATEGORIES OF EQUIPMENT AND OTHER MATERIAL

(INFCIRC/209/Mod.2 — February 1984)

1. The Director General has received letters from the Resident Representatives to the Agency of the following Member States concerning the commitments of these Member States under Article III, paragraph 2, of the Treaty on the Non-Proliferation of Nuclear Weapons (the date of each letter is given in parentheses after the name of the Member State):

Australia (25th January 1984)
Canada (13th February 1984)
Czechoslovakia (24th January 1984)
Finland (24th January 1984)
German Democratic Republic (24th January 1984)
Federal Republic of Germany (24th January 1984)
Ireland (24th January 1984)
Japan (24th January 1984)
Netherlands (24th January 1984)
Norway (24th January 1984)
Poland (24th January 1984)
Sweden (31st January 1984)
Union of Soviet Socialist Republics (24th January 1984)
United Kingdom of Great Britain and Northern Ireland (24th January 1984)
United States of America (24th January 1984)

2. In the light of the wish expressed at the end of each of the letters, both the text of the letters and their Annex are attached.

LETTER

"I have the honour to refer to [relevant previous communication] from the Resident Representative of [Member State] to the International Atomic Energy Agency.

"In the years since the procedures for export of nuclear material and of certain categories of equipment and other material enclosed with these notes were formulated, there have been considerable changes in nuclear technology which make it desirable, in the view of my Government, to clarify those parts of the trigger list which refer to the gas centrifuge enrichment process.

"In relation to the Annex entitled "Clarifications of Items on the Trigger List" attached to Memorandum B, I therefore wish to inform you that the items relating to this process and specified in paragraph 11 of that Annex include:

a) Assemblies and components especially designed or prepared for use in gas centrifuges. This includes rotating and static components listed in sections 1.1 and 1.2 of the Annex to this letter.
b) Especially designed or prepared auxiliary systems, equipment and components for gas centrifuge enrichment plants. This includes the items listed in section 2 of the Annex to this letter.

"Both categories of items are introduced by explanatory notes.

"As hitherto, my Government reserves to itself discretion as to the interpretation and implementation of the procedures provided in the above-mentioned notes and the right to control, if it wishes, the export of items relevant to the gas centrifuge enrichment process other than those specified in the Annex to this letter.

"I should be grateful if you would circulate the text of this letter and its Annex to all Member Governments for their information."

Annex

1.   *Assemblies and components especially designed or prepared for use in gas centrigues*

     Note:

The gas centrifuge normally consists of a thin-walled cylinder(s) of between 75 mm (3 ins) and 400 mm (16 ins) diameter contained in a vacuum environment and spun at high peripheral speed of the order of 300 m/sec or more with its central axis vertical. In order to achieve high speed the materials of construction for the rotating components have to be of a high strength to density ratio and the rotor assembly, and hence its individual components, have to be manufactured to very close tolerances in order to minimise the unbalance. In contrast to other centrifuges, the gas centrifuge for uranium enrichment is characterised by having within the rotor chamber a rotating disc-shaped baffle(s) and a stationary tube arrangement for feeding and extracting the UF_6 gas and featuring at least 3 separate channels, of which 2 are connected to scoops extending from the rotor axis towards the periphery of the rotor chamber. Also contained within the vacuum environment are a number of critical items which do not rotate and which although they are especially designed are not difficult to fabricate nor are they fabricated out of unique materials. A centrifuge facility however requires a large number of these components, so that quantities can provide an important indication of end use.

1.1. *Rotating Components*

a)   Complete Rotor Assemblies:

Thin-walled cylinders, or a number of interconnected thin-walled cylinders, manufactured from one of the high strength to density ratio materials described in the footnote to this Section.

If interconnected, the cylinders are joined together by flexible bellows or rings as described in Section 1.1(c) following. The rotor is fitted with an internal baffle(s) and end caps, as described in Section 1.1(d) and (e) following, if in final form. However the complete assembly may be delivered only partly assembled.

b)   Rotor Tubes:

Especially designed or prepared thin-walled cylinders with thickness of 12 mm (.50 in) or less, a diameter of between 75 mm (3 ins) and 400 mm (16 ins), and manufactured from one of the high strength to density ratio materials described in the footnote to this Section.

c)   Rings or Bellows:

Components especially designed or prepared to give localised support to the rotor tube or to join together a number of rotor tubes. The bellows is a short cylinder of wall thickness 3 mm (.125 ins) or less, a diameter of between 75 mm (3 ins) and 400 mm (16 ins), having a convolute, and manufactured from one of the high strength to density ratio materials described in the footnote to this Section.

d)   Baffles:

Disc-shaped components of between 75 mm (3 ins) and 400 mm (16 ins) diameter especially designed or prepared to be mounted inside the centrifuge rotor tube, in order to isolate the take-off chamber from the main separation chamber and, in some cases, to assist the UF_6 gas circulation within the main separation chamber of the rotor tube, and manufactured from one of the high strength to density ratio materials described in the footnote to this Section.

e)   Top Caps/Bottom Caps:

Disc-shaped components of between 75 mm (3 ins) and 400 mm (16 ins) diameter especially designed or prepared to fit to the ends of the rotor tube, and so contain the UF_6 within the rotor tube, and in some cases to support, retain or contain as an integrated part an element of the upper bearing (top cap) or to carry the rotating elements of the motor and lower bearing (bottom cap), and manufactured from one of the high strength to density ratio materials described in the footnote to this Section.

     Footnote

The materials used for centrifuge rotating components are:

a)   Maraging steel capable of an ultimate tensile strength of 2.050×10^9 N/m^2 (300 000 lb/in^2) or more;

b)   Aluminium alloys capable of an ultimate tensile strength of 0.460×10^9 N/m^2 (67 000 lb/in^2) or more;

c)   Filamentary materials suitable for use in composite structures and having a specific modulus of 12.3×10^6 or greater and a specific ultimate tensile strength of 0.3×10^6 or greater ('Specific Modulus' is the Young's Modulus in N/m^2 divided by the density in kg/m^3: 'Specific Ultimate Tensile Strength' is the ultimate tensile strength in N/m^2 divided by the density in kg/m^3).

1.2. *Static Components*

a)   Magnetic Suspension Bearings:

Especially designed or prepared bearing assemblies consisting of an annular magnet suspended within a housing containing a damping medium. The housing will be manufactured from a UF_6 resistant material (see footnote to Section 2). The magnet couples with a pole piece or a second magnet fitted to the top cap described in Section 1.1(e). The magnet may be ring-shaped with a relation between outer and inner diameter smaller or equal to 1.6:1. The magnet may be in a form having an initial permeability of 0.15 Henry/metre (120 000 in CGS units) or more, or a remanence of 98.5 per cent or more, or an energy product of greater than 80 000 joules/m^3 (10 x 10^6 gauss-oersteds). In addition to the usual material properties, it is a prerequisite that the deviation of the magnet axes from the geometrical axes is limited to very small tolerances (lower than 0.1 mm) or that homogeneity of the material of the magnet is specially called for.

b)   Bearings/Dampers:

Especially designed or prepared bearings comprising a pivot/cup assembly mounted on a damper. The pivot is normally a hardened steel shaft polished into a hemisphere at one end with a means of attachment to the bottom cap described in Section 1.1(e) at the other. The shaft may however have a hydrodynamic bearing attached. The cup is pellet-shaped with a hemispherical indentation in one surface. These components are often supplied separately to the damper.

c)   Molecular Pumps:

Especially designed or prepared cylinders having internally machined or extruded helical grooves and internally machined bores. Typical dimensions are as follows: 75 mm (3 ins) to 400 mm (16 ins) internal diameter, 10 mm (0.4 in) or more wall thickness, 1 to 1 length to diameter ratio. The grooves are typically rectangular in cross-section and 2 mm (0.08 in) or more in depth.

d)   Motor Stators:

Especially designed or prepared ring-shaped stators for high speed multiphase AC hysteresis (or reluctance) motors for synchronous operation within a vacuum in the frequency range of 600-2000 Hertz and a power range of 50-1000 volts amps. The stators consist of multiphase windings on a laminated low loss iron core comprised of thin layers typically 2.0 mm (0.080 in) thick or less.

2.   *Especially designed or prepared auxiliary systems, equipment and components for gas centrifuge enrichment plants*

   Note:

   The auxiliary systems, equipment and components for a gas centrifuge enrichment plant are the systems of plant needed to feed UF$_6$ to the centrifuges to link the individual centrifuges to each other to form cascades (or stages) to allow for progressively higher enrichments and to extract the 'product' and 'tails' UF$_6$ from the centrifuges, together with the equipment required to drive the centrifuges or to control the plant.

   Normally UF$_6$ is evaporated from the solid using heated autoclaves and is distributed in gaseous form to the centrifuges by way of cascade header pipework. The 'product' and 'tails' UF$_6$ gaseous streams flowing from the centrifuges are also passed by way of cascade header pipework to cold traps (operating at about -70ºC) where they are condensed prior to onward transfer into suitable containers for transportation or storage. Because an enrichment plant consists of many thousands of centrifuges arranged in cascades there are many kilometres of cascade header pipework, incorporating thousands of welds with a substantial amount of repetition of layout. The equipment, components and piping systems are fabricated to very high vacuum and cleanliness standards.

   The items listed below either come into direct contact with the UF$_6$ process gas or directly control the centrifuges and the passage of the gas from centrifuge to centrifuge and cascade to cascade.

   a)   Feed System/Product and Tails Withdrawal Systems:

   Especially designed or prepared process systems including:

   —   Feed autoclaves (or stations), used for passing UF$_6$ to the centrifuge cascades at up to 100 KN/m^2 (15 lb/in^2) and at a rate of 1 kg/hr or more;
   —   Desublimers (or cold traps) used to remove UF$_6$ from the cascades at up to 3 KN/m^2 (0.5 lb/in^2) pressure. The desublimers are capable of being chilled to -70ºC and heated to 70ºC;
   —   'Product' and 'Tails' stations used for trapping UF$_6$ into containers.

   This plant, equipment and pipework is wholly made of or lined with UF$_6$ resistant materials (see footnote to this Section) and is fabricated to very high vacuum and cleanliness standards.

   b)   Machine Header Piping Systems:

   Especially designed or prepared piping systems and header systems for handling UF$_6$ within the centrifuge cascades. This piping network is normally of the 'triple' header system with each centrifuge connected to each of the headers. There is thus a substantial amount of repetition in its form. It is wholly made of UF$_6$ resistant materials (see footnote to this Section) and is fabricated to very high vacuum and cleanliness standards.

   c)   UF$_6$ Mass Spectrometers/Ion Sources:

   Especially designed or prepared magnetic or quadrupole mass spectrometers capable of taking 'on-line' samples of feed, product or tails, from UF$_6$ gas streams and having all of the following characteristics:

   1.   Unit resolution for mass greater than 320;
   2.   Ion sources constructed of or lined with nichrome or monel or nickel plated;
   3.   Electron bombardment ionization sources;
   4.   Having a collector system suitable for isotopic analysis.

d)   Frequency Changers:

Frequency changers (also known as converters or invertors) especially designed or prepared to supply motor stators as defined under 1.2(d), or parts, components and sub-assemblies of such frequency changers having all of the following characteristics:

1.   A multiphase output of 600 Hz to 2000 Hz;
2.   High stability (with frequency control better than 0.1 per cent);
3.   Low harmonic distortion (less than 2 per cent); and
4.   An efficiency of greater than 80 per cent.

Footnote

Materials resistant to corrosion by UF_6 include stainless steel, aluminium, aluminium alloys, nickel or alloys containing 60 per cent or more nickel.

ANNEX V

COMMUNICATIONS RECEIVED FROM MEMBERS REGARDING THE EXPORT OF NUCLEAR MATERIAL AND OF CERTAIN CATEGORIES OF EQUIPMENT AND OTHER MATERIAL

(INFCIRC/209/Mod.3 — August 1985)

1. The Director General has received letters dated 1st July 1985 from the Resident Representatives to the Agency of the following Member States concerning the commitments of these Member States under Article III, paragraph 2, of the Treaty on the Non-Proliferation of Nuclear Weapons: Australia, Canada, Finland, the German Democratic Republic, the Federal Republic of Germany, Ireland, Japan, the Netherlands, Norway, Poland, Sweden, the Union of Soviet Socialist Republics, the United Kingdom of Great Britain and Northern Ireland, and the United States of America.

2. In the light of the wish expressed at the end of each of the letters, both the text of the letters and their Annex are attached.

LETTER

"I have the honour to refer to [relevant previous communication(s)] from the Resident Representative of [Member State] to the International Atomic Energy Agency.

"In the years since the procedures for export of nuclear materials and of certain categories of equipment and other material described in INFCIRC/209 were formulated, there have been considerable changes in nuclear technology which make it desirable, in the view of my Government, to clarify those parts of the Trigger List which refer to fuel reprocessing technology.

"In relation to the Annex entitled "Clarifications of Items on the Trigger List" attached to Memorandum B in INFCIRC/209, I therefore wish to inform you that the items relating to Fuel Reprocessing Plants and specified in paragraph 7 of that Annex also include the items listed in the Annex to this letter. These items are introduced by an explanatory note.

"As hitherto, my Government reserves to itself discretion as to the interpretation and implementation of the procedures provided in the above-mentioned documents and the right to control, if it wishes, the export of items relevant to fuel reprocessing plants other than those specified in INFCIRC/209 and in the Annex to this letter.

"I should be grateful if you would circulate the text of this letter and its Annex to all Member Governments for their information."

Annex

INTRODUCTORY NOTE: SPENT NUCLEAR FUEL REPROCESSING

Reprocessing irradiated nuclear fuel separates plutonium and uranium from intensely radioactive fission products and other transuranic elements. Different technical processes can accomplish this separation. However, over the years Purex has become the most commonly used and accepted process. Purex involves the dissolution of irradiated nuclear fuel in nitric acid, followed by separation of the uranium, plutonium, and fission products by solvent extraction using a mixture of tributyl phosphate in an organic diluent.

Purex facilities have process functions similar to each other, including: irradiated fuel element chopping, fuel dissolution, solvent extraction, and process liquor storage. There may also be equipment for thermal denitration of uranium nitrate, conversion of plutonium nitrate to oxide or metal, and treatment of fission product waste liquor to a form suitable for long-term storage or disposal. However, the specific type and configuration of the equipment performing these functions may differ between Purex facilities for several reasons, including the type and quantity of irradiated nuclear fuel to be reprocessed and the intended disposition of the recovered materials, and the safety and maintenance philosophy incorporated into the design of the facility.

The equipment listed below performs key reprocessing functions. Each comes into direct contact with the irradiated fuel or process liquor and operates in an environment characterised by criticality, radiation, and toxicity hazards. These make remote control of the process essential.

1) *Fuel element chopping*

This equipment breaches the cladding of the fuel to expose the irradiated nuclear material to dissolution. Especially designed metal cutting shears are the most commonly employed, although advanced equipment, such as lasers, may be used.

2) *Dissolvers*

Dissolvers normally receive the chopped up spent fuel. In these critically safe vessels, the irradiated nuclear material is dissolved in nitric acid and the remaining hulls removed from the process stream.

3) *Solvent extractors*

Solvent extractors both receive the solution of irradiated fuel from the dissolvers and the organic solution which separates the uranium, plutonium, and fission products. Solvent extraction equipment is normally designed to meet strict operating parameters, such as long operating lifetimes with no maintenance requirements or adaptability to easy replacement, simplicity of operation and control, and flexibility for variations in process conditions.

4) *Holding or storage vessels*

Three main process liquor streams result from the solvent extraction step. Holding or storage vessels are used in the further processing of all three streams, as follows:

a) The pure uranium nitrate solution is concentrated by evaporation and passed to a denitration process where it is converted to uranium oxide. This oxide is reused in the nuclear fuel cycle.
b) The intensely radioactive fission products solution is normally concentrated by evaporation and stored as a liquor concentrate. This concentrate may be subsequently evaporated and converted to a form suitable for storage or disposal.
c) The pure plutonium nitrate solution is concentrated and stored pending its transfer to further process steps. In particular, holding or storage vessels for plutonium solutions are designed to avoid criticality problems resulting from changes in concentration and form of this stream.

5) *Plutonium nitrate to oxide conversion system*

In most reprocessing facilities, this final process involves the conversion of the plutonium nitrate solution to plutonium dioxide. The main functions involved in this process are: process feed storage and adjustment, precipitation and solid/liquor separation, calcination, product handling, ventilation, waste management, and process control.

6) *Plutonium oxide to metal conversion system*

This process, which could be related to a reprocessing facility, involves the fluorination of plutonium dioxide, normally with highly corrosive hydrogen fluoride, to produce plutonium fluoride which is subsequently reduced using high purity calcium metal to produce metallic plutonium and a calcium fluoride slag. The main functions involved in this process are: fluorination (e.g. involving equipment fabricated or lined with a precious metal), metal reduction (e.g. employing ceramic crucibles), slag recovery, product handling, ventilation, waste management and process control.

These processes, including the complete systems for plutonium conversion and plutonium metal production, may be identified by the measures taken to avoid criticality (e.g. by geometry), radiation exposure (e.g. by shielding), and toxicity hazards (e.g. by containment).

DEFINITIONS FOR REPROCESSING

A.   *Solvent extraction equipment* — Especially designed or prepared solvent extractors such as packed or pulse columns, mixer settlers or centrifugal contractors for use in a plant for the reprocessing of irradiated fuel. Solvent extractors must be resistant to the corrosive effect of nitric acid. Solvent extractors are normally fabricated to extremely high standards (including special welding and inspection and quality assurance and quality control techniques) out of low carbon stainless steels, titanium, zirconium or other high quality materials.

B.   *Chemical holding or storage vessels* — Especially designed or prepared holding or storage vessels for use in a plant for the reprocessing of irradiated fuel. The holding or storage vessels must be resistant to the corrosive effect of nitric acid. The holding or storage vessels are normally fabricated of materials such as low carbon stainless steels, titanium or zirconium, or other high quality materials. Holding or storage vessels may be designed for remote operation and maintenance and may have the following features for control of nuclear criticality:

1) walls or internal structures with a boron equivalent of at least two per cent, or
2) a maximum diameter of 7 inches (17.78 cm) for cylindrical vessels, or
3) a maximum width of 3 inches (7.62 cm) for either a slab or annular vessel.

C.   *Plutonium nitrate to plutonium oxide conversion systems* — Complete systems especially designed or prepared for the conversion of plutonium nitrate to plutonium oxide, in particular adapted so as to avoid criticality and radiation effects and to minimise toxicity hazards.

D.   *Plutonium metal production system* — Complete systems especially designed or prepared for the production of plutonium metal, in particular adapted so as to avoid criticality and radiation effects and to minimise toxicity hazards.

ANNEX VI

STATES HAVING UNDERTAKEN COMMITMENTS FOLLOWING CONSULTATIONS IN THE LONDON CLUB

	Reservations		
	Adjustment of domestic regulations	Common Market	Additional requirements
I. Original Members			
* Canada			
France			
* Federal Republic of Germany		x	
* Japan	x		
* United Kingdom		x	
* United States			+Full-scope safeguards
* USSR			
II. Members since 1976			
Belgium	x	x	
* Czechoslovakia			+Full-scope safeguards
* German Democratic Republic			+Full-scope safeguards
Italy		x	+Other rights and obligations stemming from agreements, in particular Article IV of NPT
* Netherlands		x	
* Poland			+Full-scope safeguards
* Sweden			
Switzerland	x		
III. States having decided to conform to the Guidelines after 1978 and without having participated in the work			
* Australia (21.2.1978)			+Other requirements
Bulgaria (14.12.1984)			
* Denmark (13.8.1984)		x	
* Finland (28.1.1980)	x		+Full-scope safeguards (request)
* Greece (19.9.1984)		x	
* Hungary (2.5.1985)			
* Ireland (19.11.1984)		x	
* Luxembourg (13.11.1984)	x	x	
Portugal (10.1.1986)		x	

* Also member of Zangger Committee.
  It should be noted that Belgium, Bulgaria, France (not an NPT signatory), Greece, Italy, Portugal and Switzerland particpated in the work of the London Club or subsequently adopted its conclusions without having signed the letters concluding the work of the Zangger Committee. On the other hand, Austria and Norway took part in the Zangger Committee but not in the London Club.

COMMUNICATIONS RECEIVED FROM CERTAIN MEMBER STATES REGARDING GUIDELINES FOR THE EXPORT OF NUCLEAR MATERIAL, EQUIPMENT OR TECHNOLOGY

(INFCIRC/254 — February 1978)

1.   On 11th January 1978, the Director General received similar letters, all of that date, from the Resident Representatives to the Agency of Czechoslovakia, France, the German Democratic Republic, Japan, Poland, Switzerland, the Union of Soviet Socialist Republics and the United States of America, relating to the export of nuclear material, equipment or technology. In the light of the request at the end of each of those letters, the text is reproduced below as Letter I.

2.   On the same day, the Resident Representatives to the Agency of Canada and Sweden also addressed analogous letters to the Director General. In the light of the request expressed at the end of each of those letters, their texts are reproduced below as Letter II and Letter III respectively.

3.   On the same day, the Director General received similar letters from the Resident Representatives to the Agency of Belgium, the Federal Republic of Germany, the Netherlands and the United Kingdom of Great Britain and Northern Ireland, Members of the European Communities, relating to the export of nuclear material, equipment or technology. In the light of the request expressed at the end of each of those letters, the text is reproduced below as Letter IV.

4.   On 11th January 1978 the Resident Representative to the Agency of Italy, a Member of the European Communities, addressed a letter to the Director General relating to the same subject, the text of which is reproduced below as Letter V.

5.   On 11th January 1978 the Director General received complementary letters, all of that date, from the Resident Representatives to the Agency of Belgium, Czechoslovakia, the German Democratic Republic, Japan, Poland, Switzerland and the Union of Soviet Socialist Republics, the texts of which are reproduced belows as Letters VI, VII, VIII, IX, X, XI and XII respectively.

6.   The attachments to Letters I-V, which are in every case identical, setting forth the Guidelines for Nuclear Transfers with their Annexes, are reproduced in the Appendix.

LETTER I

The Permanent Mission of .... presents its compliments to the Director General of the International Atomic Energy Agency and has the honour to enclose copies of three documents which have been the subject of discussion between the Government of .... and a number of other Governments.

The Government of .... has decided that, when considering the export of nuclear material, equipment or technology, it will act in accordance with the principles contained in the attached documents.

In reaching this decision, the Government of .... is fully aware of the need to contribute to the development of nuclear power in order to meet world energy requirements, while avoiding contributing in any way to the dangers of a proliferation of nuclear weapons or other nuclear explosive devices, and of the need to remove safeguards and non-proliferation assurances from the field of commercial competition.

The Government of .... hopes that other Governments may also decide to base their own nuclear export policies upon these documents.

The Government of .... requests that the Director General of the International Atomic Energy Agency should circulate the texts of this note and its enclosures to all Member Governments for their information and as a demonstration of support by the Government of .... for the Agency's non-proliferation objectives and safeguards activities.

The Permanent Mission of .... avails itself of this opportunity to renew to the Director General of the International Atomic Energy Agency the assurances of its highest consideration.

LETTER II

The Permanent Mission of Canada to the IAEA presents its compliments to the Director General and has the honour to enclose copies of three documents that have been the subject of discussion between the Government of Canada and a number of other Governments.

The Government of Canada has decided that, when considering the export of nuclear material, equipment or technology, it will act in accordance with the principles contained in the attached documents as well as other principles considered pertinent by it.

In reaching this decision, the Government of Canada is fully aware of the need to contribute to the development of nuclear power in order to meet world energy requirements, while avoiding contributing in any way to the dangers of a proliferation of nuclear weapons or other nuclear explosive devices, and of the need to remove safeguards and non-proliferation assurances from the field of commercial competition.

The Government of Canada hopes that other Governments may also decide to base their own nuclear export policies upon these documents and such further principles as may be agreed upon.

The Government of Canada requests that the Director General of the International Atomic Energy Agency should circulate the text of this Note and its enclosures to all Member Governments for their information and as a demonstration of support by the Government of Canada for the Agency's non-proliferation objectives and safeguard activities.

The Permanent Mission of Canada to the IAEA avails itself of this opportunity to renew to the Director General the assurances of its highest consideration.

LETTER III

The Permanent Mission of Sweden present their compliments to the Director General of the International Atomic Energy Agency and have the honour to enclose copies of three documents which have been the subject of discussion between the Government of Sweden and a number of other Governments.

The Government of Sweden have decided that, when considering the export of nuclear material, equipment or technology, they will act in accordance with the principles contained in the attached documents.

In reaching this decision, the Government of Sweden are fully aware of the need to avoid contributing in any way to the dangers of a proliferation of nuclear weapons or other nuclear explosive devices, and of the need to remove safeguards and non-proliferation assurances from the field of commercial competition.

The Government of Sweden hope that other Governments may also decide to base their own nuclear export policies upon these documents.

The Government of Sweden request that the Director General of the International Atomic Energy Agency should circulate the text of this Note and its enclosures to all Member Governments for their information and as a demonstration of support by the Government of Sweden for the Agency's non-proliferation objectives and safeguards activities.

The Permanent Mission of Sweden take this opportunity to renew to the Director General of the International Atomic Energy Agency the assurances of their highest consideration.

LETTER IV

The Permanent Mission of .... to the International Organisations in Vienna presents its compliments to the Director General of the International Atomic Energy Agency and has the honour to enclose copies of three documents which have been the subject of discussion between the .... and a number of other Governments.

The Government of .... has decided that, when considering the export of nuclear material, equipment or technology, it will act in accordance with the principles contained in the attached documents.

In reaching this decision, the Government of .... is fully aware of the need to contribute to the development of nuclear power in order to meet world energy requirements, while avoiding contributing in any way to the danger of a proliferation of nuclear weapons or other nuclear explosive devices, and of the need to remove safeguards and non-proliferation assurances from the field of commercial competition.

As a Member of the European Community, the Government of .... so far as trade within the Community is concerned, will implement these documents in the light of its commitments under the Treaties of Rome where necessary.

The Government of .... hopes that other Governments may also decide to base their own nuclear export policies upon these documents.

The Government of .... requests that the Director General of the International Atomic Energy Agency should circulate the texts of this Note and its enclosures to all Member Governments for their information and as a demonstration of support by the Government of .... for the Agency's non-proliferation objectives and safeguards activities.

The Permanent Mission of .... to the International Organisations in Vienna avails itself of this opportunity to renew to the Director General of the International Atomic Energy Agency the assurances of its highest consideration.

LETTER V

Permanent Mission of Italy present their compliments and have the honour to enclose copies of three documents which have been the subject of discussion between the Government of Italy and a number of other Governments.

The Government of Italy have decided that, when considering the export of nuclear material, equipment or technology, they will act in accordance with the principles contained in the attached documents.

In reaching this decision, the Government of Italy are fully aware of the need to contribute to the development of nuclear power in order to meet world energy requirements, while avoiding contributing in any way to dangers of a proliferation of nuclear weapons or other nuclear explosive devices, and of the need to remove safeguards and non-proliferation assurances from the field of commercial competition.

The Italian Government underline that the undertaking referred to cannot limit in any way the rights and obligations arising for Italy out of agreements to which she is a Party, and in particular those arising out of Article IV of the Non-Proliferation Treaty.

As a Member of the European Community, the Government of Italy, so far as trade within the Community is concerned, will implement these documents in the light of their commitments under the Treaties of Rome where necessary.

The Government of Italy hope that other Governments may also decide to base their own nuclear export policies upon these documents.

The Government of Italy request that the Director General of the International Atomic Energy Agency should circulate the texts of this Note and its enclosures to all Member Governments for their information and as a demonstration of support by the Government of Italy for the Agency's non-proliferation objectives and safeguards activities.

LETTER VI

The Permanent Mission of Belgium presents its compliments to the Director General of the IAEA and, in addition to its Note P 10-92/24 of 11th January 1978, would like to draw the attention to the following.

The Government of Belgium at present are not in a position to implement fully the principles for technology transfer set out in the documents attached to the above-mentioned Note because of the lack of appropriate laws and regulations. However, the Government of Belgium intend to implement these principles fully when appropriate laws and regulations for this purpose are put into force as necessary.

The Government of Belgium request that the Director General of the IAEA should circulate the text of this Note to all Member Governments for their information.

The Permanent Mission of Belgium takes this opportunity to renew to the Director General of the IAEA the assurance of its highest consideration.

LETTER VII

The Permanent Mission of the Czechoslovak Socialist Republic to the International Organizations presents its compliments to the Director General of the International Atomic Energy Agency and has the honour to refer to its Note No. 1036/78 regarding standards of the nuclear export policies which have been adopted by the members of the Nuclear Suppliers Group.

The Government of the Czechoslovak Socialist Republic greatly appreciates the role of the International Atomic Energy Agency in the sphere of control of the provisions of the Non-Proliferation Treaty. This activity has been an important instrument of preventing proliferation of nuclear weapons. Sharing the opinion that further strengthening of safeguards lies in the interest of universal peace, the Government of the Czechoslovak Socialist Republic has decided that it would deliver nuclear material, equipment and technology defined in a trigger list, to any non-nuclear-weapon State only in a case when the whole nuclear activity of a recipient country, and not only material, equipment and technology being transferred, are subject to the Agency's safeguards.

The Government of the Czechoslovak Socialist Republic expresses its opinion that this principle, if observed by all the States — nuclear suppliers, could have made a great contribution toward strengthening and universality of the Non-Proliferation Treaty.

The Permanent Mission of the Czechoslovak Socialist Republic to the International Organizations avails itself of this opportunity to renew to the Director General of the International Atomic Energy Agency the assurances of its highest consideration.

LETTER VIII

The Permanent Mission of the German Democratic Republic to the International Organizations in Vienna presents its compliments to the Director General of the International Atomic Energy Agency and has the honour, in connection with Note No. 2/78-III addressed to the Director General of the IAEA on 11th January 1978, to state the following: in the view of the

Government of the German Democratic Republic, the guidelines for nuclear exports are such as to strengthen the regime of non-proliferation of nuclear weapons and the IAEA safeguards system. The German Democratic Republic will also in future advocate agreements to the effect that nuclear exports under the trigger list mentioned in the above Note should go only to those non-nuclear-weapon States that accept IAEA safeguards for all of their nuclear activities.

The Government of the German Democratic Republic is convinced that any reinforcement of the regime of non-proliferation of nuclear weapons will promote the peaceful uses of nuclear energy and international co-operation in this area.

The Permanent Mission requests that the present text be circulated as an official document of the International Atomic Energy Agency.

The Permanent Mission of the German Democratic Republic to the International Organizations in Vienna avails itself of this opportunity to renew to the Director General of the International Atomic Energy Agency the assurances of its highest consideration.

LETTER IX

The Embassy of Japan presents its compliments to the International Atomic Energy Agency and, in reference to its Note No. J.M. 78/21 of 11th January 1978, has the honour to inform the International Atomic Energy Agency of the following.

The Government of Japan at present is not in a position to implement fully the Principles for Technology Transfers set out in the documents attached to the above-mentioned Note because of the lack of appropriate laws and regulations.

However, the Government of Japan intends to implement these principles fully when appropriate laws and regulations for this purpose are put into force as necessary.

The Government of Japan requests that the Director General of the International Atomic Energy Agency be good enough to circulate the texts of this Note to all Member Governments for their information.

The Embassy of Japan avails itself of this opportunity to renew to the International Atomic Energy Agency the assurances of its highest consideration.

LETTER X

The Permanent Mission of the Polish People's Republic to the International Atomic Energy Agency presents its compliments to the Director General of the IAEA and has the honour to refer to its Note No. 10-96/77 regarding standards of the nuclear export policies which have been adopted by the members of the Nuclear Suppliers Group.

The Government of the Polish People's Republic greatly appreciates the role of the International Atomic Energy Agency in the sphere of control of the provisions of the Non-Proliferation Treaty. This activity has been an important instrument of preventing proliferation of nuclear weapons. Sharing the opinion that further strengthening of safeguards lies in the interest of universal peace, the Government of the Polish People's Republic has decided that it would deliver nuclear material, equipment and technology defined in a trigger list, to any non-nuclear-weapon State only in a case when the whole nuclear activity of a recipient country, and not only material, equipment and technology being transferred, are subject to the Agency's safeguards.

The Government of the Polish People's Republic expresses its opinion that this principle, if observed by all the States — nuclear suppliers, could have made a great contribution toward strengthening and universality of the Non-Proliferation Treaty.

The Government of the Polish People's Republic requests that the Director General of the IAEA should circulate the text of this Note to all Member Governments.

The Permanent Mission of the Polish People's Republic to the International Atomic Energy Agency avails itself of this opportunity to renew to the Director General of the IAEA the assurances of its highest consideration.

LETTER XI

The Permanent Mission of Switzerland presents its compliments to the Director General of the International Atomic Energy Agency and, with reference to its today's Note No. 003, has the honour to emphasize the following.

The Government of Switzerland at present is not in a position to implement fully the Principles for Technology Transfers set out in the documents attached to the above-mentioned Note because of the lack of appropriate laws and regulations. However, the Government of Switzerland intends to implement these principles fully when appropriate laws and regulations for this purpose are put into force as necessary.

The Government of Switzerland requests that the Director General of the International Atomic Energy Agency should circulate the text of this Note to all Member Governments for their information.

The Permanent Mission of Switzerland avails itself of this opportunity to renew to the Director General of the International Atomic Energy Agency the assurances of its highest consideration.

With reference to Note Verbale No. 1 from the Permanent Mission of the USSR, dated 11th January 1978, I have the honour to send you the following Declaration of the Government of the USSR:

"The Government of the Union of Soviet Socialist Republics emphasizes its determination to continue its efforts to secure agreement between countries supplying nuclear materials, equipment and technology on the principle that IAEA safeguards must be applied to all nuclear activities of non-nuclear-weapon States when those States receive any of the items mentioned in the initial list referred to in the above-mentioned Note Verbale. In this connection the Government of the USSR takes the view that the principle of full control is a necessary condition for ensuring effective safeguards which can prevent nuclear materials, equipment and technology from being used for manufacturing nuclear weapons or other nuclear explosive devices."

The Government requests that the text of the present letter be distributed as an official document of the IAEA.

Appendix

GUIDELINES FOR NUCLEAR TRANSFERS

1.   The following fundamental principles for safeguards and export controls should apply to nuclear transfers to any non-nuclear-weapon State for peaceful purposes. In this connection, suppliers have defined an export trigger list and agreed on common criteria for technology transfers.

Prohibition on nuclear explosives

2.   Suppliers should authorise transfer of items identified in the trigger list only upon formal governmental assurances from recipients explicitly excluding uses which would result in any nuclear explosive device.

Physical protection

3.   a)   All nuclear materials and facilities identified by the agreed trigger list should be placed under effective physical protection to prevent unauthorised use and handling. The levels of physical protection to be ensured in relation to the type of materials, equipment and facilities, have been agreed by suppliers, taking account of international recommendations.

   b)   The implementation of measures of physical protection in the recipient country is the responsibility of the Government of that country. However, in order to implement the terms agreed upon amongst suppliers, the levels of physical protection on which these measures have to be based should be the subject of an agreement between supplier and recipient.

   c)   In each case special arrangements should be made for a clear definition of responsibilities for the transport of trigger list items.

Safeguards

4.   Suppliers should transfer trigger list items only when covered by IAEA safeguards, with duration and coverage provisions in conformance with the GOV/1621 guidelines. Exceptions should be made only after consultation with the parties to this understanding.

5.   Suppliers will jointly reconsider their common safeguards requirements, whenever appropriate.

Safeguards triggered by the transfer of certain technology

6.   a)   The requirements of paragraphs 2, 3 and 4 above should also apply to facilities for reprocessing, enrichment, or heavy-water production, utilising technology directly transferred by the supplier or derived from transferred facilities, or major critical components thereof.

   b)   The transfer of such facilities, or major critical components thereof, or related technology, should require an undertaking (1) that IAEA safeguards apply to any facilities of the same type (i.e. if the design, construction or operating processes are based on the same or similar physical or chemical processes, as defined in the trigger list) constructed during an agreed period in the recipient country and (2) that there should at all times be in effect a safeguards agreement permitting the IAEA to apply Agency safeguards with respect to such facilities identified by the recipient, or by the supplier in consultation with the recipient, as using transferred technology.

Special controls on sensitive exports

7.   Suppliers should exercise restraint in the transfer of sensitive facilities, technology and weapons-usable materials. If enrichment or reprocessing facilities, equipment or technology are to be transferred, suppliers should encourage recipients to accept, as an alternative to national plants, supplier involvement and/or other appropriate multinational participation in resulting facilities. Suppliers should also promote international (including IAEA) activities concerned with multinational regional fuel cycle centres.

Special controls on export of enrichment facilities, equipment and technology

8.   For a transfer of an enrichment facility, or technology therefor, the recipient nation should agree that neither the transferred facility, nor any facility based on such technology, will be designed or operated for the production of greater than 20 per cent enriched uranium without the consent of the supplier nation, of which the IAEA should be advised.

Controls on supplied or derived weapons-usable material

9.   Suppliers recognise the importance, in order to advance the objectives of these guidelines and to provide opportunities further to reduce the risks of proliferation, of including in agreements on supply of nuclear materials or of facilities which produce weapons-usable material, provisions calling for mutual agreement between the supplier and the recipient on arrangements for reprocessing, storage, alteration, use, transfer or retransfer of any weapons-usable material involved. Suppliers should endeavour to include such provisions whenever appropriate and practicable.

Controls on retransfer

10.   *a)*   Suppliers should transfer trigger list items, including technology defined under paragraph 6, only upon the recipient's assurance that in the case of:

        (1)   retransfer of such items or,
        (2)   transfer of trigger list items derived from facilities originally transferred by the supplier, for with the help of equipment or technology originally transferred by the supplier;

        the recipient of the retransfer or transfer will have provided the same assurances as those required by the supplier for the original transfer.

     b)   In addition the supplier's consent should be required for: (1) any retransfer of the facilities, major critical components, or technology described in paragraph 6; (2) any transfer of facilities or major critical components derived from those items; (3) any retransfer of heavy water or weapons-usable material.

SUPPORTING ACTIVITIES

Physical security

11.   Suppliers should promote international co-operation on the exchange of physical security information, protection of nuclear materials in transit, and recovery of stolen nuclear materials and equipment.

Support for effective IAEA safeguards

12.   Suppliers should make special efforts in support of effective implementation of IAEA safeguards. Suppliers should also support the Agency's efforts to assist Member States in the improvement of their national systems of accounting and control of nuclear material and to increase the technical effectiveness of safeguards.

    Similarly, they should make every effort to support the IAEA in increasing further the adequacy of safeguards in the light of technical developments and the rapidly growing number of nuclear facilities, and to support appropriate initiatives aimed at improving the effectiveness of IAEA safeguards.

Sensitive plant design features

13.   Suppliers should encourage the designers and makers of sensitive equipment to construct it in such a way as to facilitate the application of safeguards.

Consultations

14.   *a)*   Suppliers should maintain contact and consult through regular channels on matters connected with the implementation of these guidelines.

b) Suppliers should consult, as each deems appropriate, with other Governments concerned on specific sensitive cases, to ensure that any transfer does not contribute to risks of conflict or instability.

c) In the event that one or more suppliers believe that there has been a violation of supplier/recipient understandings resulting from these guidelines, particularly in the case of an explosion of a nuclear device, or illegal termination or violation of IAEA safeguards by a recipient, suppliers should consult promptly through diplomatic channels in order to determine and assess the reality and extent of the alleged violation.

Pending the early outcome of such consultations, suppliers will not act in a manner that could prejudice any measure that may be adopted by other suppliers concerning their current contacts with that recipient.

Upon the findings of such consultations, the suppliers, bearing in mind Article XII of the IAEA Statute, should agree on an appropriate response and possible action which could include the termination of nuclear transfers to that recipient.

15. In considering transfers, each supplier should exercise prudence having regard to all the circumstances of each case, including any risk that technology transfers not covered by paragraph 6, or subsequent retransfers, might result in unsafeguarded nuclear materials.

16. Unanimous consent is required for any changes in these guidelines, including any which might result from the reconsideration mentioned in paragraph 5.

Annex A

TRIGGER LIST REFERRED TO IN GUIDELINES

Part A. MATERIAL AND EQUIPMENT

1. Source or special fissionable material as defined in Article XX of the Statute of the International Atomic Energy Agency; provided that items specified in sub-paragraph *(a)* below, and exports of source or special fissionable material to a given recipient country, within a period of 12 months, below the limits specified in sub-paragraph *(b)* below, shall not be included:

a) Plutonium with an isotopic concentration of plutonium 238 exceeding 80 per cent.

Special fissionable material when used in gram quantities or less as a sensing component in instruments; and

Source material which the Government is satisfied is to be used only in non-nuclear activities, such as the production of alloys or ceramics;

b) Special fissionable material   50 effective grams;
Natural uranium   500 kilograms;
Depleted uranium   1 000 kilograms; and
Thorium   1 000 kilograms.

2.1. *Reactors and equipment therefor:*

2.1.1. Nuclear reactors capable of operation so as to maintain a controlled self-sustaining fission chain reaction, excluding zero energy reactors, the latter being defined as reactors with a designed maximum rate of production of plutonium not exceeding 100 grams per year.

2.1.2. Reactor pressure vessels:

Metal vessels, as complete units or as major shop-fabricated parts therefor, which are especially designed or prepared to contain the core of a nuclear reactor as defined in paragraph 2.1.1 above and are capable of withstanding the operating pressure of the primary coolant.

2.1.3. Reactor fuel charging and discharging machines:
Manipulative equipment especially designed or prepared for inserting or removing fuel in a nuclear reactor as defined in paragraph 2.1.1 above capable of on-load operation or employing technically sophisticated positioning or alignment features to allow complex off-load fuelling operations such as those in which direct viewing of or access to the fuel is not normally available.

2.1.4. Reactor control rods:

Rods especially designed or prepared for the control of the reaction rate in a nuclear reactor as defined in paragraph 2.1.1 above.

2.1.5. Reactor pressure tubes:

Tubes which are especially designed or prepared to contain fuel elements and the primary coolant in a reactor as defined in paragraph 2.1.1 above at an operating pressure in excess of 50 atmospheres.

2.1.6. Zirconium tubes:

Zirconium metal and alloys in the form of tubes or assemblies of tubes, and in quantities exceeding 500 kg per year, especially designed or prepared for use in a reactor as defined in paragraph 2.1.1 above, and in which the relationship of hafnium to zirconium is less than 1:500 parts by weight.

2.1.7. Primary coolant pumps:

Pumps especially designed or prepared for circulating liquid metal as primary coolant for nuclear reactors as defined in paragraph 2.1.1 above.

2.2. *Non-nuclear materials for reactors:*

2.2.1. Deuterium and heavy water:

Deuterium and any deuterium compound in which the ratio of deuterium to hydrogen exceeds 1:5000 for use in a nuclear reactor as defined in paragraph 2.1.1 above in quantities exceeding 200 kg of deuterium atoms for any one recipient country in any period of 12 months.

2.2.2. Nuclear grade graphite:

Graphite having a purity level better than 5 parts per million boron equivalent and with a density greater than 1.50 grams per cubic centimetre in quantities exceeding 30 metric tons for any one recipient country in any period of 12 months.

2.3.1. Plants for the reprocessing of irradiated fuel elements, and equipment especially designed or prepared therefor.

2.4.1. Plants for the fabrication of fuel elements.

2.5.1. Equipment, other than analytical instruments, especially designed or prepared for the separation of isotopes of uranium.

2.6.1. Plants for the production of heavy water, deuterium and deuterium compounds and equipment especially designed or prepared therefor.

Clarifications of certain of the items on the above list are annexed.

Part B. COMMON CRITERIA FOR TECHNOLOGY TRANSFERS UNDER
PARAGRAPH 6 OF THE GUIDELINES

(1) "Technology" means technical data in physical form designated by the supplying country as important to the design, construction, operation, or maintenance of enrichment, reprocessing, or heavy water production facilities or major critical components thereof, but excluding data available to the public, for example, in published books and periodicals, or that which has been made available internationally without restrictions upon its further dissemination.

(2) "Major critical components" are:
   a) in the case of an isotope separation plant of the gaseous diffusion type: *diffusion barrier*
   b) in the case of an isotope separation plant of the gas centrifuge type: *gas centrifuge assemblies, corrosion-resistant to UF$_6$*;
   c) in the case of an isotope separation plant of the jet nozzle type: *nozzle units*;
   d) in the case of an isotope separation plant of the vortex type: the *vortex units*.

(3) For facilities covered by paragraph 6 of the Guidelines for which no major critical component is described in paragraph 2 above, if a supplier nation should transfer in the aggregate a significant fraction of the items essential to the operation of such a facility, together with the know-how for construction and operation of that facility, that transfer should be deemed to be a transfer of "facilities or major critical components thereof".

(4) The definitions in the preceding paragraphs are solely for the purposes of paragraph 6 of the Guidelines and this Part B, which differ from those applicable to part A of this Trigger List, which should not be interpreted as limited by such definition.

(5) For the purposes of implementing paragraph 6 of the Guidelines, the following facilities should be deemed to be "of the same type (i.e. if their design, construction or operating processes are based on the same or similar physical or chemical processes)":

Where the technology transferred is such as to make possible the construction in the recipient State of a facility of the following type, or major critical components thereof:

The following will be deemed to be facilities of the same type:

a)	an isotope separation plant of the gaseous diffusion type	any other isotope separation plant using the gaseous diffusion process.
b)	an isotope separation plant of the gas centrifuge type	any other isotope separation plant using the gas centrifuge process.
c)	an isotope separation plant of the jet nozzle type	any other isotope separation plant using the jet nozzle process.
d)	an isotope separation plant of the vortex type	any other isotope separation plant using the vortex process.
e)	a fuel reprocessing plant using the solvent extraction process	any other fuel reprocessing plant using the solvent extraction process.
f)	a heavy water plant using the exchange process..........	any other heavy water plant using the exchange process.
g)	a heavy water plant using the electrolytic process	any other heavy water plant using the electrolytic process.
h)	a heavy water plant using the hydrogen distillation process	any other heavy water plant using the hydrogen distillation process.

Note: In the case of reprocessing, enrichment, and heavy water facilities whose design, construction, or operation processes are based on physical or chemical processes other than those enumerated above, a similar approach would be applied to define facilities "of the same type", and a need to define major critical components of such facilities might arise.

(6) The reference in paragraph 6(b) of the Guidelines to "any facilities of the same type constructed during an agreed period in the recipient's country" is understood to refer to such facilities (or major critical components thereof), the first operation of which commences within a period of at least 20 years from the date of the first operation of (1) a facility which has been transferred or incorporates transferred major critical components or of (2) a facility of the same type built after the transfer of technology. It is understood that during that period there would be a conclusive presumption that any facility of the same type utilised transferred technology. But the agreed period is not intended to limit the duration of the safeguards imposed or the duration of the right to identify facilities as being constructed or operated on the basis of or by the use of transferred technology in accordance with paragraph 6(b)(2) of the Guidelines.

Annex

CLARIFICATIONS OF ITEMS ON THE TRIGGER LIST

A. *Complete nuclear reactors*
(Item 2.1.1 of the Trigger List)

1. A "nuclear reactor" basically includes the items within or attached directly to the reactor vessel, the equipment which controls the level of power in the core, and the components which normally contain or come in direct contact with or control the primary coolant of the reactor core.

2. The export of the whole set of major items within the boundary will take place only in accordance with the procedures of the Guidelines. Those individual items within this functionally defined boundary which will be exported only in accordance with the procedures of the Guidelines are listed in paragraphs 2.1.1 to 2.1.5.

The Government reserves to itself the right to apply the procedures of the Guidelines to other items within the functionally defined boundary.

3. It is not intended to exclude reactors which could reasonably be capable of modification to produce significantly more than 100 grams of plutonium per year. Reactors designed for sustained operation at significant power levels, regardless of their capacity for plutonium production, are not considered as "zero energy reactors".

B. *Pressure vessels*
(Item 2.1.2 of the Trigger List)

4. A top plate for a reactor pressure vessel is covered by item 2.1.1 as a major shop-fabricated part of a pressure vessel.

5. Reactor internals (e.g. support columns and plates for the core and other vessel internals, control rod guide tubes, thermal shields, baffles, core grid plates, diffuser plates, etc.) are normally supplied by the reactor supplier. In some cases, certain internal

support components are included in the fabrication of the pressure vessel. These items are sufficiently critical to the safety and reliability of the operation of the reactor (and, therefore, to the guarantees and liability of the reactor supplier), so that their supply, outside the basic supply arrangement for the reactor itself, would not be common practice. Therefore, although the separate supply of these unique, especially designed and prepared, critical, large and expensive items would not necessarily be considered as falling outside the area of concern, such a mode of supply is considered unlikely.

C. *Reactor control rods*
    (Item 2.1.4 of the Trigger List)

6.   This item includes, in addition to the neutron absorbing part, the support or suspension structures therefor if supplied separately.

D. *Fuel reprocessing plants*
    (Item 2.3.1 of the Trigger List)

7.   A "plant for the reprocessing of irradiated fuel elements" includes the equipment and components which normally come in direct contact with and directly control the irradiated fuel and the major nuclear material and fission product processing streams. The export of the whole set of major items within this boundary will take place only in accordance with the procedures of the Guidelines. In the present state of technology, the following items of equipment are considered to fall within the meaning of the phrase "and equipment especially designed or prepared therefor":

    a) Irradiated fuel element chopping machines: remotely operated equipment especially designed or prepared for use in a reprocessing plant as identified above and intended to cut, chop or shear irradiated nuclear fuel assemblies, bundles or rods; and

    b) Critically safe tanks (e.g. small diameter, annular or slab tanks) especially designed or prepared for use in a reprocessing plant as identified above, intended for dissolution of irradiated nuclear fuel and which are capable of withstanding hot, highly corrosive liquid, and which can be remotely loaded and maintained.

8.   The Government reserves to itself the right to apply the procedures of the Guidelines to other items within the functionally defined boundary.

E. *Fuel fabrication plants*
    (Item 2.4.1 of the Trigger List)

9.   A "plant for the fabrication of fuel elements" includes the equipment:

    a) Which normally comes in direct contact with, or directly processes, or controls, the production flow of nuclear material, or

    b) Which seals the nuclear material within the cladding.

10.   The export of the whole set of items for the foregoing operations will take place only in accordance with the procedures of the Guidelines. The Government will also give consideration to application of the procedures of the Guidelines to individual items intended for any of the foregoing operations, as well as for other fuel fabrication operations such as checking the integrity of the cladding or the seal, and the finish treatment to the sealed fuel.

F. *Isotope separation plant equipment*
    (Item 2.5.1 of the Trigger List)

11.   "Equipment, other than analytical instruments, especially designed or prepared for the separation of isotopes of uranium" includes each of the major items of equipment especially designed or prepared for the separation process. Such items include:

— gaseous diffusion barriers,
— gaseous diffuser housings,
— gas centrifuge assemblies, corrosion-resistant to UF_6,
— jet nozzle separation units,
— vortex separation units,
— large UF_6 corrosion-resistant axial or centrifugal compressors,
— special compressor seals for such compressors.

Annex B

CRITERIA FOR LEVELS OF PHYSICAL PROTECTION

1.   The purpose of physical protection of nuclear material is to prevent unauthorised use and handling of materials. Paragraph 3(a) of the Guidelines document calls for agreement among suppliers on the levels of protection to be ensured in relation to the type of materials, and equipment and facilities containing these materials, taking account of international recommendations.

2.   Paragraph 3(b) of the Guidelines document states that implementation of measures of physical protection in the recipient country is the responsibility of the Government of that country. However, the levels of physical protection on which these measures have to be based should be the subject of an agreement between supplier and recipient. In this context these requirements should apply to all States.

3.   The document INFCIRC/225 of the International Atomic Energy Agency entitled "The Physical Protection of Nuclear Material" and similar documents which from time to time are prepared by international groups of experts and updated as appropriate to account for changes in the state of the art and state of knowledge with regard to physical protection of nuclear material are a useful basis for guiding recipient States in designing a system of physical protection measures and procedures.

4.   The categorisation of nuclear material presented in the attached table or as it may be updated from time to time by mutual agreement of suppliers shall serve as the agreed basis for designating specific levels of physical protection in relation to the type of materials, and equipment and facilities containing these materials, pursuant to paragraph 3(a) and 3(b) of the Guidelines document.

5.   The agreed levels of physical protection to be ensured by the competent national authorities in the use, storage and transportation of the materials listed in the attached table shall as a minimum include protection characteristics as follows:

CATEGORY III

Use and Storage within an area to which access is controlled.

Transportation under special precautions including prior arrangements among sender, recipient and carrier, and prior agreement between entities subject to the jurisdiction and regulation of supplier and recipient States, respectively, in case of international transport specifying time, place and procedures for transferring transport responsibility.

CATEGORY II

Use and Storage within a protected area to which access is controlled, i.e. an area under constant surveillance by guards or electronic devices, surrounded by a physical barrier with a limited number of points of entry under appropriate control, or any area with an equivalent level of physical protection.

Transportation under special precautions including prior arrangements among sender, recipient and carrier, and prior agreement between entities subject to the jurisdiction and regulation of supplier and recipient States, respectively, in case of international transport, specifying time, place and procedures for transferring transport responsibility.

CATEGORY I

Materials in this Category shall be protected with highly reliable systems against unauthorised use as follows:

Use and Storage within a highly protected area, i.e. a protected area as defined for Category II above, to which, in addition, access is restricted to persons whose trustworthiness has been determined, and which is under surveillance by guards who are in close communication with appropriate response forces. Specific measures taken in this context should have as their objective the detection and prevention of any assault, unauthorised access or unauthorised removal of material.

Transportation under special precautions as identified above for transportation of Category II and III materials and, in addition, under constant surveillance by escorts and under conditions which assure close communication with appropriate response forces.

6.   Suppliers should request identification by recipients of those agencies or authorities having responsibility for ensuring that levels of protection are adequately met and having responsibility for internally co-ordinating response/recovery operations in the event of unauthorised use or handling of protected materials. Suppliers and recipients should also designate points of contact within their national authorities to co-operate on matters of out-of-country transportation and other matters of mutual concern.

*
**

        The table: Categorization of Nuclear Material which supplements Annex B, is reproduced in annex II to chapter seven of the Study (Note by the Secretariat).

ANNEX VII-B

FURTHER COMMUNICATIONS RECEIVED FROM CERTAIN MEMBER STATES REGARDING GUIDELINES FOR THE EXPORT OF NUCLEAR MATERIAL, EQUIPMENT OR TECHNOLOGY

Communication from Australia (INFCIRC/254/Add.1, March 1978)

1.   On 21st February 1978 the Director General received a letter dated the same day from the Governor from Australia on the Board of Governors of the Agency, enclosing a Note concerning the communications received from certain Member States, circulated in document INFCIRC/254, regarding guidelines for the export of nuclear material, equipment or technology.

2.   In accordance with the request made by the Governor from Australia in his letter, the text of the attached Note is reproduced below for the information of all Members.

The Permanent Mission of Australia to the International Atomic Energy Agency presents its compliments to the Director General.

The Government of Australia has noted the communications dated 11th January 1978[1] to the Director General of the IAEA by 15 Member States regarding common principles and guidelines which these States have agreed to follow in considering nuclear exports.

Australia welcomes this action, and the IAEA's intention to circulate the text of the guidelines to all Member States. This marks a significant contribution to the development of international arrangements under which nuclear energy can be developed to help meet world energy requirements while avoiding the dangers of nuclear proliferation.

Australia also shares the commitment of the countries concerned to support the effective implementation of IAEA safeguards and the need to remove safeguards and non-proliferation assurances from the field of commercial competition.

The States concerned have expressed the hope that other Governments will decide to follow similar nuclear export policies. In Australia's case, under the nuclear safeguards policy announced by the Government on 24th May 1977, Australia will be applying export criteria which satisfy the guidelines and have certain additional requirements

Finally, the Government of Australia would like to record its views that it is desirable to keep nuclear export policies and guidelines under review to take account of new developments and to ensure their continued appropriateness as a framework for the development of nuclear energy for peaceful purposes consistent with shared non-proliferation objectives.

The Government of Australia requests that the Director General of the IAEA should circulate the text of this Note to all Member Governments for their information.

The Permanent Mission of Australia to the International Atomic Energy Agency avails itself of this opportunity to renew to the Director General the assurances of its highest consideration.

[1]   See document INFCIRC/254.

Communication from Finland (INFCIRC/254/Add.2, March 1980)

1.   On 1st February 1980 the Director General received a letter dated 28th January 1980 from the Resident Representative of Finland to the Agency with a Note concerning communications received from 15 Member States and circulated in document INFCIRC/254, about guidelines for the export of nuclear material, equipment or technology.

2.   In accordance with the request made by the Resident Representative of Finland in his letter, the text of the Note is reproduced below for the information of all Members.

The Permanent Mission of Finland to the International Atomic Energy Agency presents its compliments to the Director General.

The Government of Finland has noted the communications of 15 Member States in document INFCIRC/254 concerning principles and guidelines which these countries will follow in their nuclear export policies. The adoption of these guidelines is in the view of the Finnish Government a significant contribution to the efforts to create an appropriate framework for the further development of peaceful uses of nuclear energy under effective safeguards.

In the communications mentioned above the States concerned have expressed the hope that other Governments may decide to follow similar nuclear export policies. In view of this and in accordance with its continuous support to the strengthening of the existing non-proliferation regime, the Government of Finland has decided to base its nuclear export policy on the criteria set forth in the guidelines. As the lack of appropriate legislation does not allow for a full and immediate implementation of all the criteria, the Government of Finland has equally decided to effect the requisite changes in the legislation to allow for a full implementation of the guidelines in the future.

In making this decision the Government of Finland is fully aware that at present international exchanges in the field of nuclear energy are to an increasing extent characterised by a complex system of safeguards requirements and overlapping controls, which have brought along various administrative and technical difficulties for importing countries. It would therefore be desirable that efforts be made towards the simplification of the existing safeguards procedures in a way that would not affect their efficiency. This would best be achieved by a system of uniform safeguards requirements based on the IAEA safeguards system and applied to all nuclear activities in all non-nuclear-weapon States.

The Government of Finland requests the Director General to circulate the text of this Note to all Member States for information.

The Permanent Mission of Finland to the International Atomic Energy Agency avails itself of this opportunity to renew to the Director General the assurances of its highest consideration.

Communication from Denmark (INFCIRC/254/Add.3, September 1984)

On 20th August 1984 the Director General received a letter dated 13th August 1984 from the Permanent Mission of Denmark to the Agency in the same terms as the letter and its attachments reproduced as letter IV in document INFCIRC/254. That document deals with communications received from certain Members regarding guidelines for the export of nuclear material, equipment or technology.

Communication from Greece (INFCIRC/254/Add.4, November 1984)

On 21st September 1984 the Director General received a letter dated 19th September 1984 from the Permanent Mission of Greece to the Agency in the same terms as the letter and its attachments reproduced as letter IV in document INFCIRC/254. That document deals with communications received from certain Members regarding guidelines for the export of nuclear material, equipment or technology.

Communication from Luxembourg (INFCIRC/254/Add/5, December 1984)

1.   On 13th November 1984, the Director General received a letter of the same date from the Permanent Mission of the Grand Duchy of Luxembourg to the International Organizations in Vienna, relating to the export of nuclear material, equipment or technology. In the light of the request at the end of this letter, the text is reproduced below as Letter I.

2.   On the same day, the Permanent Mission of the Grand Duchy of Luxembourg to the International Organizations in Vienna addressed a letter to the Director General relating to the full implementation of the principles for technology transfers referred to in Letter I. In the light of the request expressed at the end of this letter, the text is reproduced below as Letter II.

LETTER I

The Permanent Mission of the Grand Duchy of Luxembourg to the International Organizations in Vienna presents its compliments to the Director General of the International Atomic Energy Agency and, on instructions from its Government, has the honour to state the following:

The Government of Luxembourg has taken note of the communications addressed to the Director General of the IAEA on 11th January 1978 by 15 Member States and relating to the common principles and guidelines that those States have decided to apply in the export of nuclear materials, equipment and technology, and which are contained in document INFCIRC/254.

The Government of Luxembourg approves this initiative and has decided to act in accordance with the principles and guidelines contained therein.

In reaching this decision the Luxembourg Government is fully aware of the need to contribute to the development of nuclear power in order to meet world energy requirements while avoiding contributing in any way to dangers of a proliferation of nuclear weapons or other explosive nuclear devices, and the need to remove safeguards and assurances of non-proliferation from the field of world competition.

As a Member of the European Community the Government of Luxembourg, so far as trade within the Community is concerned, will implement these documents in the light of its commitments under the Treaties of Rome, where necessary.

The Government of Luxembourg requests the Director General of the IAEA to communicate the text of this Note to the Member Governments for their information and as a demonstration of support by the Government of Luxembourg for the Agency's non-proliferation objectives and safeguards activities.

The Permanent Mission of the Grand Duchy of Luxembourg to the International Organizations in Vienna takes this opportunity to renew to the Director General of the International Atomic Energy Agency the assurances of its highest esteem.

LETTER II

The Permanent Mission of the Grand Duchy of Luxembourg to the International Organizations in Vienna presents its compliments to the Director General of the International Atomic Energy Agency and, further to its Note 689/84 of 13th November 1984, has the honour to draw his attention to the following:

The Government of Luxembourg is at present not in a position to implement fully the Principles for Technology Transfers referred to in the above-mentioned Note, because of the lack of appropriate regulations.

However, the Government of Luxembourg intends to implement these principles fully when appropriate regulations are put into force.

The Government of Luxembourg requests the Director General of the IAEA to circulate this Note to all Member Governments for their information.

The Permanent Mission of the Grand Duchy of Luxembourg to the International Organizations in Vienna takes this opportunity to renew to the Director General of the International Atomic Energy Agency the assurances of its highest esteem.

Communication from Ireland (INFCIRC/254/Add.6, December 1984)

On 20th November 1984 the Director General received a letter dated 14th November 1984 from the Permanent Mission of Ireland to the International Organizations in Vienna in the same terms as the letter and its attachments reproduced as letter IV in document INFCIRC/254. That document deals with communications received from certain Members regarding guidelines for the export of nuclear material, equipment or technology.

Communication from Bulgaria (INFCIRC/254/Add.7, January 1985)

1.   On 18th December 1984, the Director General received a letter dated 14th December 1984 from the Permanent Mission of the People's Republic of Bulgaria to the International Organizations in Vienna relating to the export of nuclear material, equipment or technology.

2.   As requested in the letter, its text is reproduced below for the information of all Member States.

The Permanent Mission of the People's Republic of Bulgaria to the International Organizations in Vienna presents its compliments to the Director General of the International Atomic Energy Agency and has the honour to inform him that the Government of the People's Republic of Bulgaria has decided that, when considering the export of nuclear material, equipment or technology, it will act in accordance with the principles, contained in the document INFCIRC/254.

In taking this decision the Government of the People's Republic of Bulgaria is convinced that the strengthening of the non-proliferation regime will promote the peaceful uses of nuclear energy and international co-operation in this area.

The Government of the People's Republic of Bulgaria proposes to the Director General of the International Atomic Energy Agency to circulate the text of this Note to all Member Governments for their information as a demonstration of the support by the Government of the People's Republic of Bulgaria for the Agency's non-proliferation objectives and safeguards activities.

The Permanent Mission of the People's Republic of Bulgaria to the International Organizations in Vienna avails itself of this opportunity to renew to the Director General of the International Atomic Energy Agency the assurances of its highest consideration.

Communication from Hungary (INFCIRC/254/Add.8, May 1985)

1.   On 23rd May 1985 the Director General received a letter dated 2nd May 1985 from the Minister of Foreign Affairs of the Hungarian People's Republic relating to the export of nuclear material, equipment or technology.

2.   As requested in the letter, its text is reproduced below for the information of all Member States.

I have the honour to inform you that the Government of the Hungarian People's Republic has decided that, when considering the export of nuclear material, equipment or technology, it will act in accordance with the principles and guidelines contained in document INFCIRC/254.

In reaching this decision the Government of the Hungarian People's Republic is convinced that the strengthening of the non-proliferation regime will promote the peaceful uses of nuclear energy and international co-operation in this field.

I shall be grateful if you will be kind enough to communicate the text of this letter to all Members of the Agency for their information and as a demonstration of support by the Government of the Hungarian People's Republic for the non-proliferation objectives and safeguards activities of the International Atomic Energy Agency.

Communication from Portugal (INFCIRC/254/Add.9, February 1986)

1.   On 15th January 1986 the Director General received a Note Verbale dated 10th January 1986 from the Permanent Mission of Portugal to the International Atomic Energy Agency relating to the export of nuclear material, equipment or technology.

2.   As requested in a subsequent letter dated 22nd January 1986, the text of the letter of 10th January 1986 is reproduced below for the information of all Member States.

———————

The Permanent Mission of Portugal to the International Atomic Energy Agency presents its compliments to the Director General and has the honour to enclose copies of the documents which have been the subject of discussion between the Government of the Portuguese Republic and a number of other Governments.

The Government of the Portuguese Republic has decided that, when considering the export of nuclear material, equipment or technology, it will act in accordance with the principles contained in the attached documents.[1]

In reaching this decision, the Government of the Portuguese Republic is fully aware of the need to contribute to the development of nuclear power in order to meet world energy requirements, while avoiding contributing in any way to the dangers of a proliferation of nuclear weapons or other nuclear explosive devices, and of the need to remove safeguards and non-proliferation assurances from the field of commercial competition.

As a Member of the European Community, the Government of the Portuguese Republic, so far as trade within the Community is concerned, will implement these documents in the light of its commitments under the Treaties of Rome, as necessary.

The Government of the Portuguese Republic hopes that other Governments may also decide to base their own nuclear export policies upon these documents.

The Permanent Mission of Portugal avails itself of this opportunity to renew to the Director General of the International Atomic Energy Agency the assurance of its highest consideration.

Communication from Switzerland (INFCIRC/254/Add.10, September 1987)

1.   On 6th August 1987 the Director General received a letter dated 6th August 1987 from the Permanent Mission of Switzerland to the International Organizations in Vienna relating to the export of nuclear material, equipment or technology.

2.   As requested, the text of the letter is reproduced below for the information of all Member States.

———————

The Permanent Mission of Switzerland presents its compliments to the Director General of the International Atomic Energy Agency and, with reference to the Mission's letter No. 004 of 11th January 1978, reproduced under the heading "Letter XI" in document INFCIRC/254 of February 1978, has the honour to state that, consequent upon a revision of national legislation, the principles governing transfer of technology set forth in the documents appended to the Mission's letter No. 003 of 11th January 1978 have been fully applied in Switzerland since 1st April 1987.

The Permanent Mission would be grateful to the Director General if he could communicate the text of this letter to the Governments of all States Members of the IAEA.

The Permanent Mission of Switzerland takes this opportunity to reiterate to the Director General of the International Atomic Energy Agency the assurances of its highest esteem.

[1]   Reproduced as Appendix with Annexes in document INFCIRC/254.

ANNEX TO THE MEMORANDUM BY THE IAEA DIRECTOR GENERAL OF 20TH AUGUST 1973 ON THE FORMULATION OF CERTAIN PROVISIONS IN AGREEMENTS UNDER THE AGENCY'S SAFEGUARDS SYSTEM (INFCIRC/66/Rev.2)

1.   In the case of receipt by a State of source or special fissionable material, equipment, facilities or non-nuclear material from a supplier outside that State, the duration of the relevant agreement under the Agency's Safeguards System[1] would be related to the actual use in the recipient State of the material or items supplied. This may be accomplished by requiring, in accordance with present practice, that the material or items supplied be listed in the inventory called for by the agreement.

2.   The primary effect of termination of the agreement, either by act of the parties or effluxion of time, would be that no further supplied nuclear material, equipment, facilities or non-nuclear material could be added to the inventory. On the other hand, the rights and obligations of the parties, as provided for in the agreement, would continue to apply in connection with any supplied material or items and with any special fissionable material produced, processed or used in or in connection with any supplied material or items which had been included in the inventory, until such material or items had been removed from the inventory.

3.   With respect to nuclear material, conditions for removal are those set out in paragraph 26 or 27 of the Agency's Safeguards System; with respect to equipment, facilities and non-nuclear material, conditions for removal could be based on paragraph 26. A number of agreements already concluded have prescribed such conditions in part, by providing for deletion from the inventory of nuclear material, equipment and facilities which are returned to the supplying State or transferred (under safeguards) to a third State. The additional provisions contemplated would stipulate that items or non-nuclear material could be removed from the purview of the agreement if they had been consumed, were no longer usable for any nuclear activity relevant from the point of view of safeguards or had become practically irrecoverable.

4.   The effect of reflecting the two concepts in agreements would be that special fissionable material which had been produced, processed or used in or in connection with supplied material or items before they were removed from the scope of the agreement, would remain or be listed in the inventory, and such special fissionable material, together with any supplied nuclear material remaining in the inventory, would be subject to safeguards until the Agency had terminated safeguards on that special fissionable and nuclear material in accordance with the provisions of the Agency's Safeguards System. Thus, the actual termination of the operation of the provisions of the agreement would take place only when everything had been removed from the inventory.

[1]   The Agency's Safeguards System (1965, as Provisionally Extended in 1966 and 1968) set forth in document INFCIRC/66/Rev.2.

ANNEX IX

DECLARATION OF COMMON POLICY ADOPTED ON 20TH NOVEMBER 1984 BY THE MINISTERS FOR FOREIGN AFFAIRS OF THE MEMBER STATES OF THE EUROPEAN COMMUNITY

The Ten (hereinafter called "the Member States"), united within the framework of European political co-operation:

a) *Recalling* the rights and obligations deriving from their membership of the European Atomic Energy Community,

b) *Emphasizing* their support for the objective of non-proliferation of nuclear weapons,

c) *Referring* to the various undertakings relating to the peaceful utilisation of nuclear energy and the safeguarding thereof to which they have respectively subscribed, in particular the Treaty on the Non-Proliferation of Nuclear Weapons and the agreements concluded between the Member States, the European Atomic Energy Community and the International Atomic Energy Agency for the application of safeguards within the Community, and

d) *Taking note* of the adoption by all the Member States of the Guidelines for the Export of Nuclear Material, Equipment or Technology set forth in document INFCIRC/254 of the International Atomic Energy Agency (hereinafter called "the Guidelines"),

1.   *State* that the principles contained in the Guidelines constitute a common, fundamental set of rules for all the Member States in relation to their nuclear exports,

2.   *Declare* that, provided the provisions of the Treaties of Rome and the competence of the Member States are respected, transfers of nuclear material, equipment and technology may be made without restriction between the Member States, subject to the following additional arrangements:

2.1.   Until such time as they are used, separated plutonium and uranium enriched to more than 20 per cent will be stored by the Member States at the place of separation or enrichment to more than 20 per cent, at the places of fabrication of fuels containing plutonium or uranium enriched to more than 20 per cent, *or* in a store established and administered by a Member State, *or* in a place to be determined by common agreement between the Member States concerned.

2.1.1.   Plutonium and uranium enriched to more than 20 per cent will be transferred by the Member States upon receipt of a certificate from the consignee (see the model form annexed hereto) specifying the final destination, the quantities, the approximate date of delivery, the timetable for utilisation, the form in which delivery is to take place and the allocation of the material to one or other of the following uses:

— fuel supply for any power or research reactor in operation or under construction on the territory of a Member State or under its jurisdiction;

— fabrication on the territory of a Member State or under its jurisdiction for purposes of fuel supply to the reactors specified above or, subject to the terms of paragraph 2.1.3, for purposes of fuel supply to any reactor situated on the territory of a third-party State;

— research and development in any laboratory situated on the territory of a Member State or under its jurisdiction. Subject to the terms of paragraph 2.1.2, the materials may also be transferred to a third-party State under a co-operation agreement relating to research and development;

— utilisation in any other installation connected with an energy programme or a research and development programme and situated on the territory of a Member State or subject to its jurisdiction,

including any intermediate storage required for satisfactory implementation of the above-mentioned operations.

2.1.2.   The Government of the Member State to which the consignee belongs will confirm the correctness of the information given in the certificate referred to in paragraph 2.1.1 above.

2.1.3.   Plutonium and uranium enriched to more than 20 per cent will not be retransferred to a third State without mutual agreement between the Member State that has separated the plutonium or enriched the uranium to more than 20 per cent and the Member State desiring to effect the retransfer, without prejudice to any other rights of prior consent that may exist.

2.1.4. Paragraphs 2.1.1, 2.1.2 and 2.1.3 above do not apply to:
- plutonium having an isotopic concentration of plutonium 238 above 80 per cent;
- special fissile materials used in quantities of the order of a gramme or less as a component of sensitive measuring instruments;
- transfers to a given Member State not exceeding 50 effective grammes in the course of a year;
- retransfers to a given third State not exceeding 50 grammes in the course of a year, without prejudice to any other rights of prior consent that may exist.

2.1.5. The above arrangements will be reconsidered by the Member States in the event that an international plutonium store is set up under the aegis of the International Atomic Energy Agency.

2.2. Installations and technology relating to reprocessing, enrichment and the production of heavy water, or other installations created on the basis of such technology, may be transferred in the light of the nature and the degree of development of the nuclear programme in the recipient Member States.

2.3. No enrichment facility transferred from a Member State nor any installation created on the basis of the technology derived from such a facility may be designed or operated for the production of uranium enriched to more than 20 per cent without the agreement of the Member State supplying the facility.

2.4. In making transfers of sensitive equipment or technology, the Member States will observe the provisions relating to the protection of secret information.

2.5. The prior agreement of the supplying State will be required for any retransfer of installations, principal components of crucial importance, reprocessing or enrichment technology or the technology of heavy water production, as well as for any transfer of installations or principal components of crucial importance derived therefrom. Such retransfers and transfers between Member States may take place in consultation with the originating Member State in the light of the nature and the degree of development of the nuclear programme of the receiving Member State;

3. *State* that the Member States will apply to the nuclear materials under their jurisdiction measures of physical protection at least equal to the levels established in the Guidelines; and

4. *State* finally that, in the above-mentioned conditions, transfers between the Member States of nuclear materials, equipment and technology will be carried out in a manner compatible with the requirements of non-proliferation and free movement of goods.

MODEL FORM

REQUEST FOR TRANSFER OF PLUTONIUM OR URANIUM
ENRICHED TO MORE THAN 20 %

1. *Enrichment or reprocessing facility*

   1.1. Name or trade name of firm
   1.2. Address

2. *Consignee*

   2.1. Name or trade name of firm
   2.2. Address
   2.3. Principal activity

3. *Description of shipment*

   3.1. Total weight of material
   3.2. Weight of fissile plutonium (or uranium enriched to more than 20 per cent)
   3.3. Form of material
   3.4. Approximate date of delivery

4. *Use of the material*

   4.1. Fuel fabrication

      4.1.1. Nature of fabrication
      4.1.2. Name, trade name and address of fabrication plant
      4.1.3. Timetable for fuel fabrication

   4.2. Other uses

      4.2.1. Nature of the use
      4.2.2. Name, trade name and address of the user
      4.2.3. Timetable for use

   4.3 Final destination

      4.3.1. Nature of final use
      4.3.2. Designation of facility
      4.3.3. Name, trade name and address of final user
      4.3.4. Timetable for final use

I the undersigned certify that the information given in this form is authentic and truthful.

Date and place of signature

Signature

Name and office of signer

ANNEX X-A

NOTIFICATION TO THE AGENCY OF EXPORTS AND IMPORTS
OF NUCLEAR MATERIAL

(INFCIRC/207 — 26th July 1974)

On 11th July 1974 the Director General received letters dated 10th July from the Resident Representatives to the Agency of the Union of Soviet Socialist Republics, the United Kingdom of Great Britain and Northern Ireland and the United States of America informing him that in the interest of assisting the Agency in its safeguards activities, the Governments of these three Members had decided to provide it henceforth with information on exports and imports of nuclear material. In the light of the wish expressed at the end of these letters their text is reproduced below.

I am pleased to inform you that my Government, in the interest of assisting the IAEA in its safeguards activities, has decided that the Agency should be provided on a continuing basis with the following information:

1) With respect to the anticipated export of nuclear material (excluding exports of source material for non-nuclear purposes), in an amount exceeding one effective kilogram, for peaceful purposes to any non-nuclear-weapon State:

   a) The organization or company which will prepare the nuclear material for export;

   b) The description, and if possible the expected composition and quantity, of nuclear material in the anticipated export;

   c) The State and organization or company to which the nuclear material is to be exported and, where applicable (i.e. in those cases in which nuclear material is processed further in a second State before retransfer to a third State), the State and organization or company of ultimate destination.

   The foregoing information will be provided normally at least ten days prior to export of the material from my country; confirmation of each export, including actual quantity and composition and date of shipment, will be provided promptly after shipment;

2) With respect to each import, in an amount greater than one effective kilogram, of nuclear material which, immediately prior to export, is subject to safeguards, under an agreement with the IAEA, in the State from which the material is imported:

   a) The State and organization or company from which the nuclear material is received;

   b) The description, composition and quantity of nuclear material in the shipment.

   The information described above will be provided as soon as possible after receipt of the material.

It is intended that the provision of the information described above will be initiated as soon as possible and not later than 1st October 1974. The details of these arrangements can be discussed with you or your staff whenever convenient.

My Government would appreciate this information being brought to the attention of all Members of the Agency by means of an information circular.

ANNEX X-B

NOTIFICATION TO THE AGENCY OF EXPORTS AND IMPORTS
OF NUCLEAR MATERIAL

(INFCIRC/207/Add.1 — March 1984)

On 16th February 1984 the Director General received a letter dated 7th February 1984 from the Governor from France on the Agency's Board of Governors informing him that, in the interest of assisting the Agency in its safeguards activities, the Government of France had decided to provide it henceforth with information on exports and imports of nuclear material. In the light of the request made in this letter, its text is reproduced below.

I am pleased to inform you that my Government, in the interest of assisting the IAEA in its safeguards activities, has decided that the Agency should be provided on a continuing basis with the following information:

1) With respect to the anticipated export of nuclear material (excluding exports of source material for non-nuclear purposes), in an amount exceeding one effective kilogram, for peaceful purposes to any non-nuclear-weapon State:

   a) The organization or company which will prepare the nuclear material for export;
   b) The description, and if possible the expected composition and quantity, of nuclear material in the anticipated export;
   c) The State and organization or company to which the nuclear material is to be exported and, where applicable (i.e. in those cases in which nuclear material is processed further in a second State before retransfer to a third State), the State and organization or company of ultimate destination.

The foregoing information will be provided normally at least ten days prior to export of the nuclear material by my country; confirmation of each export, including actual quantity and composition and date of shipment, will be provided promptly after shipment.

2) With respect to each import, in an amount greater than one effective kilogram, of nuclear material which, immediately prior to export, is subject to safeguards, under an agreement with the IAEA, in the State from which the material is imported:

   a) The State and organization or company from which the nuclear material is received;
   b) The description, composition and quantity of nuclear material in the shipment.

The information described above will be provided as soon as possible after receipt of the material.

It is intended that the provision of the information described above will be initiated as soon as possible. The details of these arrangements can be discussed with you or your staff whenever convenient.

My Government would appreciate this information being brought to the attention of all Members of the Agency by means of an information circular.

ANNEX XI

EXCERPT FROM THE FINAL DECLARATION OF
THE CONFERENCE OF FOREIGN MINISTERS OF NON-ALIGNED COUNTRIES
HELD IN LUANDA FROM 4TH-7TH SEPTEMBER 1985

(Attachment to INFCIRC/332)

VII. PEACEFUL USES OF NUCLEAR ENERGY

65.   The Ministers, recalling the decisions of the Non-Aligned Countries, stressed the exceptional importance of international co-operation among the Non-Aligned and other developing countries in the field of peaceful uses of nuclear energy. This co-operation is of special significance in fields where these countries can achieve a greater degree of self-sufficiency.

66.   The Ministers affirmed the inalienable right of all States to apply and develop their programmes for peaceful uses of nuclear energy for economic and social development in conformity with their priorities, interests and needs. All States should have unhindered access to and be free to acquire technology, equipment and materials on a non-discriminatory basis for peaceful uses of nuclear energy, taking into account the particular needs of the developing countries. They deplored the pressures and threats against developing countries aimed at preventing them from pursuing their programmes for the development of nuclear energy for peaceful purposes.

67.   They also expressed their concern in this respect regarding the obstacles which the developed countries place in the way of transfer of technologies related to the peaceful uses of atomic energy by fixing conditions which are incompatible with the sovereignty of the developing countries. Each country's choices and decisions in the field of peaceful uses of nuclear energy should be respected without jeopardizing fuel cycle policies or international agreements and contracts for the peaceful uses of nuclear energy.

68.   The Ministers also stressed the need for observance of the principles of non-discrimination and free access to nuclear technology and reaffirmed the right of each country to develop programmes for the use of nuclear energy for peaceful purposes in conformity with their own freely determined priorities and needs.

69.   In this connection, the Meeting reiterated that non-proliferation should not be made a pretext for preventing States from exercising their full rights to acquire and develop nuclear technology for peaceful purposes geared to economic and social development in accordance with their priorities, interests and needs, determined in a sovereign manner.

70.   The Ministers expressed their satisfaction at the progress in the preparations for the United Nations Conference for the Promotion of International Co-operation in the Peaceful Uses of Nuclear Energy for Social and Economic Development, in the interest of developing countries and the international community as a whole. They expressed their satisfaction at the results of the work of the Preparatory Committee of the Conference and underlined the necessity for continuing detailed preparations, with the active participation of all countries, in order to fully realise the goals of the Conference. Proceeding from the positions of the Seventh Summit of Non-Aligned Countries, the participants in the Conference reaffirmed their conviction that the results of the Conference should contribute to free and unhampered access on a just and non-discriminatory basis to nuclear technology, equipment and materials needed for the development of national programmes of peaceful uses of nuclear energy.

CHAPTER SIX

IMPACT OF BILATERAL AGREEMENTS ON NUCLEAR TRADE

A. INTRODUCTION

International trade in nuclear materials and equipment is usually conducted within a framework of bilateral agreements between States, or between States and international organisations, agreements which therefore form part of public international law. The actual implementation of such agreements often takes the form of private law contracts concluded between public or private companies from the exporting or importing country. These bilateral agreements, like the commercial contracts pursuant thereto, are binding only on the parties to them, and create no rights or obligations for third parties. However, for most countries with important positions as suppliers, the export of nuclear materials and supply of services such as enrichment can be carried out legally only within the framework of bilateral agreements; this — together with the extremely similar terms of such agreements and the problems posed by their revision — means that these agreements are a key element of the international rules on nuclear trade. In fact these agreements, on par with domestic regulations on international trade, constitute the means by which most exporting States fulfil their obligations, essentially as regards non-proliferation and physical protection, undertaken in an international legal (for example the Non-Proliferation Treaty), or political framework (unilateral commitments undertaken following the work of the Zangger Committee or the London Club).

The provisions in these agreements determine the rights and obligations of the Governments Parties as regards non-proliferation guarantees; they may also further complicate nuclear trade. This is true for prior consent clauses for certain sensitive operations such as enrichment and reprocessing — especially if consent is required on a case-by-case basis, as well as for retransfers to third countries. By way of example, labelling practices may be mentioned: materials are labelled according to their country of origin for that country to keep track of the materials it owns. That situation becomes even more complex when the materials are transformed successively, in different territories, with several countries using a multiple labelling procedure[1].

Thus the United States, which has played a historical role in developing this legal technique, has concluded a whole series of bilateral agreements which, together with the problems raised by their evolution, have undoubtedly served as a reference both for the drafting of international agreements and for the determination by other countries of their own system of bilateral agreements. Section B will therefore examine the system in the United States, paying particular attention to the agreements with EURATOM. Consideration will then be given to the extent to which the bilateral agreements concluded by other countries exporting nuclear materials or equipment and services follow or differ from the original model (Sections C to G).

B. THE SYSTEM OF AGREEMENTS CONCLUDED BY THE UNITED STATES

I. THE ATOMIC ENERGY ACT OF 1954

Reversing the previous policy of secrecy and refusal to export nuclear supplies, this Act (amended on several occasions since its adoption in 1954) had as a purpose to make available to "co-operating nations" the benefits of the peaceful applications of atomic energy, as widely as developing technology and considerations of the common defence and security would permit (Section 3e of the Act).

These "considerations" led to export controls. Thus, in the case of the export of special nuclear material, facilities for producing and utilising such material, or the communication of confidential information, the Atomic

Energy Commission (today the Nuclear Regulatory Commission) was obliged not to deliver such material, communicate such information or grant facility export licences except within the framework of and in accordance with a co-operation agreement concluded with a foreign State or group of States. Such agreements were required to be approved by the President and not opposed by Congress, in addition to which the Act (Section 123) specified that the foreign Party had, in particular, to undertake to maintain the security control (safeguards) and standards set out in the agreement, not to use the material transferred pursuant to the agreement for research on or the manufacture or development of atomic weapons or for any other military purpose, and not to transfer the material or confidential information to unauthorised persons or beyond the jurisdiction of the Party concerned, except as specified in the agreement.

In the years separating the 1954 Atomic Energy Act and the Nuclear Non-Proliferation Act of 1978, the contents of these co-operation agreements changed in two ways:

— first, the setting up of the IAEA safeguards system meant that new co-operation agreements, or amendments to co-operation agreements already in force, were required to substitute IAEA control for that exercised by the United States, a substitution effected in three-cornered agreements between the United States, the IAEA and the other State receiving supplies under the co-operation agreement;
— secondly, there was a compensatory movement to strengthen the control exercised by the United States over retransfers and to add a right of approval for the reprocessing of exported material.

II. SPECIAL CASE OF THE EURATOM/ UNITED STATES AGREEMENTS

Under the EURATOM Treaty, Member States conferred wide powers on the European Atomic Energy Community (EURATOM) in the field of the dissemination of information, supplies, safeguards and property rights. Thus, the Treaty provided that co-operation agreements concluded by a Member State before the entry into force of the Treaty should be renegotiated in order to ensure that, as far as possible, the rights and obligations arising out of such agreements are assumed by the Community (Article 106).

In addition, Article 101 conferred upon EURATOM the same powers for external relations as within the Community; accordingly it may conclude agreements for exchanging information (Article 29), supplying ores, source materials and special fissile materials coming from outside the Community (Articles 52.2(b) and 64), exporting special fissile materials (Articles 59(b) and 62), or for contract work (Article 75). According to the circumstances, this power may be exclusive or allow parallel powers to Member States to conclude bilateral agreements subject to control by the Commission of their conformity with the Treaty (Article 103).

In view of the support accorded by the United States to the Community enterprise, an Agreement for Co-operation between the United States and EURATOM was signed on 8th November 1958, in implementation of the outline Agreement signed on 29th May and 19th June 1958. This Agreement set up a "Joint Program":

— to bring into operation, within the Community, "large-scale power plants using nuclear reactors of types on which research and development have been carried to an advanced stage in the United States";
— to initiate a joint research and development programme centred on these types of reactors.

To help implement this programme, it was agreed that the United States Atomic Energy Commission would sell a given quantity of contained U-235. Provision was also made for the possibility of selling additional quantities of special nuclear material, but this was made subject to a new agreement. The Community was to retain ownership of all the special nuclear material supplied by the United States, but was entitled to distribute such material to authorised users within the Community.

It was agreed that, failing written notice of non-availability of the service, the United States would undertake reprocessing, and return to the Community any material recovered after this operation. In exchange, the Community promised to comply with the three main conditions laid down in Section 123 of the Atomic Energy Act: not to use materials for military purposes, no transfer without prior agreement, and safeguards. However, in the case of the transfer of materials, equipment or devices, the Community, one of whose aims is the creation of a common market, succeeded in having itself considered as a single territory, within which transfers to authorised persons would not be subject to approval by the United States Government. Thus, only transfers outside the jurisdiction of the Community were made subject to such approval. At the same time, the EURATOM safeguards provided for under Chapter VII of the Treaty were implicitly held to be effective and satisfactory.

An amendment was made to this Agreement, on 21st-22nd May 1962, to make it possible to lease, as well as sell, the quantity of contained U-235 in uranium provided for by the 1958 Agreement and to fix the termination date of the Agreement at 31st December 1985. (Another amendment was signed on 22nd-27th August 1963 to extend to 31st December 1995 this termination date.) An Additional Agreement, supplementing the first, had been signed on 11th June 1960 (and amended on 21st-22nd May 1962 and 22nd-27th August 1963), providing for the supply of materials intended for experimental reactors and experimental reprocessing facilities. The Additional Agreement stipulated that retention by the United States of ownership of the special fissile material leased was not incompatible with the Treaty since the Community had power and authority over such material while it was within the Community. This Additional Agreement was concluded for a period of ten years. Along the lines established by the Additional Agreement, the 1962 and 1963 amendments provided for the supply of the special nuclear material required to implement research and energy production programmes not included in the Joint Program of the Co-operation Agreement, with limits as to weights. Charging amounts against these weights was made easier regarding materials returned to the United States or transferred, with the latter's agreement, to another country. This provision therefore permitted conversion or fabrication services to be provided by the Community to third countries.

The amendment of 20th September 1972 made more significant changes to the Additional Agreement, giving up predetermined weight limits in favour of a more flexible formula but leaving the fixing of transfer limits to the discretion of the United States.

III.   THE NUCLEAR NON-PROLIFERATION ACT OF 1978 AND ITS IMPACT ON CO-OPERATION AGREEMENTS

The Nuclear Non-Proliferation Act (NNPA) of 10th March 1978 contains a number of sections affecting co-operation agreements. Thus, Section 407 states that the President of the United States shall endeavour to provide, in co-operation agreements, for co-operation in protecting the international environment from radioactive, chemical or thermal contamination arising from peaceful nuclear activities. However, the essential provision with regard to agreements to be concluded after the promulgation of this Act is Section 401 (inserted as Section 123 in the Atomic Energy Act of 1954).

This Section provides that future co-operation agreements should meet nine criteria, a number which is reduced to seven when the other Contracting Party is a nuclear-weapon state. Provision is also made for certain exceptions when the agreement concerns defence and is concluded pursuant to Sections 91c., 144b. or c. of the Atomic Energy Act of 1954.

These criteria apply to nuclear material, equipment and/or installations ("production or utilization facilities") and, in certain cases, to confidential information or sensitive nuclear technology transferred to the other Party in implementation of the Agreement, and also to special nuclear material used or produced in or by using the material, equipment, facilities, etc., transferred.

The criteria relate to the following :

i)   *Continuing safeguards*
Safeguards with respect to all the materials listed above must be maintained so long as the materials, etc. remain under the jurisdiction of the co-operating party, even if the Agreement is terminated or cancelled for any reason. This applies in particular to special nuclear material used in or produced through the use of "nuclear material and equipment".

ii)   *Full-scope safeguards*
If the co-operating party is a non-nuclear-weapon state, IAEA safeguards must be maintained with respect to all nuclear material within the territory of such state or transported outside its territory but still under its control.

For non-nuclear-weapon states having ratified the NPT, this condition does not involve any obligations in addition to those applying under this Treaty. For states not bound by the NPT, on the other hand, this condition brings about a result equivalent to ratification, at least for application of safeguards on their territory.

iii)   *Ban on use for any nuclear explosive device*
The co-operating party must guarantee that no nuclear material and equipment or sensitive nuclear technology to which these criteria apply will be used to manufacture any nuclear explosive device, or for research on or development of any such device or for any other military purpose.

iv)   *Return of nuclear material and equipment*
Agreements, other than for defence, with non-nuclear-weapon states must contain a clause giving the United States the right to require the return of any nuclear material and equipment transferred and any special nuclear material produced through the use thereof, if the co-operating party detonates a nuclear explosive device or terminates or cancels a safeguards agreement with the IAEA.

The United States, which cannot exercise perpetual control over the materials or equipment supplied by a third party state or prevent the manufacture of nuclear explosive devices by means of materials supplied by a third party state, thus ensures, by this possibility of recovering its supplies, that it can exercise dissuasive pressure and, if necessary, penalise a co-operating party which has not complied with the conditions of the criteria mentioned under *(ii)* and *(iii)*.

v)   *Retransfers*
The co-operating party must guarantee that any material, restricted data or production or utilisation facility to which these criteria apply will not be retransferred to unauthorised persons or beyond the jurisdiction or control of the co-operating party without the consent of the United States.

The Act systematizes the rule established in previous co-operation agreements regarding the special nuclear material produced by any facility or through the use of any material retransferred.

vi) *Physical protection*

The co-operating party must guarantee that adequate physical security will be maintained with respect to any nuclear material to which the criteria apply and also to any production or utilisation facility transferred.

vii) *Reprocessing, enrichment or other alteration*

The co-operating party must guarantee that no material to which the the criteria apply will be reprocessed, enriched or (in the case plutonium, uranium 233, or uranium enriched to more than 20 per cent in the isotope 235, or other nuclear materials which have been irradiated) otherwise altered in form or content without the prior approval of the United States.

Again, the new element here as compared to previous practice lies in the extension to material used or produced through the use of any nuclear material or equipment supplied by the United States.

It will be noted that when, as happens frequently, these operations are performed in a third party state, the prior approval provided for under this criterion is necessary in addition to the transfer consent required under criterion *(v)*.

viii) *Storage*

The co-operating party must guarantee that plutonium, uranium 233 or uranium enriched to more than 20 per cent, to which these criteria apply, will be stored in a facility approved in advance by the United States.

ix) *Sensitive nuclear technology*

The co-operating party must guarantee that any special nuclear material, production facility or utilisation facility produced or constructed under the jurisdiction of the co-operating party through the use of any sensitive nuclear technology transferred pursuant to the agreement, will be subject to all the requirements specified above.

In addition, Section 402 of the Nuclear Non-Proliferation Act, which has not been inserted in the Atomic Energy Act provides that :

— no source or special nuclear material may be exported for enrichment without the prior consent of the United States for such enrichment, unless this has been expressly provided for in a new or amended agreement for co-operation ;
— no major critical component of any enrichment, reprocessing or heavy water production facility may be exported under any agreement for co-operation unless expressly provided for in the agreement.

It was decided that the policy defined in the Nuclear Non-Proliferation Act would first be implemented in the framework of new agreements for co-operation, completed :

— by "subsequent arrangements" entered into by the United States competent authorities with the co-operating party. These arrangements, which concern the granting of United States consent to the transfer, storage, or alteration of materials originating in the United States or produced by means of United States supplies (for example, reactors or the major components of reactors), are subject to a special procedure ;
— by export licences, the granting of which implies that the above-mentioned criteria have been met.

However, the Act could not terminate co-operation set up under agreements for co-operation entered into prior to the date of its enactment (Section 405). Thus, it provided that the President should initiate a programme to renegotiate existing agreements to bring them into line with the new requirements of the Act and, in particular, the nine criteria listed above.

Despite the rigid and restrictive character of the Nuclear Non-Proliferation Act, United States policy retains a certain amount of flexibility. Thus, Section 401 gives the President of the United States power to exempt a proposed agreement for co-operation from any of the criteria laid down in the Act, if he determines that its inclusion would be seriously prejudicial to the achievement of non-proliferation objectives or otherwise jeopardise the common defence and security of the United States. In the Nuclear Non-Proliferation Policy Statement of 16th June 1981, the President of the United States, while opposed to transfers of reprocessing technology and the transfer of plutonium, specified that he did not intend to create obstacles to civil reprocessing and the development of fast breeder reactors in countries with an advanced nuclear programme.

Details of this policy were given in a "Plutonium Use Paper", approved by the President on 4th June 1982, which defined countries with an advanced nuclear programme as the Member States of EURATOM and Japan. These countries could hope to receive long-term approval for reprocessing spent fuel of United States origin and for using the plutonium extracted. Other countries subjected to full-scope IAEA safeguards could receive prior consent to have fuel reprocessed in a nuclear-weapon State, but not to receive back and use the plutonium recovered.

This increased flexibility is reflected in the regulations on the export of nuclear materials and equipment promulgated by the Nuclear Regulatory Commission (NRC) on 14th April 1986 (51 FR 12598) which extend the general licence to export by-product material to include americium 241 contained in industrial process control equipment, take account of new adherents to the NPT, to modify the list of countries subject to particular restrictions, and of the prior consent of Canada to the re-export of materials and equipment originating in that country. An amendment of 4th August 1986 (51 FR 27825), on the other hand, imposes more restrictive limits on the general licence for the export of tritium contained in radioluminescent sources and other items.

IV. NEW AGREEMENTS NEGOTIATED UNDER THE NNPA

Amendements to existing agreements, or new agreements have been concluded with Australia, Canada, Sweden, Norway, and Egypt and more recently, with Japan.

The Agreement of 5th July 1979 with *Australia* provides for IAEA safeguards with respect to all Australian nuclear activities and to nuclear materials of Australian origin in the United States. The right of prior consent given to the United States, for example in relation to retransfers and reprocessing, only applies to the proportion of fuel of United States origin contained in a reactor not supplied by the United States. General approval is given to enrichment at less than 20 per cent, and agreement to storage is not required if the plutonium or uranium 233 have not been separated from the fuel elements.

An Exchange of Notes of 2nd August 1985 specified the responsibilities of each of the Parties with respect to safeguards, physical protection and retransfers.

The Protocol of 23rd April 1980 amending the Agreement for Co-operation with *Canada* also provides for reciprocal safeguards.

An example of flexibility, while at the same time complying with the NNPA criteria, is provided by the agreements concluded by the United States with *Sweden* on 19th December 1983 and *Norway* on 12th January 1984[2].

Thus, no transfer of materials, equipment or components from one country to the other, whether directly or through a third country, may be made unless confirmation is given by the competent authorities in the receiving country that the transfer is subject to the terms of the agreement.

The non-proliferation guarantees are perpetual in the sense that they continue to apply for as long as the materials transferred or produced through the use of materials, equipment or components transferred remain usable for a nuclear activity relevant from the risk of proliferation viewpoint, even when, for whatever reason, the agreements for co-operation were suspended, terminated or expired. As a rule, the IAEA safeguards contained in the safeguards agreements signed by the United States and by each of the two States mentioned above, apply. If, for whatever reason, these safeguards were no longer applied, the parties undertake to conclude an agreement to ensure the effective continuation of the safeguards in compliance with the safeguards principles of the Agency and in accordance with a procedure specified in an agreed interpretation of the agreement (Agreed Minute).

Since Sweden and Norway are both NNW (non-nuclear-weapon) States within the meaning of the NPT, the safeguards are "full-scope" and apply to all nuclear activities conducted on their territory or under their jurisdiction or control. The undertaking not to use material transferred for military purposes or for research on or the development or use of nuclear explosive devices, is reciprocal. In the Agreed Minute, Sweden has also had the scope of the term "military purpose" clarified (see chapter nine of this Study - Definitions) and the production of tritium using material, equipment or components transferred, prohibited.

If either Party fails to comply with the relevant provisions of the agreement for co-operation or of the safeguards agreement with the IAEA, the other Party may require the return of any items transferred or special nuclear material produced through the use thereof. Hypothetically, a detonation by Norway or Sweden of a nuclear explosive device would also entitle the United States to exercise this right of recovery. However, even in the event of such breaches, provision is made for payment at a fair market price.

The NNPA criteria concerning storage and retransfers are to be found together in a single Article of these Agreements. Formally, the agreement of the other Party is required for the storage or retransfer of items received under the Agreement or special nuclear material produced through the use thereof. However, the United States gave its approval in principle for retransfers for reprocessing in the facilities at Sellafield (United Kingdom) or La Hague (France), on condition that the United States be informed of such shipments, that the material transferred be subject to the EURATOM/ United States Agreement for Co-operation, and above all, that Sweden or Norway retain their rights over the plutonium extracted such that the United States may exercise its right to prior consent with regard to any retransfer of this plutonium to Sweden (or Norway) or to any other country and for whatever purpose. In the case of Sweden, this agreement in principle extends, according to the Agreed Minute, to the conversion and fabrication of fuel outside Sweden, and takes account of Swedish projects for the storage and conditioning of spent fuel.

These Agreements provide that reprocessing, enrichment to more than 20 per cent and alteration require the prior consent of the other Party. However, the Agreed Minute with Sweden exempts from the prior consent requirement the alteration of limited quantities of spent fuels necessary for testing or analysis, and refers to processes involving compaction and encapsulation of such fuels. Moreover, these two Agreements take into consideration the fact that safeguards may also be exercised, for example over a particular facility or fuel load, under agreements concluded with third countries or a group of third countries. Under the Agreed Minute, a conflict between different rights to exercise control is solved by means of a rule of proportionality established between the quantity of transferred material used in the production of special nuclear material as compared to the total amount of new fuel, and similarly for subsequent generations.

The provisions of these Agreements dealing specifically with physical protection will be examined in chapter eight of this Study.

A difference between the two Agreements studied may be observed with regard to sensitive nuclear technology. The Agreement with Sweden provides that restricted data and sensitive nuclear technology may be transferred if provided for by an amendment to the principal Agreement or by a separate Agreement.

The Agreement with Norway (see *appendix 9* to the Study), on the other hand, provides for the transfer of "information" other than "Restricted Data", as defined in the Atomic Energy Act. The transfer of sensitive nuclear technology requires an amendment to the Agreement.

Lastly, account was taken to a certain extent of the concern that the complexity and slowness of the procedures for granting prior consent would affect the continuity of supply, by an undertaking that the physical protection and safeguards provisions would be implemented in such a way as to be consistent with prudent management practices required for the economic and safe conduct of the Swedish and Norwegian nuclear programmes.

The Agreement concluded with *Japan* on 4th November 1987 is intended to replace the previous co-operation agreement between both countries. It gives yet another example of reconciling the criteria fixed by the NNPA with the needs of the nuclear industry. By inserting the concept of programmatic approval, the Agreement, which is supplemented by an Implementing Agreement does away with — under certain conditions — the *ad hoc* prior consent requirement for subsequent conversion and reprocessing operations regarding the exported materials (in practice in British and French facilities). Also transfers of unirradiated source materials and low enriched uranium to countries designated in writing may be the subject of an agreement in principle. A special feature of the Agreement is that it contains specific instructions for the air transport of plutonium recovered following reprocessing.

The European Community on the other hand, has not agreed to renegotiating the EURATOM/United States Agreement which will expire at the end of 1995. However, discussions on the future of this Agreement are in progress.

C. AGREEMENTS CONCLUDED BY AUSTRALIA

Large amounts of uranium ore are found in Australia which has concluded several export agreements in this connection. Given that its position as an exporter developed relatively late, while non-proliferation and physical protection considerations were already given high priority at national level, it is not surprising that such considerations are given pride of place in the agreements.

By way of example, the Agreement concluded on 20th July 1978 with *Finland* provides that nuclear material subject to the Agreement shall not be diverted for use in connection with nuclear weapons or other nuclear explosive devices, with research on or the development of nuclear weapons or other nuclear explosive devices, or used for any military purpose. Such material shall be subject to the safeguards applied by the IAEA under a Non-Proliferation Treaty safeguards agreement or equivalent, or, failing which, to control exercised by the Party supplying the material.

The Agreement applies not only to nuclear material transferred between the two Parties, whether directly or through a third country, but also to material derived from the material transferred and even to subsequent generations of such material. However, in these latter two cases, and in accordance with a principle of proportionality with the transferred nuclear material used for their production :

— safeguards apply for as long as the materials subject to the Agreement remain usable, for any nuclear activity relevant from the point of view of safeguards, under the jurisdiction of the receiver Party ;
— the transfer of such material beyond the jurisdiction of the receiver Party, enrichment to more than 20 per cent, and reprocessing are subject to the prior consent of the supplier Party. Examination of a request for prior consent must, however, take into account not only non-proliferation considerations, but also energy requirements and the need of the receiver Party to ensure proper management of spent nuclear fuel and the disposal of nuclear waste. As a minimum, measures of physical protection which satisfy the requirements of the IAEA recommendations are to be applied.

Similar Agreements were concluded between Australia and the Philippines (8th August 1978), South Korea

(2nd May 1979), Sweden (18th March 1981, amended on 12th July 1982) and Switzerland (28th January 1986).

A Co-operation Agreement concluded in 1972 with *Japan* was revised on 5th March 1982 along similar lines. It also applies, however, to equipment transferred between the two countries and, where relevant, through the intermediary of a third country. This Agreement with Japan should be read together with the Agreement concluded between Australia and France on 30th October 1980 concerning the conversion and/or enrichment in France of nuclear material of Australian origin supplied to Japan under a contract of sale concluded prior to 2nd December 1972.

Negotiating agreements with the *European Community* and two of its Member States, the *United Kingdom* and *France* was a lengthy process, in large measure due to internal Community reasons. In 1978, the United Kingdom entered into negotiations with Australia to conclude a safeguards agreement covering the supply of uranium by Australia, and France expressed the wish to conclude a similar agreement. However, following a ruling by the Court of Justice of the European Communities on 1st November 1978 relating to the Physical Protection Convention, the Commission of the European Communities decided that it also had to be a Party to any agreement affecting supplies and the nuclear common market. Agreement was reached on co-ordinating the bilateral Agreements and the one concluded by the Community, giving rise to the three connected Agreements :

— Agreement with the United Kingdom concerning nuclear transfers between Australia and the United Kingdom, signed on 24th July 1979 (and entering into force on the same date) ;
— Agreement with France concerning nuclear transfers, signed on 7th January 1981 (and entering into force on 12th September 1981)[3] ;
— Agreement with EURATOM concerning transfers of nuclear material from Australia to the European Atomic Energy Community, signed on 21st September 1981 (and entering into force on 15th January 1982). The provisions of this Agreement are considered as supplementary to those of the bilateral Agreements with the United Kingdom and France and in some instances replace them.

The scope of the Australia/EURATOM Agreement is limited to nuclear material transferred from Australia to the Community as well as, in accordance with the principle of proportionality used in the Agreement with Finland, to all forms of nuclear material obtained by chemical or physical processes or isotopic separation and to all generations of nuclear material produced by isotopic irradiation from material of Australian origin.

The Australia/EURATOM Agreement contains the same provisions as those noted in the Agreement with Finland (ban on diversion for military purposes, IAEA safeguards), but includes a reference to the so-called Verification Agreement of 5th April 1973 (see chapter

four of this Study) and the tripartite Agreements, United Kingdom/France, the Community and the Agency of 6th September 1976 and 27th July 1978. The physical protection provisions refer to documents INFCIRC/254 Annex B, and INFCIRC/225/Rev.1).

As concerns the retransfer of nuclear material to which the Agreement applies, the territory of the Community is considered as forming a whole, and the prior consent of Australia is only required for a transfer beyond the Community to any other country. However, Australia drawing the consequences from the policy it applied to its commercial partners, stated in an accompanying letter (No. 2) that it did not require consent on a case-by-case basis for transfers of nuclear material for fuel cycle operations such as enrichment to below 20 per cent when such transfers are to third countries which have an agreement with Australia. The Community must, however, notify the transfers to Australia, which reserves the right to tell the Community that it considers it necessary to suspend, cancel or abstain from effecting them.

Enrichment beyond 20 per cent in the isotope uranium 235 is not envisaged "in the present circumstances" and the Parties will, if appropriate, discuss the conditions to be applied (Annex B to the Agreement).

Although Australia was at first not enthusiastic about the principle of reprocessing, it nevertheless acknowledged, during negotiations, that it could be legitimate if performed in accordance with agreed conditions set out in Annex C to the Agreement. Reprocessing may be effected only to promote the efficient use of energy resources or for the management of materials contained in spent fuel in accordance with a nuclear fuel cycle programme described in an Implementing Arrangement to be concluded between the Parties. The plutonium separated will be stored and used in accordance with this programme. Both the reprocessing and the storage of the separated plutonium remain subject to IAEA safeguards. The reprocessing and use of the separated plutonium for other peaceful non-explosive purposes, including research, will be undertaken only on the conditions agreed between the Parties.

Similar clauses on reprocessing have been included in the Agreements concluded with Sweden, on 18th March 1981, Japan, on 5th March 1982, and Switzerland, on 28th January 1986.

The scope of the bilateral Agreements between Australia and the United Kingdom or France, on the other hand, is wider and extends to non-nuclear material intended for use in reactors, and to equipment and technology. The Agreement with France, for example, takes into consideration the fact that both Australia and France apply in this respect the Guidelines laid down by the London Club (INFCIRC/254).

Agreements were also concluded with the *United States*, on 5th July 1979, and *Canada*, on 9th March 1981. The procedure of prior consent is also used for retransfers of source materials or irradiated material enriched to less than 20 per cent, to countries named in a list drawn up by Australia.

More recently, on 28th January 1986, an Agreement was concluded between Australia and *Switzerland* and supplemented by an exchange of letters on the same day. It should be noted that in the letters, a general agreement is given by the supplier countries, under certain conditions, for re-export to approved third countries of the nuclear material delivered and for proceeding with the various fuel cycle operations. The Agreement is reproduced in *appendix 15* to this Study.

D.   AGREEMENTS CONCLUDED BY CANADA

In an initial period, Canada adopted an export policy involving both the supply of natural uranium and the construction in India, Romania and Argentina of CANDU reactors, a policy implemented by Agreements for Co-operation concluded, in particular, with Pakistan, Japan, Australia, EURATOM, Sweden, Spain and Iran. Under these Agreements, supplies could be delivered only within the framework provided for by each Agreement, and could not exceed the legitimate needs of the recipient for jointly defined purposes. Provision was made for bilateral safeguards with the option of substituting those of the IAEA, but these were limited to the right to examine the design of facilities, the keeping of materials accounts, progress reports, approval of reprocessing methods, and a right of inspection. As regards the Agreement with EURATOM, the safeguards applied were considered sufficient and subsequently, like the EURATOM/United States Agreement, no provision was made for bilateral control.

However, the Agreement concluded with *India* in December 1963 and amended in 1966 concerning the power plants of Rajasthan in India and Douglas Point in Canada had no safeguards clause, only an undertaking by India to use the CANDU reactor supplied and the plutonium it would produce exclusively for peaceful purposes. The setting off, on 18th May 1974, of a nuclear explosive device — although India maintained that since this explosion was for peaceful purposes, it was not in breach of its commitments — led Canada to make important changes in its export policy[4].

The Co-operation Agreements concluded with *South Korea* on 26th January 1976, and *Argentina* on 30th January 1976, reflect this new policy which concerns, in particular, the following points :

— the undertaking not to use the items in question for military purposes is extended to cover the development and manufacture of nuclear explosive devices, the consequences of any breach being the stopping of supplies and the return of items already delivered. Moreover, this undertaking, like the rights and obligations listed below, concerns not only nuclear material, other relevant materials and equipment and facilities supplied in implementation of the Agreement, but also the special nuclear material produced through the use of the items supplied;

— conclusion by the Contracting Party of a safeguards agreement with the IAEA so as to ensure that the materials and items listed above are subject to safeguards;
— the transfer of these items outside the jurisdiction of the Contracting Party is subject to the prior consent of the supplier State;
— the reprocessing of the materials supplied or obtained through use of the materials or items supplied is prohibited unless authorised by the supplier State.

The Agreement of 6th October 1959 with EURATOM remained in force until amended by an Interim Arrangement concluded by an exchange of letters on 16th January 1978. This demonstrated the desire to reach a compromise following an embargo on supplies decided by Canada in 1977. This Interim Arrangement provided, *inter alia*, for reprocessing and for enrichment beyond 20 per cent :

— the European Community undertook not to use material for the manufacture of nuclear weapons or any other nuclear explosive device or for any other military purpose. It accepted verification of compliance with this undertaking by means of the safeguards provided for in the EURATOM Treaty and the IAEA/EURATOM/non-nuclear-weapon Member States Verification Agreement, or under trilateral agreements between the IAEA, the European Community and the United Kingdom or France (although the first of these had not yet entered into force and the second was still being negotiated). For its part, France facilitated conclusion of this arrangement by negotiating the trilateral Agreement mentioned above and by a statement that material to which the Canada/EURATOM Agreement applied would not be used in France before the entry into force of the trilateral Agreement ;
— for its part, Canada accepted that reprocessing, enrichment beyond 20 per cent in uranium 235, and the storage of plutonium and uranium enriched beyond 20 per cent should simply require notification by EURATOM and consultation as to the appropriate safeguards for the operation envisaged, without prejudicing the commercial or industrial policy of either Party.

Canada also agreed to allow nuclear material of Canadian origin to be delivered directly or transferred to France for enrichment or reprocessing there before the entry into force of the above-mentioned EURATOM/France/IAEA Agreement; this was subject to the condition that such material leave French territory after the normal period required for such operations.

The problem of mixing Canadian material with that from elsewhere was also solved by means of a rule of proportionality.

A new stage was reached in 1981, after completion of the INFCE study. The Agreement was again amended with an exchange of letters which replaced, *inter alia*, the Interim Arrangement of 1978[5]. Provision was made for a firm commitment to non-proliferation through the application of IAEA safeguards and adequate physical protection measures, and the case-by-case agreement on reprocessing, storage and the use of plutonium was abandoned. From then on, reprocessing and the storage of plutonium was allowed whenever agreement had been reached between the Parties that such operations formed an integral part of a nuclear energy programme described in detail to the other Party. The question of uranium enriched beyond 20 per cent was left open.

A further harmonisation of the Canadian and Community points of view was achieved in 1984, and implemented by another amendment to the Agreement in the form of an exchange of letters dated 21st June 1985 (see *appendix 13* to the Study). This amendment prolongs the Agreement of 1959 for a further twenty years, with tacit renewal every five years, and makes the retransfer procedure appreciably more flexible with a view to establishing a network of partner countries amongst which Canadian-origin material can circulate as freely as possible.

Consequently, the retransfer by the Community to such third parties of natural uranium, depleted uranium, uranium enriched to 20 per cent or less in the isotope U-235 and heavy water is generally allowed for the future, provided that such third parties have been identified by Canada and that procedures acceptable to both the Contracting Parties have been established. In a Protocol attached to the 1985 Agreement, Canada drew up a list of such third party countries in the light of requests submitted by the Community to protect the industrial and commercial interests of its Member States. As concerns procedures, Canada stated that it would endeavour gradually to simplify, as far as possible and in ways compatible with its non-proliferation policy, notification and other retransfer procedures.

The retransfer over a period of twelve months of "small quantities of materials" to all third countries which have signed the Non-Proliferation Treaty is also authorised in accordance with simplified procedures detailed in administrative arrangements. The retransfer of material or equipment to third countries other than those mentioned above, on the other hand, remains subject to the prior written consent of Canada.

It is also envisaged to introduce a procedure whereby inclusion in or withdrawal of material from the framework of the Agreement could be authorised, which gives support, at least in principle, to the practice of the exchange of non-proliferation obligations, commonly called "flag swap" (see *annex* to this chapter).

Canada resumed its pre-1974 practice by concluding, on 18th June 1985, an Agreement for Co-operation with *Turkey* containing provisions applicable to the transfer of nuclear facilities, material and technology between the Parties. Naturally, activities to which the Agreement applies must be conducted exclusively for peaceful purposes and subject to IAEA safeguards and recommendations as to physical protection.

To conclude, mention may be made of the Free Trade Agreement between Canada and the *United States* (planned entry into force on 2nd January 1989) which should facilitate uranium exports to American industry.

E. AGREEMENTS CONCLUDED BY THE FEDERAL REPUBLIC OF GERMANY

In support of the export efforts made by its nuclear industry, the Federal Republic of Germany has concluded scientific and technical co-operation agreements with numerous countries[6]. The Federal Republic of Germany fully adopts non-proliferation objectives, but rather than the unilateral imposition of standard rules by exporting countries, prefers co-operation arrangements decided on a case-by-case basis and strengthening non-proliferation policy by establishing links of co-operation (Einbindung). However, in order to avoid being placed in an unfavourable position as compared to its competitors, the Federal Republic of Germany supports efforts at harmonisation between supplier countries on condition that the restrictions on international trade remain reasonable.

These co-operation agreements define the spheres covered and the procedures for their implementation. In accordance with the case-by-case policy, the fields covered vary and are more or less extensive in different agreements. This illustrates the pragmatic approach adopted by the Federal Republic of Germany, as confirmed by the procedures for implementing the agreements. The exchange of staff, the participation of scientific experts and engineers in the other Party's projects, advisory services and exchanges of information or documentation may or may not be accompanied by the communication of the technological information relating to the installations supplied. However, in terms of a statement of 17th June 1977, the Federal Government decided, for the time being at least, no longer to authorise the export of reprocessing installations or the transfer of the technology related thereto[7].

Moreover, these agreements remain framework agreements, leaving the definition of the scope, the solving of the practical difficulties raised by their application and adaptation and the financial arrangements to be dealt with in special agreements drafted by mixed committees.

As concerns non-proliferation, the Federal Republic of Germany has refused to impose "full-scope safeguards". Its Agreement with Brazil concluded on 27th June 1975, however, reflects the changes in the international consensus following the Indian explosion; it tends to treat "nuclear explosive devices" as equivalent to nuclear weapons, extends the application of the IAEA safeguards to material produced or processed in transferred installations or through the use of the related technological information, and restricts the export of sensitive equipment and technology[8].

In all the agreements, including that of 9th May 1984 with the *People's Republic of China*, supplies and technology transfers are made conditional upon use for peaceful purposes and the conclusion of a safeguards agreement with the IAEA (a three-cornered agreement in the case of Brazil). In addition, the agreement with *Brazil*[9] prohibits retransfers to third, non-nuclear-weapon States unless the State concerned is itself bound by a safeguards agreement with the Agency. Moreover, the retransfer to any third State of sensitive nuclear material, equipment, installations and related technological information remains subject to the prior consent of the Contracting Party which supplied them. However, approval by the People's Republic of China of retransfers by the Federal Republic of Germany to other Community Member States is assumed to have been given. The Agreement with *Iran* of 3rd July 1976 does not include this requirement, doubtless because it did not envisage the supply of sensitive materials, equipment or technology to this country.

In the Agreement with the People's Republic of China[10], on the other hand, the retransfer to another State of materials supplied requires the authorisation of both Parties.

F.   AGREEMENTS CONCLUDED BY FRANCE

France occupies an important position in international trade in both fuel cycle services and nuclear installations, especially nuclear reactors.

From the non-proliferation viewpoint, French policy has evolved in stages[11]. In agreement with the London Club countries, France wishes to prevent the proliferation of nuclear weapons and ensure that the export of material and equipment which is sensitive from the proliferation viewpoint does not become the subject of commercial competition. While recognising that certain technologies present more proliferation risks than others, France feels that the guarantee of continuity in the supply of nuclear material or fuel cycle services and the satisfaction of legitimate requests for technology transfers are favourable to the policy of non-proliferation.

French law does not require that nuclear material, services or equipment be exported within the framework of a co-operation agreement. However, in addition to domestic regulations on foreign trade (licensing regime), the control exercised by the French Government over the public enterprises and main industrial companies operating in the nuclear sector gives it an effective means of ensuring that these enterprises and companies comply with its policy. Moreover, in many cases, scientific and technical co-operation results from agreements concluded directly by the Atomic Energy Commission (Commissariat à l'Energie Atomique) rather than from agreements signed by the Government.

The objective of a number of co-operation agreements concluded by France has been limited to an exchange of scientific and technical information relating to various spheres of the peaceful use of nuclear energy. Thus, the Agreement concluded on 3rd April 1969 with *Indonesia* related to uranium prospecting, the use of radionuclides and radiation, activation analysis, radiation protection, nuclear engineering, etc., and was to be implemented essentially by the communication of scientific and technical documentation and by the basic and advanced training in France of Indonesian personnel. Provisions regulated the taking out of patents for invention and the granting of irrevocable and non-exclusive licences with the option of ceding sub-licences to the other Party.

Of recent agreements of this type, mention may be made of the Agreements for Co-operation in the use of nuclear energy for peaceful purposes concluded, on 29th August 1980 with *Bangladesh*, and on 6th March 1980 with the *United Arab Emirates*.

Certain of these Co-operation Agreements, such as that of 14th May 1970 with *Switzerland*, also provided for the supply of nuclear fuel by France and the reprocessing in France of spent fuel sent from Switzerland.

More interesting are the Co-operation Agreements concluded in conjunction with the export of nuclear reactors. Two such agreements in particular — that with *Iraq* of 18th November 1975, and that with *Pakistan* of 17th March 1976 — gave rise to a certain amount of controversy.

The objectives of the Agreement with Iraq were:

— to promote trade in installations, equipment and other supplies, materials (especially source and special fissile materials) and provision of information not falling within the classification of security, and of services. It was provided that the conditions to which such operations were to be made subject would be laid down, on a case-by-case basis, in an agreement between the Contracting Parties and/or the public or private bodies concerned; and
— to promote the development of co-operation between the competent bodies by exchanging specialists, research workers and technical experts.

Provision was made for promoting the exchange of scientific and technical information and documentation in all spheres of relevance to the use of nuclear energy for peaceful purposes. However, the information and documents exchanged were not to be communicated to public or private third parties without the prior written consent of the Party supplying the document or information.

As concerns non-proliferation, each Party undertook that materials, equipment and installations received in the context of the Agreement, and nuclear material produced with the help of the material or equipment supplied or exchanged in the context of the Agreement:

— would not be used for military purposes to manufacture nuclear explosive devices;
— would not be transferred to persons not authorised to possess them or to bodies beyond the jurisdiction of one of the Contracting Parties without the prior consent in writing of the other Party;
— would be submitted to IAEA safeguards on the territory of the Contracting Party receiving the said material and equipment.

Moreover, each Party undertook that no source or special fissile material received pursuant to the Agreement, nor any special fissile material recovered or obtained as a by-product would be transferred to another State without being made subject to the IAEA safeguards or, failing which, without the prior consent of the supplier Contracting Party.

Agreements of a more recent type, for example the Agreement with *South Korea* of 4th April 1981 and the Agreement with *Egypt* of 27th March 1981, also provide for implementation by means of specific arrangements between the competent public bodies or by contract, with respect to industrial projects, the supply of nuclear materials, equipment and installations and the transfer of technological information. Provision is made for setting up liaison groups or joint co-ordination committees.

With regard to the exchange of information, the Agreement with South Korea specifies that the Parties are not obliged to transfer information, documents or equipment of a confidential nature, the transmission or supply of which was not provided for under the implementing arrangements or contracts.

As concerns non-proliferation, the following are expressly made subject to the IAEA safeguards (example of the Agreement with South Korea):

— nuclear installations and specialised equipment designed, constructed or operated on the basis of or using specific information supplied by France;
— special fissile material or other nuclear material, including subsequent generations of special fissile material obtained, processed or used on the basis or by means of an item supplied by France or on the basis of specific information supplied by France.

For its part, France undertook not to use the fissile material obtained and returned to France to manufacture nuclear weapons or for any other military purpose or to manufacture any nuclear explosive device.

The Agreement with Egypt specified in addition that in the event that the IAEA safeguards could no longer be applied on the territory of one or both Contracting Parties, the Parties undertook to devise and introduce as soon as possible a method of control as effective and with the same scope as the IAEA system. These two Agreements also include a physical protection clause. More recently, a nuclear power plant was constructed in the *People's Republic of China* on the basis of a "gentleman's agreement" concluded on 5th May 1983. The Parties agreed on a gradual transfer of technology from France to China so as to enable the Chinese nuclear industry to supply nearly all the necessary equipment as from the sixth nuclear unit. The Parties also agreed to co-operate in manufacturing fuel assemblies. It was specified that the technology transferred by France to China would be used exclusively for peaceful purposes and that should China export nuclear material or components produced by means of this technology and transferred towards a third party non-nuclear-weapon State, the receiver State should clearly undertake to use such material and components for peaceful purposes only. The rules governing these guarantees and undertakings are to be specified at the appropriate time. A co-operation agreement was also concluded between the nuclear safety agencies of both countries.

Another aspect of French policy with regard to nuclear trade agreements can be seen from the exchanges of letters between the French Government and the governments of countries using the services offered by France for the reprocessing of spent fuel. By way of example, the exchange of letters dated 10th July 1979 with *Sweden* specified that both Governments would apply the London Club Guidelines. Moreover, rules governing possession, storage, transfer and use when plutonium is being re-exported from France are, at the appropriate time, to be laid down in a specific agreement between the two Governments. Provision is also made for an undertaking to use material recovered after reprocessing exclusively for peaceful and non-explosive purposes. Transfers to third countries or purchases by France are subject to the same condition. Reprocessing contracts must provide for the radioactive waste resulting from reprocessing to be sent back to Sweden. More recently, (exchange of letters dated 30th March 1983), details were agreed on physical protection measures during transport to France of the spent fuel. The spent fuel for reprocessing and any material derived therefrom are subject in France to the Safeguards Agreement between France, EURATOM and the IAEA, concluded on 27th July 1978 (confirmed by an exchange of letters dated 16th May 1983). A new exchange of letters on 21st November 1986 (see *appendix 10* to the Study) deals with the reprocessing in France of spent fuel from Sweden and with the use of recovered plutonium.

G.   AGREEMENTS CONCLUDED BY ARGENTINA

Certain countries which did not participate in the work of the Zangger Committee or the London Club have nevertheless agreed between themselves on scientific and technical co-operation in the field of nuclear energy. Argentina will be used here as an example of such countries. While Argentina continues to require a supply of material such as heavy water, or nuclear equipment, mainly from countries having adopted the export policies defined by the Zangger Committee or the London Club, it has nevertheless succeeded in achieving its own export capability and is able to supply technical assistance in certain fields.

Argentina's policy with regard to scientific and technical co-operation is directed principally at the other Latin American countries and, to a lesser degree, countries such as India, which share its reservations about the conditions imposed by the main exporting countries and the People's Republic of China (15th April 1985)[12].

The technique adopted by the negotiators of these agreements was fairly conventional : a framework agreement whose provisions were applied by means of an implementing agreement (for example the Technical Co-operation Agreement with *Venezuela* of 29th February 1972, followed by an Implementing Agreement of 8th August 1979, or the Co-operation Agreement with *Chile* of 3rd November 1976, the implementation of which was entrusted to the respective Atomic Commissions).

The scope of these agreements varies from case to case, and extends from the exchange of scientific and technical information and of personnel to assistance with regard to prospecting for ores and the production of radionuclides, and even to the supply of equipment and complete installations (reactors and research laboratories, pilot units for processing uranium, etc.).

The Agreement concluded on 17th May 1980 with *Brazil*[13], a country whose nuclear development is comparable to that of Argentina, is of particular interest.

In the preamble to the Agreement, emphasis is laid :

— on the right of all countries to have access to nuclear energy and the related technology, so as to be able to use this form of energy to promote their economic and social development ; and

— on the need to prevent the proliferation of nuclear weapons by non-discriminatory measures which impose restrictions designed to guarantee general and complete nuclear disarmament under strict international control.

Co-operation between Argentina and Brazil, as defined in this Agreement, covers a vast area : reactor technology, fuel cycle, the manufacture of equipment and provision of services, radioisotopes, radiation protection and nuclear safety, physical protection, and basic and applied research.

Implementing agreements are to organise joint technical meetings to study programmes and set up mixed bodies responsible for the technical and economic management of mutually agreed programmes and projects.

In the absence of restrictions or reservations, the Parties may make free use of the information exchanged and shall facilitate the mutual supply of the nuclear material, equipment and services required to implement the joint programmes and also national development plans.

Both countries have undertaken to use material or equipment supplied by the other Party, or material obtained through the use of such material or equipment, and material used in equipment supplied under the Agreement, exclusively for peaceful purposes. However, safeguard procedures are to be implemented after consultations between the Parties which, "if necessary" are to conclude safeguards agreements with the IAEA.

NOTES AND REFERENCES

1. For further details, see *Bilateral Agreements and the Evolution of the International Safeguards System*, Uranium Institute, London, September 1981.

2. The text of the Agreement concluded with Norway is reproduced in appendix 9 to this Study.

3. As a protest against the nuclear tests in Polynesia, an embargo on deliveries of uranium to France was imposed in June 1983, and the performance of contracts concluded was suspended. Performance did not recommence until 1986. For its part, the European Community limited its purchases in Australia.

4. Declaration to the Canadian House of Commons on 22nd December 1976 — INFCIRC/243, 20th January 1977.

5. This Agreement was concluded in the form of an exchange of letters on 18th December 1981; it is reproduced in appendix 12 to this Study.

6. Agreements whose purpose is to meet the Federal Republic of Germany's own needs in source materials (cf. EURATOM) or which concern fuel cycle services have not been included, nor have agreements concluded with countries of comparable technical development and concerning, for example, high flux or fast breeder reactors, the separation of isotopes by centrifuging, or nuclear safety.

7. Concerning German policy in the field of nuclear exports, see Gerhard Meyer-Wölse, *Rechtsfragen des Exports von Kernanlagen in Nichtkernwaffenstaaten*, Institüt für Völkerrecht der Universität Göttingen, Band 62, Carl Heymann Verlag, 1979. See also Erwin Häckel, *The Politics of Nuclear Exports in West Germany*, in Nuclear Exports and World Politics — Policy and Regime, Macmillan Press, 1983.

8. Werner Boulanger, *Das Deutsch-Brasilianische Abkommen über Zusammenarbeit auf dem Gebiet der friedlichen Nutzung der Kernenergie* of 27th June 1975.

9. The text of this Agreement is reproduced in the OECD/NEA Nuclear Law Bulletin, No. 16, November 1975.

10. The text of this Agreement is reproduced in appendix 11 to this Study.

11. The main stages are:

    1961: Limitation of co-operation with Israel;
    1968: Declaration to the General Assembly of the United Nations that although France had decided not to sign the Non-Proliferation Treaty, it intended to respect the spirit of the Treaty;
    1976: Communiqués of October and December by the External Nuclear Policy Council, outlining French non-proliferation policy;
    1978: The undertaking of commitments following the London Club consultations.

12. Regarding the bilateral agreements entered into by the People's Republic of China see *La coopération nucléaire internationale de la République populaire de Chine*, by Simone Courteix in Annuaire français de droit international, XXXII, CNRS, Paris, 1986.

13. The text of this Agreement is reproduced in the OECD/NEA Nuclear Law Bulletin, No. 27, June 1981. See also the Joint Nuclear Policy Statement by the Presidents of both countries in Nuclear Law Bulletin No. 37, June 1986.

145

ANNEX

SWAP AGREEMENTS

To save the cost and avoid the risks of transport as well as customs or exchange formalities, nuclear operators often exchange nuclear material with one another. This is because uranium is fungible in nature if considered as a source material in particular, if the swap relates to uranium of the same physico-chemical form, i.e. at the same stage of the fuel cycle (for example, uranium oxide or non-enriched hexafluoride).

The term "swap" has been generally adopted by the nuclear industry to designate the exchange of nuclear material, and more especially, an exchange without such material being physically moved.

Such swaps may take the form of a delivery in kind, for example to help out in the event of an emergency. In such a case, account must be taken of the technical constraints inherent in container capacities, and in devices for filling or emptying tanks, which make it impossible to determine accurately in advance the amounts transferred, and agreement must therefore be reached on tolerance margins and criteria for adjusting quantities and any balances.

More generally, such swaps are carried out at the request of customers by means of a simple transfer in the records kept by the operators or those holding material at the various stages of the fuel cycle (book or accounting transfer). Such operators hold material belonging to their various customers, which is stored in their installations pending or subsequent to processing. The transfer of ownership takes effect when the transfer is recorded in the books of the operator.

It is also possible to swap material with different enrichment levels or at different stages of the cycle (for example natural uranium oxide for enriched uranium hexafluoride). The respective economic value of the batches swapped (market price plus, where appropriate, cost of the conversion or enrichment services carried out) is used to determine equivalence.

However, the fungible nature of nuclear material which, in a manner of speaking, is intrinsic or natural, is limited in practice by policy constraints which result from the rights which certain States supplying material or services reserve for themselves. These include, for example, the right to give prior consent to any retransfer, a right which, in certain cases, is exercised in relation to subsequent generations of material and which may enter into conflict with similar rights retained by the States on whose territory operations such as isotopic separation are carried out. Then, in the administration of security control systems (for example the EURATOM system), batches of material are given a code of origin indicating the specific control imposed in implementation of Article 77(b) of the Treaty of Rome, a code which may be likened to a flag; this explains the term "international flag swap", normally used to designate an exchange of codes of batches of nuclear material. Such an exchange of codes thus symbolises an exchange of non-proliferation and safeguards obligations. It is usually a necessary condition for an exchange in kind or an accounting transfer, a related transaction which meets the above-mentioned goal, namely, it saves costs and avoids the risk of transport. Similar concerns may be felt by the owner of a stock of material who wishes to effect permutations between batches he possesses of different origins.

It might be thought that it is of little importance to a State, which subjects supplies and services to particular requirements, whether these are applied to the actual batch of material which is the object of the transaction or to another batch of the same quantity and physical nature. In fact, States have certain reservations with regard to such transactions and are careful to retain the right to prohibit them by provisions in bilateral agreements relating to "authorised persons" or "international transfers" or to prior consent to enrichment or reprocessing operations.

There seems at present to be a compromise developing between the concern of certain supplier States to follow closely what happens to material supplied, and the financial concerns of the users; the purpose is to procure greater legal security for these swap operations by shortening delays and making government authorisations less discretionary in nature.

An early result of this measure can be found in the exchange of letters between Canada and EURATOM, dated 21st June 1985, which (in paragraph 5) provides for the setting up of mechanisms to regulate the subjecting of materials to, and the withdrawal of other materials from the Agreement (in this connection, see the EURATOM Supply Agency's Annual Report for 1986).

It seems that the United States also has fewer objections to this type of transaction, provided that the swaps remain compatible with its policy objectives and export licence procedure and that they are approved by all the Governments concerned*.

Hopefully, the rapid spread of such transactions will lead to the establishment of specific criteria providing guarantees that they are effected in total conformity with non-proliferation requirements.

---

* For further details, see *Les accords de SWAP dans le domaine nucléaire* by Renata Broll and Claudio Sartorelli, Nuclear Inter Jura '87, Congress Proceedings, INLA, Antwerp (to be published).

CHAPTER SEVEN

PHYSICAL PROTECTION OF NUCLEAR MATERIAL

A. INTRODUCTION

At the beginning of the 1970s it was generally acknowledged that the precautions taken to ensure that information about the manufacture of nuclear weapons was kept secret might well have been to no avail. That a group of determined and technically competent individuals could find enough relevant information in technical publications and scientific journals to manufacture a nuclear explosive device was viewed as a possibility. Any attempt to effect such manufacture might begin with the theft of plutonium or highly enriched uranium. There was also reason to fear that radioactive materials might be stolen for the purpose of causing radioactive contamination or simply to be used as blackmail. Lastly, the sabotage of a nuclear installation or container of nuclear materials during transport could lead to the public's being exposed to a radiation risk. However, radical groups in various countries in the world were then demonstrating that they had no scruples about resorting to terrorist acts, and certain of these were directed against nuclear installations.

Several of the measures taken under the various systems of safeguards to limit the risk of "horizontal" proliferation by individual States also served to prevent, and especially detect *a posteriori* the theft or misappropriation of nuclear material by individuals (for example, accounting procedures for nuclear material). However, they seemed insufficient and the need for supplementary measures was felt, to counter more effectively this new risk, at the level of installation design (introduction of

protected areas, alarms, etc.) and operation (surveillance and guarding), and also the organisation of the transport of nuclear material.

It was rapidly agreed at international level that an effort should be made to increase the effectiveness and homogeneity of such measures. These measures, varying in accordance with the degree and nature of the risk, were to be applied by the State or States concerned, i.e. the State which authorised the nuclear materials to be held on its territory, or the States involved in an international transfer of such materials. The intention was to prevent certain countries, due to laxity, from becoming preferential targets for theft or sabotage since this would not only be contrary to the interests of the countries concerned, but also to those of the international community as a whole.

A first step in strengthening co-operation was the publication of various technical and organisational recommendations made by the IAEA to its Member States. Thereafter, certain States decided to make compliance with an appropriate level of physical protection a precondition for the conclusion of agreements to supply nuclear materials. Subsequently, an international convention, modelled on other conventions to prevent and curb certain types of offence and crime recognised as harmful to the international community, was adopted providing for various measures of mutual police and judicial assistance.

B. IAEA RECOMMENDATIONS

In 1971, the IAEA referred the matter in question to a group of experts whose work led, in 1972, to Recommendations relating to the physical protection of nuclear material. After some amendments by another group of experts, these Recommendations were published by the IAEA in 1975 in the INFCIRC series, under the document number 225.

After this venture was approved by the IAEA General Conference in 1975, the Agency set up an advisory group on the physical protection of nuclear material which, in February 1977, submitted a corrected version of these Recommendations, entitled "The Physical Protection of Nuclear Material" (*INFCIRC/225/Rev.1*). This version (published in June 1977), takes into account studies conducted in certain countries and the conclusions of an IAEA Symposium on this topic in 1975. The Vienna Agency has also been asked to update these Recommendations regularly in the light of the developing technology.

These Recommendations, which deal with requirements in the field of the transport, use and storage of nuclear material, are not binding on States. The latter retain exclusive jurisdiction to create, implement and maintain in working order a system of physical protection within their national boundaries, responsibility for the practical implementation of physical protection measures lying with the operators duly authorised to possess nuclear material. In cases of the international transfer of such material, responsibility for physical protection must be regulated in an agreement between the States concerned.

The Agency is not empowered to set up an international physical protection system, nor to supervise or administer any such system; its role in this sphere is in general limited to drafting recommendations and, on the request of a State, giving advice. It may also informally communicate to a State the results of observations it has made during its normal safeguards activities. For safeguards purposes, the Agency must be informed, by special report, of any loss of nuclear material within the territory of a State or during international transfers.

The Agency thus recommends that States promulgate detailed regulations, and make compliance a precondition for granting licences for activities involving the need for physical protection. States are invited to co-ordinate action by their competent administrative authorities and by delegation if necessary, enforce strict compliance with the requirements using, where appropriate, the threat of sanctions.

According to the IAEA Recommendations, national physical protection regulations should :
— determine conditions to minimise the possibilities for unauthorised removal of nuclear material or for sabotage, for example by limiting access to protected areas to persons whose trustworthiness has been established, through the use of security devices and surveillance ;

— enable the State to take rapid and comprehensive measures to locate and recover missing nuclear material.

The system recommended by the Agency is based on the categorisation of nuclear material with reference to the potential hazard of the material concerned ; this depends on the type of material and its fissile isotope content, physical and chemical form, radiation level and quantity.

The requirements recommended for physical protection of nuclear material in use and storage are classified, in order of complexity and decreasing strictness, from Category I to Category III (see the table in *annex I* to this chapter).

The same applies to nuclear material during transport. Thus, for Category I material, advance notice of the shipment must be given to the receiver who should confirm his readiness to accept delivery. In certain cases, advance authorisation is granted on the basis of the performance of a security survey prior to the application. Provision should also be made for supplying locks and seals, searching vehicles, and providing for escorts or guards, means of communication and emergency action measures[1].

In the case of international transport, a contract between the sender and the receiver should stipulate clearly the point at which physical protection responsibilities are transferred from one to the other, and give any information which would assist the receiver to take adequate measures.

Furthermore, an agreement should be concluded between the sending and receiving States and, where relevant, with the States whose territory is being used for the transit of the materials, to fix the details of their co-operation including in particular, the maintenance of communications enabling verification of the continuing integrity of the shipment and recovery actions to be taken in the event of loss of the materials.

It may be noted in passing that, for quite understandable reasons, document INFCIRC/225/Rev.1 recommends that vehicles used for transporting protected nuclear material should not be marked in any special way; this, of course, is the contrary of requirements under the regulations on the international transport of dangerous goods.

C. MEASURES PRESCRIBED BY THE GUIDELINES OF THE NUCLEAR SUPPLIERS GROUP

As a general rule, States have based their domestic physical protection legislation on the above-mentioned Recommendations. The strengthening of physical protection measures was also discussed in the negotiations between supplier countries of nuclear material and equipment. Thus, the communication presented on behalf of the European Community by Italy (INFCIRC/322 of April 1985), notes that Member States apply to nuclear material under their jurisdiction, physical protection measures at least as strict as those laid down in the Guidelines contained in document INFCIRC/254 (Guidelines of the Nuclear Suppliers Group — so-called London Club).

At the level of international law, whereas the statements of 22nd April 1974 arising from the work of the Zangger Committee (INFCIRC/209) were restricted to specifying the application of the IAEA safeguards, those which followed the Nuclear Suppliers Group's work, on the other hand, contained provisions relating to physical protection. The Appendix entitled "Guidelines for Nuclear Transfers", attached to the Communications gathered together in document INFCIRC/254 of February 1978 (see chapter five of this Study), provides that :

"*(a)* All nuclear materials and facilities identified by the agreed *trigger list* should be placed under effective physical protection to prevent unauthorized use and handling. The levels of physical protection to be ensured in relation to the type of materials, equipment and facilities, have been agreed by suppliers, taking account of international recommendations.

(b) The implementation of measures of physical protection in the recipient country is the responsibility of the Government of that country. However, in order to implement the terms agreed upon amongst suppliers, the levels of physical protection on which these measures have to be based should be the subject of an agreement between supplier and recipient.

(c) In each case special arrangements should be made for a clear definition of responsibilities for the transport of trigger list items."

With further details given in Annex B to document INFCIRC/254 (Criteria for Levels of Physical Protection), these provisions reflect the agreement of London Club supplier countries :

— on a classification of nuclear material for physical protection purposes (with regular updating), and on the main guidelines for protection measures applicable to each category of nuclear material ;

— on the principle that the levels of physical protection on which these measures must be based should be the subject of an agreement between the supplier and the receiver.

This text recognises that responsibility for implementing physical protection measures lies, within its own territory, with the receiver State, and invites it to use document INFCIRC/225 as a model for this purpose. However, these Guidelines, by making transfers of nuclear articles dependent upon compliance with the fundamental principles they contain — and thus, in particular, on an agreement concerning the levels of physical protection — implicitly give the supplier State power to assess whether the measures envisaged by the receiver State are satisfactory. This is all the more true in that the protection characteristics provided for in Annex B for each category are expressly stated to be minimum requirements.

The classification of nuclear material annexed to the London Club Guidelines is modelled on that contained in document INFCIRC/225/Rev.1. It does, however, recommend, without making this an obligation, that low-enriched fuel (less than 10 per cent fissile content), and natural or depleted uranium and thorium not included in INFCIRC/225, should be placed in Category II (see the Table reproduced in *annex II* to this chapter).

From the viewpoint of physical protection levels, the protection characteristics of transport operations, identical for Categories II and III, are as follows :

"*Transportation* under special precautions including prior arrangements among sender, recipient and carrier, and prior agreement between entities subject to the jurisdiction and regulation of supplier and recipient States, respectively, in case of international transport specifying time, place and procedures for transferring transport responsibility."

Transportation of Category I nuclear material, on the other hand, must in addition be carried out "under constant surveillance by escorts and under conditions which assure close communication with appropriate response forces".

It is also provided that :

"Suppliers should request identification by recipients of those agencies or authorities having responsibility for ensuring that levels of protection are adequately met and having responsibility for internally co-ordinating response/recovery operations in the event of unauthorized use or handling of protected materials. Suppliers and recipients should also designate points of contact within their national authorities to co-operate on matters of out-of-country transportation and other matters of mutual concern".

D. CONVENTION ON THE PHYSICAL PROTECTION OF NUCLEAR MATERIAL

Following meetings held at the IAEA Headquarters between 1977 and 1979 and attended by the representatives of 58 States and the European Community[2], the

Convention on the Physical Protection of Nuclear Material was opened for signature on 3rd March 1980[3].

The Convention entered into force on 8th February

1987 after being ratified, accepted or approved by twenty-one States. (The list of Contracting Parties is reproduced in appendix 8 to this Study).

The Convention applies, in its entirety, to nuclear material used for peaceful purposes in course of international transport. Those of its provisions dealing with administrative co-operation and mutual judicial assistance also apply to nuclear material used for peaceful purposes while in domestic use, storage and transport.

It will be noted that the phrase "used for peaceful purposes" limits the scope of the Convention. For, the nuclear-weapon States did not agree to its being extended to cover materials used for military purposes, stressing that, for these latter, they applied much stricter protection measures than those required under the Convention. Reference is made to this fact in the Preamble.

The Convention applies to nuclear material and not to the facilities or means of transport in which such material may be held. However, some of the protection measures required under the Convention help, in an indirect or ancillary fashion, to ensure the protection of facilities or means of transport.

International nuclear transport is defined as "the carriage of a consignment of nuclear material by any means of transportation intended to go beyond the territory of the State where the shipment originates beginning with the departure from a facility of the shipper in that State and ending with the arrival at a facility of the receiver within the State of ultimate destination". The term "facility" is not defined. It therefore does not necessarily mean a nuclear facility. It could be an airport or a port. The result, however, is that the levels of protection specified in Annex I to the Convention must be applied on their respective territories by the sender and receiver States, between the border and the facility of origin or of ultimate destination.

Thus, in international nuclear transport, the States Parties concerned have a duty to ensure that the nuclear material transported is protected in compliance with the levels described in Annex I to the Convention (Levels of Physical Protection to be applied in International Transport of Nuclear Material as categorized in Annex II of the Convention; the table therein is reproduced as *annex III* to this chapter). These levels vary depending on which of the three categories described in the table in Annex II the material in question falls into. They are the same as those adopted by the members of the London Club with one exception, relating to spent fuel. In addition to the rules applicable to Category II international nuclear transport, the transport of more than 500 kilogrammes of natural uranium requires advance notification of the shipment, specifying the mode of transport, expected time of arrival and confirmation of receipt of shipment.

This direct obligation on the sender or receiver State Party to apply the levels of protection naturally covers materials on board a ship or aircraft coming under its jurisdiction (for example because it is registered in that State). Similarly, when nuclear material is transported between two facilities of a single State Party through international waters or airspace, the State Party concerned is responsible for applying the appropriate levels of protection.

However, the Convention also attempts to ensure, in an indirect fashion, the application of its levels of protection by States which are not parties. For, States Parties are forbidden to export or authorise the export of nuclear material, or import or authorise the import of such material from a State not party to the Convention unless it has received assurances that the material will be protected during the international transport at the levels described in Annex I to the Convention. The same applies when a State Party is asked to allow nuclear material to be transitted through its territory between States that are not parties to the Convention. However, this obligation is qualified by the phrase "as far as practicable", apparently to take account of a simple transit through territorial waters or straits and of overflight by aircraft.

When, on the other hand, the nuclear material in question is not in course of international nuclear transport but being *used*, transported or *stored* within a State Party's national territory, the provisions of domestic legislation alone apply, except as otherwise provided in the following paragraphs.

The second part of the Convention deals with administrative co-operation between States Parties and with mutual judicial assistance. These provisions are of general application and are not limited to cases of the international transport of nuclear material used for peaceful purposes. Article 5.2 even makes it possible for States not party to the Convention to ask States Parties for co-operation and assistance in the recovery and protection of nuclear material. States Parties, for their part, must inform "other States" concerned of any theft or other unlawful taking of nuclear material and of any "creditable threat" thereof.

More generally, the Convention provides that States Parties shall make known to each other their competent authority and point of contact with responsibility for physical protection of nuclear material and for co-ordinating recovery and response operations in the event of any unauthorised removal, use or alteration of nuclear material or in the event of any creditable threat thereof (Article 5.1). While this Convention is not the first to organise international information exchange about certain criminal activities, it will be observed that emphasis is laid on co-operation for purposes of recovering and returning stolen or missing material, this being the first priority when theft or something comparable has occurred.

The second priority of the co-operation introduced under the Convention is to ensure that offenders are punished by making it impossible for any State Party to constitute a sanctuary for offenders. In this respect, the

Convention is clearly modelled on provisions common to various international Conventions designed to curb criminal activities such as traffic in women and children, counterfeiting, drug trafficking and, in particular, agreements to prevent air piracy and threats to the safety of aircraft, or for the protection of diplomats. The rule is that each State Party must take appropriate measures (including detention) to ensure that the alleged perpetrator of any of the offences defined in the Convention can, should he be present in its territory, be prosecuted there, even if the offence was not committed within the territory concerned, or else extradited. States Parties afford one another the greatest measure possible of judicial assistance, especially with regard to supplying any evidence at their disposal.

In conclusion, it should be pointed out that the various co-operation or safeguards agreements concluded in recent years often include physical protection clauses which are based on the Convention or which refer expressly to either document INFCIRC/225/Rev.1, or to document INFCIRC/254 or reproduce its provisions in the form of an annex.

NOTES AND REFERENCES

1. As a detailed example of such measures, see the chapter on the United States in Volume II of this Study concerning national regulations.

2. The signature of this Convention by the European Community caused problems at the negotiation stage and gave rise to a ruling of the Court of Justice of the European Communities on 14th November 1978 (OJEC C 302/2 of 16th December 1978). The signature of the Convention by the European Community did not give it any separate vote in addition to that of its Member States.

3. Convention on the Physical Protection of Nuclear Material — Official Record of the Negotiation, IAEA Legal Series No. 12, Vienna 1982. The text of the Convention, in English and French, is reproduced in Nuclear Law Bulletin No. 24 (December 1979).

L.A. Herron, *Legal Aspects of International Co-operation in the Physical Protection of Nuclear Facilities and Materials*, in Nuclear Inter Jura' 81, Proceedings of INLA Congress, Palma de Mallorca, pp.293-303.

Ha Vinh Phuong, *The Physical Protection of Nuclear Material*, in Nuclear Law Bulletin, No. 35, OECD/NEA, June 1985, pp.113-119.

ANNEX I

CATEGORIZATION OF NUCLEAR MATERIAL[e]
(INFCIRC/225/Rev. 1)

Material	Form	Category		
		I	II	III
1. Plutonium[a,f]	Unirradiated[b]	2 kg or more	Less than 2 kg but more than 500 g	500 g or less[c]
2. Uranium 235[d]	Unirradiated[b]			
	— uranium enriched to 20% ^{235}U or more	5 kg or more	Less than 5 kg but more than 1 kg	1 kg or less[c]
	— uranium enriched to 10% ^{235}U but less than 20%	—	10 kg or more	Less than 10 kg[c]
	— uranium enriched above natural, but less than 10% ^{235}U	—	—	10 kg or more
3. Uranium 233	Unirradiated[b]	2 kg or more	Less than 2 kg but more than 500 g	500 g or less[c]

a) All plutonium except that with isotopic concentration exceeding 80% in plutonium 238.

b) Material not irradiated in a reactor or material irradiated in a reactor but with a radiation level equal to or less than 100 rads/hour at one metre unshielded.

c) Less than a radiologically significant quantity should be exempted.

d) Natural uranium, depleted uranium and thorium and quantities of uranium enriched to less than 10% not falling in Category III should be protected in accordance with prudent management practice.

e) Irradiated fuel should be protected as Category I, II or III nuclear material depending on the category of the fresh fuel. However, fuel which by virtue of its original fissile material content is included as Category I or II before irradiation should only be reduced one Category level, while the radiation level from the fuel exceeds 100 rads/h at one metre unshielded.

f) The State's competent authority should determine if there is a credible threat to disperse plutonium malevolently. The State should then apply physical protection requirements for Category I, II or III of nuclear material, as it deems appropriate and without regard to the plutonium quantity specified under each category herein, to the plutonium isotopes in those quantities and forms determined by the State to fall within the scope of the credible dispersal threat.

ANNEX II

CATEGORIZATION OF NUCLEAR MATERIAL
(INFCIRC/254/Annex B)

Material	Form	Category		
		I	II	III
1. Plutonium[a]	Unirradiated[b]	2 kg or more	Less than 2 kg but more than 500 g	500 g or less[c]
2. Uranium 235	Unirradiated[b] — uranium enriched to 20% ^{235}U or more	5 kg or more	Less than 5 kg but more than 1 kg	1 kg or less
	— uranium enriched to 10% ^{235}U but less than 20%	—	10 kg or more	Less than 10 kg[c]
	— uranium enriched above natural, but less than 10% ^{235}U[d]	—	—	10 kg or more
3. Uranium 233	Unirradiated[b]	2 kg or more	Less than 2 kg but more than 500 g	500 g or less
4. Irradiated fuel			Depleted or natural uranium, thorium or low-enriched fuel (less than 10 % fissile content)[e,f]	

a) As identified in the Trigger List.

b) Material not irradiated in a reactor or material irradiated in a reactor but with a radiation level equal to or less than 100 rads/hour at one metre unshielded.

c) Less than a radiologically significant quantity should be exempted.

d) Natural uranium, depleted uranium and thorium and quantities of uranium enriched to less than 10% not falling in Category III should be protected in accordance with prudent management practice.

e) Although this level of protection is recommended, it would be open to States, upon evaluation of the specific circumstances, to assign a different category of physical protection.

f) Other fuel which by virtue of its original fissile material content is classified as Category I or II before irradiation may be reduced one category level while the radiation level from the fuel exceeds 100 rads/hour at one metre unshielded.

ANNEX III

CATEGORIZATION OF NUCLEAR MATERIAL
(Convention on the Physical Protection of Nuclear Material)

Material	Form	Category		
		I	II	III[c]
1. Plutonium[a]	Unirradiated[b]	2 kg or more	Less than 2 kg but more than 500 g	500 g or less but more than 15 g
2. Uranium 235	Unirradiated[b] — uranium enriched to 20% ^{235}U or more	5 kg or more	Less than 5 kg but more than 1 kg	1 kg or less but more than 15 g
	— uranium enriched to 10% ^{235}U but less than 20%	—	10 kg or more	Less than 10 kg but more than 1 kg
	— uranium enriched above natural, but less than 10% ^{235}U	—	—	10 kg or more
3. Uranium 233	Unirradiated[b]	2 kg or more	Less than 2 kg but more than 500 g	500 g or less but more than 15 g
4. Irradiated fuel			Depleted or natural uranium, thorium or low-enriched fuel (less than 10 % fissile content)[d,e]	

a) All plutonium except that isotopic concentration exceeding 80% in plutonium 238.

b) Material not irradiated in a reactor or material irradiated in a reactor but with a radiation level equal to or less than 100 rads/hour at one metre unshielded.

c) Quantities not falling in Category III and natural uranium should be protected in accordance with prudent management practice.

d) Although this level of protection is recommended, it would be open to States, upon evaluation of the specific circumstances, to assign a different category of physical protection.

e) Other fuel which by virtue of its original fissile material content is classified as Category I or II before irradiation may be reduced one category level while the radiation level from the fuel exceeds 100 rads/hour at one metre unshielded.

CHAPTER EIGHT

SAFETY, RADIATION PROTECTION AND THIRD PARTY LIABILITY IN THE INTERNATIONAL TRANSPORT OF NUCLEAR MATERIAL

A. INTRODUCTION

The export and import of nuclear material and equipment is effected by transporting them from one country to another, often using international waters or airspace. Transports of packages containing radioisotopes amount, each year, to millions of shipments[1]. There are many different aspects to the regulation of the transport of radioactive materials, both in national and international law. The provisions designed to ensure nuclear safety and the protection of workers and the public against radiation are included in regulations for the transport of dangerous goods, about which a study has already been published, in 1980, by the OECD Nuclear Energy Agency (Regulations Governing the Transport of Radioactive Materials). Thus, for the purposes of the present Study, the basic structure only of the international regulations on the transport of radioactive materials will be described, attention nevertheless being drawn to certain difficulties which may result from the limits to the scope of application of the international conventions or because the implementation of some of the international regulations is left to national law (see section B below). More generally, the temporary storage which may be rendered necessary by transfers of loads is subject to national regulations designed to ensure that storage facilities meet nuclear safety and radiation protection requirements.

The purpose of the various international regulations is to protect those working in the transport sector, the public as well as goods being transported, from the risk of contamination and external irradiation, the heat generated by radioactive materials and, in the case of fissile materials, criticality and also, where necessary, from the risks related to the chemical properties of the radioactive materials. Account must also be taken of the fact that the transport operation itself involves risks (collision, falling, etc.) and specific obligations (with regard to the weight and size of packages, for example).

Essentially, protection against these risks is ensured by the packaging. Depending on the seriousness of the risk, packaging may be of the ordinary industrial type (for materials with a low specific activity) or may be special in nature. Certain types of packaging must be approved either by the competent authority of the country of origin or by the authorities of each of the countries concerned.

Moreover, the risks inherent in the radioactive properties of nuclear material have made it necessary to establish a special liability and insurance regime covering all activities in the nuclear fuel cycle with the exception of the mining and processing of ores and their transport. This chapter contains a brief description of the provisions of this special regime which are relevant from the viewpoint of the international transport of nuclear material (see section C below).

B. INTERNATIONAL REGULATIONS ON THE TRANSPORT OF RADIOACTIVE MATERIALS

A high degree of harmonisation has been achieved in the field of the transport of radioactive materials, not only between the international regulations governing each mode of transport but also between these and most of the corresponding national regulations.

This harmonisation is essentially the result of the work of two organisations within the United Nations "family" : the *Economic and Social Council* (ECOSOC) and the *International Atomic Energy Agency* (IAEA).

Since 1957, ECOSOC, assisted by a Committee of Experts based in Geneva, has published recommendations relating to the specification, classification and labelling of dangerous goods and the consignment documents required for such goods[2]. Radioactive materials are contained in Class 7.

In 1959, ECOSOC adopted a resolution that the IAEA be asked to draft recommendations on the safe transport of radioactive materials, and in 1961, the IAEA published its *Regulations for the Safe Transport of Radioactive Materials*. These Regulations have been revised several times, most recently in 1985. They are published under the reference No. 6 of the IAEA Safety Series, and a Note on certain aspects of the Regulations, a Directive with regard to their implementation and a list of the competent authorities, are also published in the same Safety Series under the numbers 7, 9 and 37, respectively.

These Regulations are of direct application and compulsory only with regard to the IAEA's own operations or to operations of assistance effected by it. In all other cases, the Regulations, like those of ECOSOC, amount to incentives only, and serve as a model for Member States when drafting their domestic regulations and for competent international organisations in regulating a given mode of transport. The influence enjoyed by the IAEA Regulations explains, however, the high degree of harmonisation evident from a reading of national regulations in this field and the standards issued by international organisations specialised in the different modes of transport. Several such organisations are empowered to regulate a given mode of the transport of dangerous goods, and insert their own regulations between the IAEA Recommendations and national regulations, adapting them to the transport mode concerned.

The procedure for issuing such rules varies depending on the Convention or international regulations in question. In certain cases, practical implementation is left to the particular country's domestic regulations which, in the case of international transport, are at least two in number (the country of origin, on whose territory loading takes place, and the country of destination) and will often be more, where goods are carried in transit through the territory of a third country (or when the ship or aircraft used is registered in a third country). Some such countries may not be bound by the regulations or any other relevant measure adopted by organisations, such as an international Convention. Even if they are Contracting Parties to such a Convention, they may have issued reservations thereto or made use of options left open by the Convention. Lastly, delays in updating some of these international or national texts in line with the most recent addition of the IAEA Regulations may also create problems for those involved when different modes of transport are used in succession.

I. AIR TRANSPORT

Responsibility for the regulation of the international transport by air of dangerous goods, including radioactive materials, today lies with the *International Civil Aviation Organisation* (ICAO), created by the Chicago Convention of 7th December 1944 with its headquarters in Montreal. Until recently, the International Air Transport Association (IATA), an Association of the main airline companies, directly transposed the IAEA Regulations, but since 1974, has simply relayed the ICAO Regulations.

Consideration should be given to three ICAO texts which, in decreasing order of importance, are as follows :

1. Chicago Convention

The Chicago Convention is based on the principle of complete and exclusive sovereignty of each State over its own territory. Consequently, the norms, recommended practices and international procedures adopted by ICAO with regard to subjects of concern to the safety of air travel, are of direct application only above the high seas (Article 12). In all other cases, including loading and unloading, Contracting States are bound only "to collaborate in securing the highest practicable degree of uniformity" (Article 37) by bringing their own regulations or practices "into full accord" with the international standards. This facilitates compliance with the various national regulations concerned, namely those of the States of origin, destination, transit and of the operator (State in which the aircraft is registered). Moreover, Contracting States are obliged to notify ICAO of any differences between their own domestic regulations and the norms and practices recommended (Article 38). Each Contracting State has, however, the right, for reasons of public order and safety, to prohibit the carriage of articles other than those already prohibited, provided that no distinction is made in this respect between its national aircraft and those of other States [Article 35(b)].

2. Annex 18, dealing with the safe transport of dangerous goods by air

As far as radioactive materials are concerned (Class 7), this Annex to the Chicago Convention, adopted by the ICAO Council on 26th June 1981, is based on the IAEA Transport Regulations and the recommendations of the ECOSOC Committee of Experts, and makes a distinction between "international standards", with regard to which any difference in national regulations must be notified by the States (Article 38 of the Convention), and "recommended practices", with which States are obliged only to try to conform. This Annex is supplemented by technical instructions prepared, kept up to date and published by ICAO[3].

Thus, while the provisions of the Annex and technical instructions apply to all types of international civil aviation operations falling within the scope of the Convention, it is merely recommended that States apply these provisions to domestic flights and surface transport to and from airports.

The Annex lays down the principle that the transport of dangerous goods is prohibited except under the conditions specified in the Annex and in accordance with the detailed provisions of the technical instructions. There is thus an absolute ban on the transport of pyrophoric radioactive materials (Article 4.3 h). On the other hand, the States concerned may, in certain circumstances and on certain conditions, agree to make exceptions to the ban on the transport of explosive radioactive materials (Article 4.2 b).

The most important aspect of Annex 18 is the division of responsibilities between the consignor and the operator of the aircraft. A consignor wishing to send a package or overpack of dangerous goods by air is responsible for ensuring that such transport is not prohibited and that the goods are classified, packaged, marked, labelled and accompanied by a transport document in accordance with the conditions specified in the Annex and technical instructions.

As for the operator of the aircraft, he must not accept any such packages or overpacks until he has seen a precise description thereof and papers certifying that the relevant provisions of the technical instructions have been complied with. He must verify that the appropriate marks and labels have been put on and that there are no leakages or other damage compromising the integrity of the package. He must establish an acceptance check list and comply with the provisions of the instructions concerning the distances radioactive materials must be kept apart from persons, animals, non-developed photographic film, and concerning any decontamination.

A supplement to the Annex lists the differences notified by certain States. Some of these differences (notified by the United States, Italy, Japan and the USSR) concern radioactive materials.

3. Technical instructions for the safe transport of dangerous goods by air

These instructions, updated annually, are essentially technical in nature and lay down, in particular, the requirements for packaging, marking and loading.

The *International Air Transport Association* (IATA) publishes, for use by its members and companies having signed the IATA Interline Traffic Agreement Cargo, regulations concerning the articles subject to restrictions. These regulations served as technical instructions until publication by ICAO of its instructions; since that time, the IATA regulations reproduce the provisions of the technical instructions but present them in the form it used previously.

II. TRANSPORT BY SEA

At the origin of regulations in this field, the 1960 International Conference for the Safety of Life at Sea accomplished the following:

— it laid down, in Chapter VII of the *International Convention for the Safety of Life at Sea* (SOLAS), a general framework for the transport by sea of dangerous goods; and

— it entrusted the *International Maritime Organization* (IMO)[4] with the task of drafting a single international Code for the transport by sea of dangerous goods (IMDG Code) and recommended Governments Parties to the Convention to adopt this Code. The above-mentioned SOLAS Convention of 17th June 1960 was repealed and replaced by a new Convention dated 1st November 1974 (which entered into force in 1980), and a revised version of Chapter VII of this Convention was adopted by the Maritime Safety Committee of IMO in 1983 (and entered into force on 1st July 1986).

There are therefore two texts which need to be considered: the SOLAS Convention and its Annexes, and the IMDG Code.

1. The SOLAS Convention

The SOLAS Convention is organised along similar lines to the Chicago Convention. It applies to ships authorised to fly the flag of a Contracting State, and the Contracting Parties undertake to issue all laws and regulations and to take any other measures necessary to give full and complete effect to the provisions of the Convention.

Chapter VII, relating to the carriage of dangerous goods, specifies that it applies to dangerous goods classified under regulation 2 (for example, Class 7, radioactive materials) which are carried in packaged form or in solid form in bulk in all ships to which the regulations apply, i.e. to ships (other than warships, troop carriers, sailing ships, fishing boats, etc.) making an "international voyage". "International voyage" means a voyage between a country to which the Convention applies and a port situated outside the country in question, or vice versa.

Lastly, it will be noted that the Contracting Parties must communicate to IMO the text of any Acts, decrees, etc. they promulgate concerning matters falling within the scope of the Convention. This obligation covers special rules established by agreement between all or only some of the Contracting Parties.

National governments may also allow the promulgation of provisions different from those prescribed by the Convention, on condition that they are at least as effective.

2. The International Maritime Dangerous Goods Code (IMDG Code)

This Code was drafted by the Maritime Safety Committee of the IMO; the first version, published in 1965, was re-issued in a new edition in 1977, which itself has been amended several times since[5].

This Code lays down the basic principles. Detailed recommendations relating to each type of material and a number of good practice recommendations are contained in the classes corresponding to these materials. The note corresponding to any given material can be found with the aid of a technical index.

The Code is presented in the form of recommendations which Contracting States are invited to adopt. As a rule, these recommendations apply only to the ships covered by the Convention. However, IMO feels it desirable for manufacturers, packers and carriers to follow the advice given as to terminology, packaging and labelling; and for road, rail and port services to adopt the provisions relating to classification and labelling and to separate the various classes of goods into loading and unloading zones on the basis of the Code's recommendations.

III. TRANSPORT BY POST

The *Universal Postal Union*, a specialised agency of the United Nations with its headquarters in Berne, has published and regularly updates, in principle every five years, Detailed Regulations (the latest edition of which is dated 27th July 1984) for implementing the Universal Postal Convention of 1894.

Article 121 of the Regulations specifies in particular the limits restricting the radioactive materials which may be sent by post and under what conditions (content and conditioning in compliance with the IAEA Regulations providing for special exceptions for certain categories of consignment, prior authorisation from the competent authorities of the country of origin, labelling as prescribed by the above-mentioned Article 121).

In Europe, other modes for the international transport of dangerous goods and, in particular radioactive materials, have been regulated on a regional basis.

IV. TRANSPORT BY RAIL

The *Central Office for International Railway Transport*, whose headquarters are in Berne, has published International Regulations concerning the Carriage of Dangerous Goods by Rail (RID), in implementation of the International Convention concerning the Carriage of Goods by Rail (CIM) of 1890, amended on 7th February 1970.

Consequently, the scope of these Regulations is the same as that of the CIM Convention. Article 1 of the Convention provides that it applies to "the carriage of goods consigned under a through consignment note made out for carriage over the territories of at least two of the

Contracting States and exclusively over lines included in the list compiled in accordance with Article 59".

It must therefore be checked that the stations sending the goods and those receiving them are in different Contracting States (West European countries but also some East European ones) and that the lines used are included in the list drawn up by the Central Office. Special provisions also exist regulating cases of transit and mixed transport (railway road/other form of transport).

RID is in principle revised every five years, and an accelerated amendment procedure is also used.

As for the Member States of the Council for Mutual Economic Assistance (COMECON), they comply with the "requirements for the transport of dangerous goods by rail" (SMGS) published by the Council. As far as radioactive materials are concerned, these requirements, like the RID, are based on the IAEA Transport Regulations, which makes it possible to be bound by both the SMGS and by RID/CIM, an option taken up by certain socialist countries. However, there may be differences in detail due, for example, to updating not being done at the same time.

V. TRANSPORT BY ROAD

In 1957, the *Economic Commission for Europe*, a United Nations body based in Geneva, sponsored the conclusion of the "European Agreement concerning the International Carriage of Dangerous Goods by Road — ADR", the annexes to which specify the materials and objects whose international carriage by road is either prohibited or authorised under certain conditions (packaging, labelling, construction, equipment and movement of the vehicle).

The ADR defines "international transport" as "any transport operation performed on the territory of at least two Contracting Parties by vehicles" (motor vehicles, articulated vehicles, trailers and semi-trailers), as defined in Article 4 of the Convention on Road Traffic of 19th September 1949, other than vehicles belonging to or under the orders of the armed forces of a Contracting Party.

Although this Agreement has been ratified by a large number of European countries, it is advisable to verify that all countries whose territory is involved in any given transport operation are in fact bound by it.

Joint meetings of experts from the Economic Commission for Europe and the Berne Central Office study the problems shared by road and rail transport with a view to avoiding conflicting provisions. The ADR may be updated by means of an accelerated amendment procedure similar to that used for the RID.

VI. TRANSPORT BY INLAND WATERWAY

In the 1960s, the Economic Commission for Europe also prepared a draft European Agreement on the International Carriage of Dangerous Goods by Inland Waterways (ADN), but no further progress was made on this Agreement.

Traffic on the Rhine, on the other hand, is regulated by the European Agreement concerning the International Carriage of Dangerous Goods on the Rhine (ADNR), drafted by the Central Commission for Navigation on the Rhine (CCNR), located in Strasbourg, and although not much use is made of the river to transport radioactive materials, some of the Agreement's provisions do expressly relate to such materials.

C. THIRD PARTY LIABILITY AND INSURANCE

International nuclear trade, whether it be the export of source materials and equipment, the international transfer of nuclear materials for various purposes such as the enrichment or reprocessing of fuel, or simply the transport of radioisotopes, gives rise to many different problems of liability and insurance. A desire to subject such activities to a uniform system of law is thus one of the main reasons behind the adoption of several international Conventions[6] on nuclear third party liability :

— The Paris Convention on Third Party Liability in the Field of Nuclear Energy, dated 29th July 1960 (amended in 1964 and 1982). The Paris Convention entered into force on 1st April 1968.

— The Convention supplementary to the Paris Convention (Brussels Supplementary Convention), dated 31st January 1963 (amended in 1964 and 1982). This Convention entered into force on 4th December 1974.

— The Vienna Convention on Civil Liability for Nuclear Damage, opened for signature on 21st May 1963. The Vienna Convention entered into force on 12th November 1977.

— Convention relating to Civil Liability in the field of Maritime Carriage of Nuclear Material, adopted in Brussels on 17th December 1971. This Convention entered into force on 15th July 1975.

The definitions in the Paris and Vienna Conventions (the former applying only to certain countries in Western Europe whereas the latter can apply to countries all over the world) of the basic principles of the nuclear operator's liability are more or less identical[7] :

— sole and exclusive liability ;
— limitation of liability in amount and time ;
— compulsory financial security ;
— unitary aspect of legal jurisdiction ;
— non-discrimination.

As for the Brussels Supplementary Convention, it introduces a system of additional compensation for nuclear incidents for which a nuclear operator is liable within the meaning of the Paris Convention, paid for out of public funds provided by the Contracting Parties to this Convention.

The 1971 Brussels Convention was adopted to solve the problems arising from the situation of duplication of liability resulting from the simultaneous application of the maritime and nuclear Conventions in the field of transport. The 1971 Convention gives priority to nuclear law over maritime law by providing that, in the event of a nuclear accident, liability will be channelled to the nuclear operator.

Considering only the case of transport, and more particularly that of the international transport of nuclear materials, it should be emphasized that the main objective of the Paris and Vienna Conventions (Article 4 of the Paris Convention and Article II of the Vienna Convention) is to avoid any gaps in the chain of liability for damage suffered on the territory of the Contracting Parties during transport operations.

The basic rule of these Conventions is that the person liable is a priori the nuclear operator sending the materials since he will have been responsible for supervising their packaging and containment and for ensuring compliance with the specific rules governing health and safety during transport. His liability ends only when the operator of another nuclear installation (normally the operator to whom the materials are being sent) assumes liability for the materials in question pursuant to the express terms of a contract in writing or, failing which, when the operator of another nuclear installation has taken charge of the materials.

When the nuclear materials have been sent to a person within the territory of a non-Contracting State, the liability of the nuclear operator, according to the provisions of the Conventions, ends only when the materials have been unloaded from the means of transport by which they arrived in the territory of the non-Contracting State. In the inverse situation, when materials are sent from a non-Contracting State to a Contracting State, liability for damage suffered in the territory of Contracting States lies with the nuclear operator to whom the materials have, with his written consent, been sent, after they have been loaded on to the means of transport concerned.

The Conventions also provide that Contracting Parties may legislate to allow a carrier, at his express request, to be considered as liable for the transport of nuclear materials in place of the operator normally liable. So far, however, this option has never been used.

One of the major practical uses of the special nuclear third party liability regime is to provide a satisfactory legal basis for insuring damage[8]. As concerns international transport, the Contracting Parties to the Paris Convention have adopted a model certificate of financial security for the transport of nuclear substances, designed to facilitate the verification of insurance when frontiers are being crossed. Presenting such a certificate does not, however, in itself, confer the right to enter the territory of a Contracting Party. Furthermore, the carriage in transit of nuclear materials may be made subject by the country concerned to certain special conditions of financial security and liability.

More generally, it should be noted that several countries which export nuclear materials or large items of nuclear equipment have, even when not themselves Parties to the Paris or Vienna Conventions, consistently adopted a policy of requiring, as a pre-condition to the conclusion of contracts of delivery, that the receiver country ratify one or other of these Conventions or, as a minimum, that it adopt legislation based on the same principles. This requirement was justified by the concern of these countries to identify clearly in advance, in compliance with the rules of the Conventions in question, the operator liable in the event of an accident which could be attributed to the materials or equipment transferred, and thus to protect themselves against any claim against the supplier. This concern demonstrates the practical advantage there would be, to facilitate international nuclear trade, in increasing the number of countries which apply this nuclear third party liability regime.

NOTES AND REFERENCES

1. In 1978, the IAEA set up a Standing Advisory Group on the Safe Transport of Radioactive Materials (SAGS-TRAM) to assist it in updating its Transport Regulations. On SAGSTRAM's recommendation, the Agency has been compiling the shipments of different types of packages containing nuclear material. A Table listing the shipments for 1981 in several OECD countries is contained in appendix 19 to this Study.

2. United Nations Committee of Experts on the Transport of Dangerous Goods. Recommendations on the transport of dangerous goods (Orange Book), Fourth Revised Edition, 1986. It should be noted that the Secretariat of this Committee is assured by the Economic Commission for Europe whose headquarters are also in Geneva.

3. Technical Instructions for the Safe Transport of Dangerous Goods (DOC 9284 — AN/905), approved and published by decision of the ICAO Council, 1987-1988 edition.

4. At the time the Intergovernmental Consultative Maritime Organization — IMCO, with its headquarters in London.

5. The first consolidated edition of the International Maritime Dangerous Goods Code (IMDG Code) was published in 1981; since then, a number of updatings have been published, the most recent being Amendment 23-86, adopted in September 1986.

6. An updated list of the Contracting Parties to these Conventions is published periodically in the OECD/NEA Nuclear Law Bulletin.

7. For further details, see "Nuclear Third Party Liability", OECD/NEA, 1976. This Analytical Study is shortly to be updated.

8. The policy issued by nuclear insurers can be either a special policy for each transport operation or an open policy. With the latter, insurers cover for a determined period (usually one year) all nuclear risks involved in transports carried out by the operator during that period.

   Although, under the Conventions, the operator must have security for each nuclear incident up to the maximum amount of liability, the ceiling of the insurers' commitment is stipulated for all the consequences of nuclear incidents occurring during the same transport operation covered (cover per transport operation).

   The insurer gives the operator a certificate for the carrier, which gives the name and address of the operator and the type and duration of the cover. The certificate also specifies the nuclear substances carried and the itinerary covered by the security (a model certificate established within NEA is reproduced in Nuclear Law Bulletin, No. 2, November 1968, page 65 *et seq.*

   A premium is calculated per transport, taking into account the nature and bulk of the substances carried, the mode of transport used (land, air or sea), the distance to be covered and the safety measures taken for the transport. The premiums thus calculated are set out in a statement of accounts and are paid at the end of each term.

   This system works out satisfactorily for all transport operations between installations located in countries having ratified the Paris Convention. However, some countries require that insurance be taken out with a company authorised to operate on their own territory. This requirement, which does not appear to conform to the generally-acknowledged interpretation of the Convention, results in duplicate insurance being taken out for the same transport operation.

   Liability covered by the policy is that of the law applicable to the nuclear incident. If the latter occurs in the State of a non-Contracting Party or if the damage is suffered on that territory, in principle, neither the Paris nor the Vienna Conventions apply. The operator may, however, be held liable for the damage caused under the law applicable in that country. In such case, the insurance policies cover non-contractual liability incurred by the operator or other insured persons designated in the contract, but only insofar as the damage is caused by a nuclear incident within the meaning of the nuclear Conventions and under the same insurance conditions (same limitation in amount and in time and same exclusions).

   In principle, the policies exclude damage to the means of transport but provide for the possibility of reincluding such damage in cases where national law makes use of the option in this respect, set out in the Convention; most countries Party to the Paris Convention have done so.

CHAPTER NINE

REVIEW OF DEFINITIONS USED IN THE REGULATION OF NUCLEAR TRADE[1]

A. INTRODUCTION

This chapter will analyse the definitions of nuclear material, installations and "sensitive" nuclear-related equipment, as well as the technology concerning these materials or equipment, insofar as they appear:

— in the statutes of international organisations or in their official acts;
— in multilateral international agreements, or for declarations of intent made in parallel by several States (the London Club, for example), in the memoranda or annexes attached to these declarations;
— in bilateral international agreements.

Definitions by way of national legislation will be examined in the monographs on the regulation of nuclear trade in Member countries of the OECD Nuclear Energy Agency (see Volume II of this Study).

From a strictly legal viewpoint, a term defined in any one of the texts mentioned above should be used, in the sense in which it is defined, only for the purposes of interpreting that particular text. Consequently these definitions should be used with caution when analysing texts which do not provide definitions. Nevertheless, at certain times and in certain areas (safeguards, for example) a consensus emerged on the acceptability of certain terms or concepts; a consensus to which it would appear legitimate to make reference in order to fill the gaps in other texts on the same subject.

On the other hand, the definitions which will be examined herein should be distinguished from the definitions sometimes given to the same terms or concepts (for example, nuclear fuel, small quantities ...) in texts concerning other fields than that which is the subject of this Study (for example, nuclear third party liability, radiation protection, transport of radioactive substances ...).

B. DEFINITIONS OF NUCLEAR MATERIAL, MODERATORS AND CLADDING MATERIALS

I. NUCLEAR MATERIAL

1. Original definitions

The texts creating international nuclear organisations or issued by them borrowed, from the United States Atomic Energy Acts of 1946 and 1954 as well as from the co-operation agreements concluded by that country, a system of definitions which relies essentially on the distinction between "source material" and "special fissionable material". The adoption of this system leads to three observations.

From a historical point of view, the consideration of natural uranium, even when elaborated in metal form, as a simple "source material", or raw material, presupposes the use of enriched uranium, which is the object elsewhere of a detailed sub-definition, in the reactor types developed in the United States (PWR or BWR)[2]. This may be compared to the competing system of definitions of British origin ("prescribed substances"), which was better adapted to reactor types using natural uranium in metal form (Magnox, UNGG)[3].

Uranium or thorium are simply included as "source material" or, are the object of a special definition, according to whether the definitions are part of a text which has as its sole or principal objective safeguards or on the contrary, like the EURATOM Treaty, are part of a text where the definitions are also applicable to a supply regime.

Finally, the United States laws mentioned above contain a definition of "by-product material" which concerns essentially radioisotopes, and is not included in the system of definitions of the international organisations.

The Table in *annex I* of this chapter presents in parallel the definitions of material contained in the IAEA Statute of 26th October 1956 (Article XX), the OEEC Convention of 20th December 1957 (Article 18)[4], and the EURATOM Treaty of 25th March 1957 (Article 197).

2. Evolution of the definitions in the field of safeguards

i) It should be recalled that neither the 1968 Treaty on the Non-Proliferation of Nuclear Weapons (NPT), nor the 1959 Antarctic Treaty nor the 1967 Treaty of Tlatelolco, contain a definition of nuclear material. The NPT (Article III.1, 2, 3) nevertheless uses the terms "source material" and "special fissionable material". These can be understood as having the same meaning as those in the Statute of the Vienna Agency since the NPT provides for the application of the IAEA safeguards system to these nuclear materials.

The application of the OEEC Convention on the Establishment of a Security Control having been suspended in 1976, there remains to be examined the evolution of the definitions in the respective systems of the IAEA and EURATOM and in the declarations resulting from the work of the Zangger Committee and the "London Club".

ii) As concerns the IAEA Safeguards System, the successive versions could only refer to the above-mentioned statutory definitions. Without its being as legally binding, the same is true for the safeguards agreements concluded by the Agency, and in particular, the 1973 Agreement between the IAEA, the European Community and its non-nuclear weapon Member States, for the application of the NPT[5].

In the successive editions of the IAEA Safeguards System prior to the NPT (INFCIRC/26 of 1961, extended in 1964 and System revised in 1965 — INFCIRC/66 — temporarily extended in 1966 and 1968), we can observe the emergence of a definition of "nuclear material" in reference to either "source material" or to "special fissionable material" within the meaning of Article XX of the Statute as well as precisions concerning the concept of "effective kilogram", "enrichment", and "improved nuclear material", and "throughput" (rate at which nuclear material is introduced in an operating facility) necessary to the implementation of the Safeguards System.

These concepts, with the exception of the last two terms, have been applied in document INFCIRC/153 the "Blue Book", entitled "The Structure and Content of Agreements between the Agency and States Required in Connection with the Treaty on the Non-Proliferation of Nuclear Weapons" and in particular, in the 1973 Community Agreement cited above (see *annex II* to this chapter).

iii) As concerns the safeguards system established by Chapter VII of the EURATOM Treaty, it is in effect governed by Commission Regulation No. 3227/76 of 19th October 1976. The system having been harmonized with that of the IAEA as a result of the above-mentioned 1973 Agreement, the definitions contained in prior Regulations are only of historical interest. The results of this harmonization are obvious from a reading of *annex III* to this chapter.

iv) For States having adhered to the declarations of intent as a result of the work of the Zangger Committee and/or the London Club, or to the Declaration of common policy of the European Community, it is necessary to keep in mind the definitions appearing in these documents (see annexes in chapter five of this Study) which define exemptions to the export procedures laid down by these documents.

3. Definitions of the EURATOM Treaty for the application of Chapter VI on supplies

i) As in the field of safeguards, it appeared advisable not to impose constraints which were not justified by the objective of the provisions in question. It also seemed expedient not to disperse available resources of the Community organisations concerned. However, since the objective of Chapter VI of the EURATOM Treaty differs from that of Chapter VII (safeguards), it is possible that the definitions of the scope of application of the procedures laid down by Chapter VI, in particular the exclusion of "small quantities", do not correspond, in certain respects to analogous definitions in the area of safeguards. The additional definitions necessary or considered useful for the application of Chapter VI are indicated hereafter.

ii) The definition of the average rate of concentration in ores provided for by Article 197(4) of the Treaty was fixed by Council Regulation No. 9 of 2nd February 1960.

Article 1 of this Regulation provides that "the average concentrations are the ratio between the weight of uranium (U) or thorium (Th) contained in any form whatsoever, in a given quantity of ore and the weight of that same quantity of ore". For uranium-bearing ores the average is 0.1 per cent or more uranium ; for thorium-bearing ores, the average (except monazites) is 3 per cent or more thorium ; for monazites the average is 10 per cent or more thorium or 0.1 per cent or more uranium.

iii) Article 74 of the Treaty authorises the Commission to exempt from the provisions of Chapter VI, "small quantities" of ores, source material or special fissile materials of the order of those normally used in research.

The definition of "small quantities" is laid down by Commission Regulation (EURATOM) No. 3137/74 of 12th December 1974 (OJEC L 333/27) which provides that :

"With regard to special fissile materials, transfers within, imports into and exports from the Community shall be exempt from the provisions of Chapter VI of the Treaty provided that the quantities involved, referred to the elemental form, do not exceed 200 grammes of uranium 235, uranium 233 or plutonium in any one transaction up to an annual limit of 1 000 grammes of any of the substances per user. In the case of imports and exports this shall apply subject to the provisions of agreements for co-operation concluded by the Community with third countries".

4. Definitions concerning the physical protection of nuclear material

i) The Convention of 3rd March 1980 on the Physical Protection of Nuclear Material contains definitions (Article 1) :

"For the purposes of this Convention :

a) *nuclear material* means plutonium except that with isotopic concentration exceeding 80 per cent in plutonium 238 ; uranium 233 ; uranium enriched in the isotopes 235 or 233 ; uranium containing the mixture of isotopes as occurring in nature other than in the form of ore or ore-residue ; any material containing one or more of the foregoing ;

b) *uranium enriched in the isotope 235 or 233* means uranium containing the isotopes 235 or 233 or both in an amount such that the abundance ratio of the sum of these isotopes to the isotope 238 is greater than the ratio of the isotope 235 to the isotope 238 occurring in nature."

It is evident from the text that the concern not to disperse efforts results, as in the case of safeguards, in excluding plutonium from the provisions of the Convention when its isotopic concentration exceeds 80 per cent in plutonium 238, as well as uranium ores and their residues.

On the other hand, in the categorisation of nuclear material (Annex II of the Convention) it is to be noted that Category III contains thresholds, expressed in mass, below which the protection measures imposed by the Convention are not, as a matter of law, applicable. A Note (C) assimilates natural uranium to materials in Category III (see chapter seven, annex III of this Study).

ii) The policy declarations resulting from the work of the London Club (INFCIRC/254) also contain a categorisation of nuclear material for the purpose of physical protection (see chapter seven, annex II of this Study). The differences between the two categorisations, the second being less precise as regards the exemption thresholds, are to be noted.

II. MODERATORS

The term "non-nuclear material for reactors", in the declarations resulting from the Zangger Committee's work (INFCIRC/209, Memorandum B, paragraph 2.2) as well as from that of the London Club (INFCIRC/254, Annex A, Trigger List) comprises the following identical definitions :

"2.2 *Non-nuclear materials for reactors* :

2.2.1 Deuterium and heavy water :

Deuterium and any deuterium compound in which the ratio of deuterium to hydrogen exceeds 1 :5000 for use in a nuclear reactor as defined in paragraph 2.1.1 above in quantities exceeding 200 kg of deuterium atoms for any one recipient country in any period of 12 months.

2.2.2 Nuclear grade graphite :

Graphite having a purity level better than 5 parts per million boron equivalent and with a density greater than 1.50 grams per cubic centimetre in quantities exceeding 30 metric tons for any one recipient country in any period of 12 months".

III. CLADDING MATERIALS

The declarations resulting from the work of the Zangger Committee include, in Memorandum B, the following paragraph :

"2.1.6 Zirconium tubes :

Zirconium metal and alloys in the form of tubes or assemblies of tubes, and in quantities exceeding 500 kg, especially designed or prepared for use in a reactor as defined in paragraph 2.1.1 above, and in which the relationship of hafnium to zirconium is less than 1 :500 parts by weight".

The Trigger List in Annex A attached to the declarations of the London Club members took over this definition but qualified the 500 kg quantity mentioned above as "per year" :

"2.1.6 Zirconium tubes :

Zirconium metal and alloys in the form of tubes or assemblies of tubes, and in quantities exceeding 500 kg *per year*, especially designed or prepared for use in a reactor as defined in paragraph 2.1.1 above, and in which the relationship of hafnium to zirconium is less than 1 :500 parts by weight".

This specification was taken up in substance by certain Member States of the IAEA in a complementary declaration within the Zangger Committee (INFCIRC/209/Mod.1).

"a. Paragraph 2.1.6 to read : Zirconium tubes : Zirconium metal and alloys in the form of tubes or assemblies of tubes, and in quantities exceeding 500 kg per year, especially designed or prepared for use in a reactor as defined in paragraph 2.1.1 above and in which the relationship of hafnium to zirconium is less than 1 :500 parts by weight."

C. DEFINITIONS CONCERNING SENSITIVE INSTALLATIONS AND MATERIALS FROM THE VIEWPOINT OF NON-PROLIFERATION

The possibility of constructing a nuclear weapon by means of an installation designed to produce substances capable of releasing a great deal of atomic energy appears in the Paris Agreements of 23rd October 1954 with regard to the termination of the regime of occupation in the Federal Republic of Germany. Apart from atomic weapons themselves, Annex II to these Agreements also covers "the means of production specially concerned for their production" unless these devices or means of production are used for civil purposes or unless they are used for scientific, medical or industrial research in the field of fundamental or applied science.

In contrast, safeguards systems instituted either by bilateral agreements between country X and the United States, Canada, the United Kingdom, or by virtue of the IAEA, OECD/NEA, or the EURATOM Treaty only apply in practice to nuclear material, even if the IAEA and NEA Statutes also mean to cover "equipment". However, the concept of a production or utilisation facility cannot be ignored when establishing safeguards on nuclear material. Knowledge of the characteristics of installations or treatment processes permits an evaluation of the production of materials subject to control. Moreover, the installation serves to keep track of the materials and is visited by inspectors.

The idea that the proliferation of nuclear weapons could directly result from the supply of "equipment and material especially designed or prepared for the processing, use or production of special fissionable material" reappears in the Non-Proliferation Treaty [Article III.2(b)] ; on the other hand, the IAEA safeguards which in non-nuclear-weapon States are a condition for supplying equipment, are only to be applied to "source material" or "special fissionable material" used or produced by this equipment.

As regards the application of safeguards to the export of installations and "sensitive" materials, see the annexes to chapter five of this Study.

D. DEFINITIONS CONCERNING TECHNOLOGY

The IAEA Statute (Article III.A.3) requires the Agency "to foster the exchange of scientific and technical information on peaceful uses of atomic energy" [see also the Statute of the OECD Nuclear Energy Agency, Article 7(c)]. Similarly, Article 2(a) of the EURATOM Treaty provides that the Community shall "promote research and ensure the dissemination of technical information". Several provisions of the Treaty (see chapter two of this Study) and regulations made pursuant thereto specify what is meant by the dissemination of knowledge. From these, we can deduce the definition of technology transfers.

The setting up of safeguards tended on the contrary to restrict these transfers. The Statute of the IAEA (Article III.A.5), still under the title "information", opened the possibility of applying its safeguards. The same possibility exists within the Statute of the NEA under the more general heading of "services" [Article 6(a)]. All the same, these provisions remain a potentiality in the first security systems put in place (INFCIRC/66/Rev.2, 24th September 1968, OEEC Convention on the Establishment of a Security Control, EURATOM Treaty, Chapter VII and implementing regulations prior to 19th October 1976). Specifications concerning the examination of installations or specialised equipment plans and especially the approval of chemical processing procedures for irradiated material, simply allowed the identification of sensitive technologies.

In effect, the purpose of the London Club negotiations was to reinforce safeguards by providing for a more effective application to transfers of certain sensitive technologies (facilities for reprocessing, enrichment or "heavy water" production, or "major critical compenents thereof"). Consequently, Annex A (Trigger List), part B of the London Club Guidelines (INFCIRC/254) contains the following definitions :

"1. *Technology* means technical data in physical form designated by the supplying country as important to the design, construction, operation, or maintenance of enrichment, reprocessing, or heavy water production facilities or major critical components thereof, but excluding data available to the public, for example, in published books and periodicals, or that which has been made available internationally without restrictions upon its further dissemination.

"2. *Major critical components* are :

  a) in the case of an isotope separation plant of the gaseous diffusion type : *diffusion barrier* ;

  b) in the case of an isotope separation plant of the gas centrifuge type : *gas centrifuge assemblies, corrosion-resistant to UF_6* ;

  c) in the case of an isotope separation plant of the jet nozzle type : the *nozzle units* ;

  d) in the case of an isotope separation plant of the vortex type : the *vortex units*.

"3. For facilities covered by paragraph 6 of the Guidelines for which no major critical component is described in paragraph 2 above, if a supplier nation should transfer in the aggregate a significant fraction of the items essential to the operation of such a facility, together with the know-how for construction and operation of that facility, that transfer should be deemed to be a transfer of *facilities or major critical components thereof*.

"4. The definitions in the preceding paragraphs are solely for the purposes of paragraph 6 of the Guidelines and this Part B, which differ from those applicable to part A of this Trigger List, which should not be interpreted as being limited by such definition".

The same list specifies what is to be understood by installations "of the same type", that is if their design, construction or operating processes are based on the same or similar physical or chemical processes (see chapter five, annex VII of this Study).

It has been mentioned above that the question of materials for isotope separation (INFCIRC/209/Mod.1), isotopic separation plants of the gas centrifuge type (INFCIRC/209/Mod.2), reprocessing plants (INFCIRC/209/Mod.3) had been developed and completed within the framework of the Zangger Committee.

E. MILITARY USES, NUCLEAR WEAPONS AND OTHER NUCLEAR EXPLOSIVE DEVICES

The OEEC Security Control Convention (Article 17) seems to be the only multilateral agreement to have attempted a definition of what is to be understood by "military uses" and, in contrast, by peaceful uses :

"A military purpose within the meaning of Article 1 includes the use of special fissionable materials in weapons of war and excludes their use in reactors for the production of electricity and heat or for propulsion".

As to the EURATOM Treaty, Article 84 provides in effect that safeguards under the Treaty are not applicable to :

"materials intended to meet defence requirements which are in course of being specially processed for this purpose or which, after being so processed, are, in accordance with an operational plan, placed or stored in a military establishment".

However the concepts of "defence requirements" or "specially processed" were not defined with more precision.

Since the Indian explosion, several other bilateral agreements are based on a triple distinction between "nuclear weapons", "other military uses" and "nuclear explosive devices" and specify what is to be understood by "military uses".

It should be noted that the NPT does not contain a definition of "nuclear weapons". On the other hand, the Treaty of Tlatelolco defines in Article 5 a nuclear weapon as : "any device which is capable of releasing nuclear energy in an uncontrolled manner and which has a group of characteristics that are appropriate for use for warlike purposes". The earlier Paris Agreements of 23rd October 1954 (Annex II) in noting the undertaking of the Federal Republic of Germany not to build "atomic weapons" on its territory gave a very broad definition :

"a) An atomic weapon is defined as a weapon which contains, or is designed to contain or utilise, nuclear fuel or radioactive isotopes and which, by explosion or other uncontrolled nuclear transformation of the nuclear fuel, or by radioactivity of the nuclear fuel or radioactive isotopes, is capable of mass destruction, mass injury or mass poisoning;

"b) Furthermore, any part, device, assembly or material especially designed for, or primarily useful in, any weapon as set forth under paragraph (a), shall be deemed to be an atomic weapon."

The differences between these definitions is to be noted. That of the Tlatelolco Treaty requires an uncontrolled release of nuclear energy whereas that of the Paris Agreements considers as an "atomic weapon" a device capable of causing massive poisoning due to the radioactivity of radioactive isotopes, even without an uncontrolled nuclear transformation.

The dominating tendency is more towards the definition of the Treaty of Tlatelolco, at least that is the definition retained in certain bilateral agreements concluded by the United States. Thus, the Agreement between the United States and the Republic of Korea of 24th November 1972, defines an "atomic weapon" (Article 1-3) in the following way :

"*Atomic weapon* means any device utilising atomic energy, exclusive of the means for transporting or propelling the device (where such means is a separable and divisible part of the device), the principal purpose of which is for use as, or for development of, a weapon, a weapon prototype, or a weapon test device".

F.   EXAMPLES OF DEFINITIONS USED IN BILATERAL AGREEMENTS

It should be noted that traditional national legislative drafting of texts and treaties may, or may not, include definitions. Agreements concluded by the Federal Republic of Germany or by Argentina do not usually contain definitions. It is not customary either in France to include definitions in a treaty; nevertheless, in this field this practice is, as we have seen earlier, generally followed in France. In contrast, the definition of the main terms used in a legal text is of general practice in common law countries.

Even more so than for the definitions in the preceding paragraphs, and for those extracted from statutes or instruments of international organisations or multilateral agreements, the definitions contained in bilateral agreements apply only for the interpretation of the agreements in which they are used. This is particularly true for definitions established by way of simple reference to Article XX of the IAEA Statute.

In this way Annex A.1 "definitions" of the co-operation Agreement between France and the Arab Republic of Egypt of 27th March 1981 provides as follows :

"Nuclear material :

Nuclear material means any source material or any special fissionable material as those terms are defined in Article XX of the Statute of the International Atomic Energy Agency.

The term source material is not to be interpreted as applying to ores or ore residues.

Any decision by the Board of Governors of the Agency under Article XX which amends the list of materials considered to be *source material* or *special fissionable material* shall not affect the terms of this Agreement unless the two Parties to this Agreement have informed each other in writing that they accept such amendment".

This same formulation is to be found in the Agreement between France and Australia concerning nuclear transfers [Agreement of 7th January 1981, Article I(e)][6] and the Agreement between Australia and EURATOM of 21st September 1981 relative to the transfer of nuclear materials [Article I(c)].

In these circumstances, it does not seem useful to give an exhaustive account, even if limited to agreements actually in force, of definitions contained therein. Some examples of definitions are provided to illustrate the orientations of this Study, and concern bilateral agreements which are referred to in chapter six of this Study.

As concerns the definition of "*nuclear material*", the introduction of a distinction between "source material" and "special nuclear material" in co-operation agreements concluded by the United States in accordance with the Atomic Energy Act of 1954, influenced, without doubt, this distinction in the statutes of international organisations and in multilateral agreements. Of more immediate interest are the definitions according to which the supplier State retains a right of control, in the form — for example — of prior consent rights for retransfers of material derived from material delivered. In this way, the Agreement of 23rd July 1985 between the United States and the People's Republic of China concerning the peaceful uses of nuclear energy contains the following definitions (Article 1) :

"5)   *material* means source material, special nuclear material or by-product material, radioisotopes other than by-product material, moderator material, or any other such substance so designated by agreement of the parties;

6) *source material* means *i)* uranium, thorium, or any other material so designated by agreement of the parties, or *ii)* ores containing one or more of the foregoing materials, in such concentrations as the parties may agree from time to time ;

7) *special nuclear material* means *i)* plutonium, uranium 233, or uranium enriched in the isotope 235, or *ii)* any other material so designated by agreement of the Parties ;

8) *by-product material* means any radioactive material (except special nuclear material) yielded in or made radioactive by exposure to the radiation incident to the process of producing or utilising special nuclear material ;

....

10) *high enriched uranium* means uranium enriched to twenty percent or greater in the isotope 235 ;

11) *low enriched uranium* means uranium enriched to less than twenty percent in the isotope 235".

In recent agreements concluded by Australia, this concern for continuity of control does not result from the definition of nuclear material but from the provisions relating to the article limiting the scope of application of the Agreement. Article II therefore of the 1981 Australia/EURATOM Agreement cited above, provides that it shall also apply to :

"*b)* All forms of nuclear material prepared by chemical or physical processes or isotopic separation provided that the quantity of nuclear material so prepared shall only be regarded as falling within the scope of this Agreement in the same proportion as the quantity of nuclear material used in its preparation, and which is subject to this Agreement, and bears to the total quantity of nuclear material so used ;

c) all generations of nuclear material produced by neutron irradiation provided that the quantity of nuclear material so produced shall only be regarded as falling within the scope of the Agreement in the same proportion as the quantity of nuclear material which is subject to this Agreement and which, used in its production, contributes to this production".

As we have seen in chapter six of this Study, this tendency to extend the scope of application of powers in the agreements is impeded by both practical and political difficulties ; certain agreements concluded by the United States, Canada and Australia, and even more often, the common interpretations given to these agreements, establish a rule of proportionality such that "such rights shall in practice be applied to that proportion of special nuclear material produced which represents the ratio of transferred material used in the production of the special nuclear material to the total amount of material so used, and similarly for subsequent generations". (Example :

Co-operation Agreement United States/Sweden, 19th December 1983 — interpretation).

France resorted to a special definition of "special nuclear material recuperated or obtained as a by-product" to achieve the same result. In this way, the 1975 Co-operation Agreement between France and the Republic of Iraq cited above (Annex g) provides that :

"*g)* by *special fissionable material recuperated or obtained as a by-product* is meant special fissionable material obtained from source material or special fissionable material supplied under this Agreement, or from one or several treatments effectuated through equipment or facilities supplied under this Agreement".

In other cases France did not resort to a particular definition but adopted the procedure for a specific pledge of non-utilisation for military purposes. The 1981 Co-operation Agreement between France and the Arab Republic of Egypt (Article VI.1.c) provides that each Contracting Party shall ensure that :

"*c.* The material, nuclear material, equipment and facilities obtained from or by elements listed in paragraph 1 A of this Article, including all successive generations of special fissile products recuperated or obtained as by-products, are not used for the purposes of designing, developing, manufacturing, acquiring or testing of nuclear weapons or other nuclear explosive devices, nor for any other military purpose, and that they are placed under IAEA safeguards".

With regard to the *definition of non-nuclear material* (moderators, cladding materials) we can distinguish two types of definitions : as to the first, in the Co-operation Agreement between EURATOM/Canada of 6th October 1959 (Article XIV) the term "material" includes nuclear material, moderators and implicitly, cladding materials :

"*e)* *material* means source material, special nuclear material, heavy water, graphite of nuclear quality, and any other substance which by reason of its nature or purity is specially suitable for use in nuclear reactors".

The definition of the term "material" which appears in the Agreements concluded by the United States is less specific as concerns moderators and cladding materials (for example, the above-mentioned 1985 Agreement with the People's Republic of China, Article 1).

In contrast, the definition of the term "material" which regroups the different categories which are the object of this Agreement, is completed by a special definition of moderators :

"9) *moderator material* means heavy water, or graphite or beryllium of a purity suitable for use in a reactor to slow down high velocity neutrons and increase the likelihood of further fission, or any other such material so designated by agreement of the Parties".

Part III of the 1981 Co-operation Agreement between France and Egypt (Annex A.I) defines moderators as :

"Deuterium and heavy water :

Deuterium and any deuterium compound in which the ratio of deuterium to hydrogen exceeds 1 :5000 for use in a nuclear reactor as defined above and supplied in quantities exceeding 200 kg of deuterium atoms for any recipient country in any period of 12 months.

Nuclear grade graphite :

Graphite having a purity level better than 5 parts per million boron equivalent and with a density greater than 1.50 grams per cubic centimetre supplied in quantities exceeding 30 metric tons for any recipient country in any period of 12 months".

The second type of definitions is more recent and exemplifies the importance accorded by the Zangger Committee or the London Club to the proliferation possibilities of moderators (see chapter five of this Study).

In any case, the Agreements concluded by the United States retain the first type of definitions in which moderators are included in the broader definition of "material" while also being the object of a sub-definition.

As concerns *definitions of sensitive installations and materials*, an evolution can be noted in the co-operation agreements concluded before the restrictions agreed to within either the Zangger Committee or the London Club. The coupled definitions — installations and equipment — had as a primary goal to help delimit the scope of application of an agreement. By way of example one can cite, for the United States, the Co-operation Agreement with Switzerland, of 30th December 1965 which distinguishes between equipment :

"E. *Equipment and devices* and *equipment or device* means any instrument, apparatus, or facility, except an atomic weapon, capable of making use of or producing special nuclear material, and component parts thereof."

and reactors :

"G. *Reactor* means an apparatus, other than an atomic weapon, in which a self-supporting fission chain reaction is maintained by utilising uranium, plutonium, or thorium, or any combination of uranium, plutonium, or thorium".

More recent agreements (United States/Australia, 5th July 1979, United States/Canada, 23rd April 1980, United States/Egypt, 29th June 1981, United States/ Norway, 12th January 1984) rely on a derived distinction in internal United States legislation between "production facility" and "utilization facility". By production facility is meant any reactor designed or used principally for the production of plutonium or uranium 233, any facility designed or used for the separation of uranium or plutonium isotopes, any facility designed or used for the reprocessing of irradiated materials containing special nuclear material or any device designated by agreement of the Parties. By utilization facility is meant any reactor other than those designed or used primarily for the production of plutonium or uranium 233.

This definition of production facility is given more specificity by the definition of "sensitive nuclear facility" which means any facility designed or used primarily for uranium enrichment, nuclear fuel reprocessing, heavy water production, or the fabrication of nuclear fuel containing plutonium.

There is no reference to the isotopic separation of plutonium but, on the other hand, the definition is extended to the production of heavy water and to the fabrication of mixed fuel containing plutonium.

It should be noted that another definition, that of "equipment" regroups production and utilization facilities as well as other additions resulting from the definition of sensitive nuclear facility — and opens the possibility of extension. "Equipment" means any production or utilization facility (including uranium enrichment facilities and those for the reprocessing of nuclear fuel) or any facility for heavy water production or for the fabrication of nuclear fuel containing plutonium or any device designated as such by agreement of the Parties.

Turning from complete facilities to the components thereof, the following points can be made :

— on the one hand, the definition of "major critical components" and simply "components". A component means a part of the "equipment" or other device so designated by agreement of the Parties, whereas "major critical component" means "any part or group of parts essential to the operation of a sensitive nuclear facility" ;
— on the other hand, the definition of the term "reactor" means any apparatus, other than a nuclear weapon or other nuclear explosive device, in which a self-sustaining fission chain reaction is maintained by utilising uranium, plutonium or any combination thereof, or any apparatus designated as such by agreement of the Parties.

The 1983 Agreement between the United States and Sweden excluded the definitions of production and utilization facilities and retained those of "sensitive nuclear facility", "major critical components", and "reactor". It provides a different definition of "equipment" and "components" from that of preceding agreements.

"Equipment" therefore means any reactor, other than those designed or used primarily for the production of plutonium or uranium 233 or any device so designated by the Parties.

As to "components", this term means a part of the equipment listed in annex or any other device so designated by the Parties. The annexed list includes pressure tubes, zirconium tubes, the "internals", the control rod mechanism and parts thereof, as well as reactor vessels, fuel loading or unloading machines, control rods, primary coolant pumps and facilities for the production of nuclear fuel not containing plutonium. It should be noted that this list is prior to that of the United States/China Agreement considered below.

For France, the Agreement concluded with Iraq on 18th March 1975 provides in an Annex that:

"*a)* by *facility* is meant all buildings and constructions specially designed and/or built for the purpose of being used in a nuclear energy programme, such as nuclear power plants and research reactors;

b) by *equipment* is meant the most important machinery, apparatus, or instruments or the major components thereof specially designed or produced for the purpose of being used in a nuclear energy programme".

In order to implement by means of and within the framework of bilateral agreements, the policies adopted by the major suppliers as a result of the work of the Zangger Committee and the London Club, the definitions of "equipment", "facilities" or "reactors" contribute towards specifying the objective of the agreement. This objective is nevertheless limited by the restrictions agreed to concerning "sensitive facilities" from the viewpoint of proliferation.

In this way, the 1985 Agreement between the United States and the People's Republic of China (Article 1) includes the following five definitions:

"12) *facility* means any reactor, other than one designed or used primarily for the formation of plutonium or uranium 233, or any other item so designated by agreement of the Parties;

13) *reactor* is defined in Annex I, which may be modified by mutual consent of the Parties;

14) *sensitive nuclear facility* means any plant designed or used primarily for uranium enrichment, reprocessing of nuclear fuel, heavy water production or fabrication of nuclear fuel containing plutonium;

15) *component* means a component part of a facility or other item, so designated by agreement of the Parties;

16) *major critical component* means any part or group of parts essential to the operation of a sensitive nuclear facility".

In the same spirit, the 1981 Agreement between Australia and France defines the term "equipment" in such a way as to include "sensitive equipment" within the meaning of the London Club Guidelines (INFCIRC/254).

The same evolution can be noted in the 1981 Agreement concluded by France with Egypt:

"*Facility* means all devices, equipment, buildings, capable of containing nuclear material or in which fissile material can be produced or treated by means of physical or chemical processes for the operation of which they have been designed".

"*Equipment* means the major elements or components specified in Part II of the Agreement".

The definitions in this Part II are based on the Trigger List of the London Club Guidelines.

For *definitions relating to technology*, bilateral agreements have moved away from concerns for the protection of patents and know-how, to the essentially political question of preventing the spread of sensitive technologies.

In this way, the above-cited Agreement between the United States and the People's Republic of China defines:

"17) *sensitive nuclear technology* means any information (including information incorporated in a facility or an important component) which is not in the public domain and which is important to the design, construction, fabrication, operation or maintenance of any sensitive nuclear facility, or such other information so designated by agreement of the Parties".

Along the same lines, the definitions contained in agreements concluded by Australia make specific reference to the London Club Guidelines. Thus, the Agreement cited above with France provides the following definition:

"*g)* *technology* means technical data in physical form, including technical drawings, photographed documents — negatives and prints, recordings, design data and technical and operating manuals, designated by the supplier Party as important for the design, production, operation or maintenance of enrichment, reprocessing or heavy water production facilities or major critical components thereof and any other technology as may be agreed between the Parties, but excluding data available to the public, for example in published books and periodicals".

This definition encompasses "sensitive technologies" mentioned in INFCIRC/254. It also appears in the agreements concluded by the United States with Australia or Norway. In other agreements, we find, on the contrary, a definition of "restricted data" which means:

"all information concerning:

i) the design, manufacture or use of nuclear weapons;

ii) the production of special nuclear material;

iii) the use of special nuclear material for the production of energy excluding information which has been declassified or removed from the category of restricted information by a Party".

The Protocol amending the United States/Canada Agreement of 23rd April 1980 includes on the one hand, two definitions relating to security and national defence. This concerns a definition of restricted data and a definition of the term "classify". This regroups restricted data as well as material and services considered confidential by the laws and regulations of the two countries. It contains on the other, a definition resulting from industrial or intellectual property law : the term "information" which applies in effect to information of a technical nature other than that falling within the public domain.

For France the same result was obtained by an implicit reference to the classifications of internal French regulations. The above-cited 1975 Agreement with Iraq contains in an Annex the following definition :

"*d*) by *knowledge not comprised by security restrictions* is meant that information which is not classified as *defence confidential* or *defence secret*".

More recent agreements, for example the 1981 Agreement with Egypt, contain a more detailed definition but one which also refers to a designation by the supplier Party, that is to say in this case, France :

"Technological information :

Technological information means technical data in material form, in particular technical plans, photographic documents, negatives and prints, recordings, project data, operating processes and instructions, designated by the supplier party as important for the design, realisation, operation and maintenance of enrichment, reprocessing or heavy water production facilities or major critical components thereof, or any other technology as agreed to by the Parties, but excluding information available to the public, for example through published books or periodicals".

With regard to *definitions of military uses* — or by reasoning *a contrario*, peaceful uses — including nuclear weapons and other nuclear explosive devices for example, the United States/China Agreement cited above, in addition to the definition of the term "reactor", gives the following definition in Article 1 :

"4) *peaceful purposes* include the use of information, technology, material, facilities and components in such fields as research, power generation, medicine, agriculture, and industry but do not include use in, research specifically on or development of any nuclear explosive device or any military purpose".

The older 1965 United States/Switzerland Agreement (and the other co-operation agreements of the same period) gives an example of the term "atomic weapon" :

"C. *Atomic weapon* means any device utilising atomic energy, exclusive of the means for transporting or propelling the device (where such means is a separable and divisible part of the device), the principal purpose of which is for use as, or for development of, a weapon, a weapon prototype, or a weapon test device".

To be noted is the more recent concern of Australia which uses a definition of "military ends" to encompass, other than nuclear weapons themselves, military nuclear propulsion. For example, the Agreement Australia/ France cited above, Article 1, by interpretation agreed to by the Parties gives the following definition :

"*d*) *military purpose* means direct military applications of nuclear energy such as nuclear weapons, military nuclear propulsion, military nuclear rocket engines or military nuclear reactors but does not include indirect uses such as power for a military base drawn from a civil power network, or production of radiosotopes to be used for diagnosis in a military hospital".

The other agreements concluded by France have not attempted to define these terms. All the same, the formulation of a specific pledge of peaceful use of the other party can be considered as equivalent. The 1981 Agreement between France and Egypt also provides :

"Article VI

1. Each Contracting Party undertakes that :
    A. The material, nuclear material, equipment, facilities and technological information transferred from one to the other shall not be utilised for the purpose of designing, developing, manufacturing, acquiring, or testing of nuclear weapons or other nuclear explosive devices or for any other nuclear military use".

In conclusion, a characteristic of the definitions contained in the bilateral agreements should also be noted ; this consists of distinguishing "*Parties*" to the agreement, that is to say, signatory States, from "*persons*" authorised to hold material, facilities, and components delivered in execution of the agreement. The Agreement between the United States and the People's Republic of China cited above therefore provides as follows (Article 1) :

"For the purposes of this Agreement :

1) *Parties* means the Government of the United States of America and the Government of the People's Republic of China ;

2) *Authorised person* means any individual or any entity under the jurisdiction of either Party and authorised by that Party to receive, possess, use or transfer material, facilities or components ;

3) *Person* means any individual or any entity subject to the jurisdiction of either Party but does not include the Parties to this agreement".

Several of these agreements, those concluded by Canada in particular (for example, the Canada/Switzerland Agreement of 6th March 1958) distinguish between "persons" and "government enterprises":

"h) *Government enterprises* means Atomic Energy of Canada Limited and Eldorado Mining and Refining Limited as for the Government of Canada, and such other enterprises under the jurisdiction of either Contracting Party as may be agreed between the Contracting Parties.

i) *Persons* means individuals, firms, corporations, companies, partnerships, associations and other entities private or governmental, and their respective agents and local representatives; but the term "persons" shall not include Government enterprises as defined in paragraph (h) of this Article".

NOTES AND REFERENCES

1. In the absence of official texts in English, the citation of definitions should be considered as unofficial translations.
2. PWR = Pressurised Water Reactor; BWR = Boiling Water Reactor.
3. Magnox = Magnesium Oxide; UNGG = natural uranium, gas, graphite.
4. Today the OECD.
5. With some modifications, these definitions are incorporated in Article 98 of the Agreement of 5th April 1973 between Belgium, Denmark, the Federal Republic of Germany, Ireland, Italy, Luxembourg, the Netherlands, the European Atomic Energy Community and the IAEA for the application of the Treaty on the Non-Proliferation of Nuclear Weapons (so-called "Verification Agreement").

   Community rules are in any case taken account of in Article 98.J a ii and b ii, in which the "receipt" or "shipment" "within the States" signatories are neither an "import" or "export" within the terms of the Agreement. Article 98.O specified, in the same vein, that the modification of the list of nuclear material will take effect only after acceptance by the Community and the States (compare with paragraphs 107 a and b and 112 of INFCIRC/153).

6. Older agreements contain more explicit definitions of "source material" and "special fissionable material" since they make no reference to the definitions contained in the Statute of the IAEA. But these definitions expressly provide for the possibility of ulterior clarifications or modifications to be agreed by the Parties (see Co-operation Agreement France/Japan of 26th February 1972, Article VIII e, f, and h; Co-operation Agreement France/Iran of 28th November 1975 Annex).

ANNEX I

NUCLEAR MATERIAL

IAEA STATUTE	OECD CONVENTION ON THE ESTABLISHMENT OF A SECURITY CONTROL IN THE FIELD OF NUCLEAR ENERGY	EURATOM TREATY

Article XX: Definitions

As used in this Statute:

1. The term "special fissionable material" means plutonium 239, uranium 233; uranium enriched in the isotopes 235 or 233; any material containing one or more of the foregoing; and such other fissionable material as the Board of Governors shall from time to time determine; but the term "special fissionable material" does not include source material.

2. The term "uranium enriched in the isotopes 235 or 233" means uranium containing the isotopes 235 or 233 or both in an amount such that the abundance ratio of the sum of these isotopes to the isotope 238 is greater than the ratio of the isotope 235 to the isotope 238 occurring in nature.

3. The term "source material" means uranium containing the mixture of isotopes occurring in nature; uranium depleted in the isotope 235; thorium; any of the foregoing in the form of metal, alloy, chemical compound, or concentrate; any other material containing one or more of the foregoing in such concentration as the Board of Governors shall from time to time determine; and such other material as the Board of Governors shall from time to time determine.

Note:

To our knowledge the Board of Governors has not used the powers granted by paragraphs 1 and 3.

Article 18

a. The term "special fissionable material" means plutonium 239; uranium 233; uranium enriched in the isotopes 235 or 233; any material containing one or more of the foregoing; and such other fissionable material as the Steering Committee shall from time to time determine; but the term "special fissionable material" does not include source material.

b. The term "uranium enriched in the isotopes 235 or 233" means uranium containing the isotopes 235 or 233 or both in an amount such that the abundance ratio of the sum of these isotopes to the isotope 238 is greater than the ratio of the isotope 235 to the isotope 238 occurring in nature.

c. The term "source material" means uranium containing the mixture of isotopes occurring in nature; uranium depleted in the isotope 235; thorium; any of the foregoing in the form of metal, alloy, chemical compound, or concentrate; any other material containing one or more of the foregoing in such concentrations as the Steering Committee shall from time to time determine; and such other material as the Steering Committee shall from time to time determine.

d. The term "material" means source material and special fissionable material.

Article 197

For the purposes of this Treaty

1. The term "special fissionable materials" shall mean plutonium 239, uranium 233; uranium enriched in the isotopes 235 or 233; any material containing one or more of the foregoing; and such other fissionable materials as shall be defined by the Council acting by means of a qualified majority vote on a proposal of the Commission; but the term "special fissionable materials" shall not include source materials.

2. The term "uranium enriched in the isotopes 235 or 233" shall mean uranium containing the isotopes 235 or 233 or both in an amount such that the abundance ratio of the sum of these isotopes to the isotope 238 is greater than the ratio of the isotope 235 to the isotope 238 occurring in nature.

3. The term "source material" shall mean uranium containing the mixture of isotopes occurring in nature; uranium depleted in the isotope 235; thorium; any of the foregoing in the form of metal, alloy, chemical compound, or concentrate; any other material containing one or more of the foregoing in such concentration as shall be defined by the Council acting by means of a qualified majority vote on a proposal of the Commission.

4. The term "ores" shall mean any ore containing, in such average concentration as shall be defined by the Council acting by means of a qualified majority vote on a proposal of the Commission, substances from which the source materials as defined above can be obtained by appropriate chemical and physical processing.

DEFINITIONS USED FOR IMPLEMENTING THE IAEA SAFEGUARDS AGREEMENTS UNDER THE NPT

(INFCIRC/153, February 1983 — "Blue Book")

Definitions

98. "Adjustment" means an entry into an accounting record or a report showing a shipper/receiver difference or material unaccounted for.

99. "Annual throughput" means, for the purposes of paragraphs 79 and 80 above, the amount of nuclear material transferred annually out of a facility working at nominal capacity.

100. "Batch" means a portion of nuclear material handled as a unit for accounting purposes at a key measurement point and for which the composition and quantity are defined by a single set of specifications or measurements. The nuclear material may be in bulk form or contained in a number of separate items.

101. "Batch data" means the total weight of each element of nuclear material and, in the case of plutonium and uranium, the isotopic composition when appropriate. The units of account shall be as follows:

   a) grams of contained plutonium;
   b) grams of total uranium and grams of contained uranium 235 plus uranium 233 for uranium enriched in these isotopes; and
   c) kilograms of contained thorium, natural uranium or depleted uranium.

   For reporting purposes the weights of individual items in the batch shall be added together before rounding to the nearest unit.

102. "Book inventory" of a material balance area means the algebraic sum of the most recent physical inventory of that material balance area and of all inventory changes that have occurred since that physical inventory was taken.

103. "Correction" means an entry into an accounting record or a report to rectify an identified mistake or to reflect an improved measurement of a quantity previously entered into the record or report. Each correction must identify the entry to which it pertains.

104. "Effective kilogram" means a special unit used in safeguarding nuclear material. The quantity in "effective kilograms" is obtained by taking:

   a) for plutonium, its weight in kilograms;
   b) for uranium with an enrichment of 0.01 (1 per cent) and above, its weight in kilograms multiplied by the square of its enrichment;
   c) for uranium with an enrichment below 0.01 (1 per cent) and above 0.005 (0.5 per cent), its weight in kilograms multiplied by 0.0001; and
   d) for depleted uranium with an enrichment of 0.005 (0.5 per cent) or below, and for thorium, its weight in kilograms multiplied by 0.00005.

105. "Enrichment" means the ratio of the combined weight of the isotopes uranium 233 and uranium 235 to that of the total uranium in question.

106. "Facility" means:

   a) a reactor, a critical facility, a conversion plant, a fabrication plant, a reprocessing plant, an isotope separation plant or a separate storage installation; or
   b) any location where nuclear material in amounts greater than one effective kilogram is customarily used.

107. "Inventory change" means an increase or decrease, in terms of batches, of nuclear material in a material balance area; such a change shall involve one of the following:

   a) Increases:
      i) import;
      ii) domestic receipt: receipts from other material balance areas, receipts from a non-safeguarded (non-peaceful) activity or receipts at the starting point of safeguards;

iii) nuclear production: production of special fissionable material in a reactor; and

iv) de-exemption: re-application of safeguards on nuclear material previously exempted therefrom on account of its use or quantity.

b) Decreases:

i) export;

ii) domestic shipment: shipments to other material balance areas or shipments for a non-safeguarded (non-peaceful) activity;

iii) nuclear loss: loss of nuclear material due to its transformation into other element(s) or isotope(s) as a result of nuclear reactions;

iv) measured discard: nuclear material which has been measured, or estimated on the basis of measurements, and disposed of in such a way that it is not suitable for further nuclear use;

v) retained waste: nuclear material generated from processing or from an operational accident, which is deemed to be unrecoverable for the time being but which is stored;

vi) exemption: exemption of nuclear material from safeguards on account of its use or quantity; and

vii) other loss: for example, accidental loss (that is, irretrievable and inadvertent loss of nuclear material as the result of an operational accident) or theft.

108. "Key measurement point" means a location where nuclear material appears in such a form that it may be measured to determine material flow or inventory. "Key measurement points" thus include, but are not limited to, the inputs and outputs (including measured discards) and storages in material balance areas.

109. "Man-year of inspection" means, for the purposes of paragraph 80 above, 300 man-days of inspection, a man-day being a day during which a single inspector has access to a facility at any time for a total of not more than eight hours.

110. "Material balance area" means an area in or outside of a facility such that:

a) the quantity of nuclear material in each transfer into or out of each "material balance area" can be determined; and

b) the physical inventory of nuclear material in each "material balance area" can be determined when necessary, in accordance with specified procedures,

in order that the material balance for Agency safeguards purposes can be established.

111. "Material unaccounted for" means the difference between book inventory and physical inventory.

112. "Nuclear material" means any source or any special fissionable material as defined in Article XX of the Statute. The term source material shall not be interpreted as applying to ore or ore residue. Any determination by the Board under Article XX of the Statute after the entry into force of this Agreement which adds to the materials considered to be source material or special fissionable material shall have effect under this Agreement only upon acceptance by the State.

113. "Physical inventory" means the sum of all the measured or derived estimates of batch quantities of nuclear material on hand at a given time within a material balance area, obtained in accordance with specified procedures.

114. "Shipper/receiver difference" means the difference between the quantity of nuclear material in a batch as stated by the shipping material balance area and as measured at the receiving material balance area.

115. "Source data" means those data, recorded during measurement or calibration or used to derive empirical relationships, which identify nuclear material and provide batch data. "Source data" may include, for example, weight of compounds, conversion factors to determine weight of element, specific gravity, element concentration, isotopic ratios, relationship between volume and mano-meter readings and relationship between plutonium produced and power generated.

116. "Strategic point" means a location selected during examination of design information where, under normal conditions and when combined with the information from all "strategic points" taken together, the information necessary and sufficient for the implementation of safeguards measures is obtained and verified; a "strategic point" may include any location where key measurements related to material balance accountancy are made and where containment and surveillance measures are executed.

ANNEX III

EXEMPTIONS FROM SAFEGUARDS

BLUE BOOK (INFCIRC/153)

TERMINATION OF SAFEGUARDS

Consumption or dilution of nuclear material

11. The Agreement should provide that safeguards shall terminate on nuclear material subject to safeguards thereunder upon determination by the Agency that it has been consumed, or has been diluted in such a way that it is no longer usable for any nuclear activity relevant from the point of view of safeguards, or has become practically irrecoverable.

   Provisions relating to nuclear material to be used in non-nuclear activities

13. The Agreement should provide that if the State wishes to use nuclear material subject to safeguards thereunder in non-nuclear activities, such as the production of alloys or ceramics, it shall agree with the Agency on the circumstances under which the safeguards on such nuclear material may be terminated.

35. The Agreement should provide that safeguards shall terminate on nuclear material subject to safeguards thereunder the conditions set forth in paragraph 11 above. Where the conditions of that paragraph are not met, but the State considers that the recovery of safeguarded nuclear material from residues is not for the time being practicable or desirable, the Agency and the State shall consult on the appropriate safeguards measures to be applied. It should further be provided that safeguards shall terminate on nuclear material subject to safeguards under the Agreement under the conditions set forth in paragraph 13 above, provided that the State and the Agency agree that such nuclear material is practicably irrecoverable.

EXEMPTIONS FROM SAFEGUARDS

36. The Agreement should provide that the Agency shall, at the request of the State, exempt nuclear material from safeguards, as follows:

   a) Special fissionable material, when it is used in gram quantities or less as a sensing component in instruments;

   b) Nuclear material, when it is used in non-nuclear activities in accordance with paragraph 13 above, if such nuclear material is recoverable; and

   c) Plutonium with an isotopic concentration of plutonium 238 exceeding 80 per cent.

EURATOM REGULATION 3227/76

DEROGATIONS AND EXEMPTIONS

Article 22

a) In order to take account of any particular circumstances in which safeguarded materials are used or produced, the Commission may, in the "particular safeguard provisions" referred to in Article 7, grant producers and users of nuclear materials a derogation from the rules governing the form and frequency of notification provided for in this Regulation.

The Commission may so decide especially in the case of installations holding only small quantities which are kept in the same state for long periods.

b) At the request of the persons or undertakings concerned in accordance with the form set out in Annex VIII, the Commission may exempt the following materials from declaration, provided that they are not processed or stored together with non-exempted nuclear materials:

   — special fissile materials which are used in quantities of the order of a gramme or less as sensing components in instruments,

   — plutonium with an isotopic concentration of plutonium 238 in excess of 80 per cent,

   — nuclear materials which are used exclusively in non-nuclear activities.

If the conditions for exemption cease to be fulfilled, the exemption shall be rescinded. The person or undertaking concerned shall inform the Commission in accordance with the form set out in Annex IX that the conditions for exemption no longer exist.

Article 23

This Regulation shall not apply to holders of finished products used for non-nuclear purposes which incorporate nuclear materials that are virtually irrecoverable.

37. The Agreement should provide that nuclear material that would otherwise be subject to safeguards shall be exempted from safeguards at the request of the State, provided that nuclear material so exempted in the State may not at any time exceed:

    a) One kilogram in total of special fissionable material, which may consist of one or more of the following:

        i) Plutonium;

        ii) Uranium with an enrichment of 0.2 (20 per cent) and above, taken account of by multiplying its weight by its enrichment; and

        iii) Uranium with an enrichment below 0.2 (20 per cent) and above that of natural uranium, taken account of by multiplying its weight by five time the square of its enrichment;

    b) Ten metric tons in total of natural uranium and depleted uranium with an enrichment above 0.005 (0.5 per cent);

    c) Twenty metric tons of depleted uranium with an enrichment of 0.005 (0.5 per cent) or below; and

    d) Twenty metric tons of thorium;

or such greater amounts as may be specified by the Board of Governors for uniform application.

Appendices 1 to 8

STATUS OF RELEVANT MULTILATERAL AGREEMENTS

APPENDIX 1

TREATY ON THE NON-PROLIFERATION OF NUCLEAR WEAPONS

Depositaries: United Kingdom, United States, USSR
Date of adoption: 1st July 1968
Date of entry into force: 5th March 1970

Contracting Parties	Date of Ratification Accession/Succession	Contracting Parties	Date of Ratification Accession/Succession
Afghanistan	4th February 1970	Germany, Federal Republic of	2nd May 1975
Antigua and Barbuda (succ.)	1st November 1981	Ghana	5th May 1970
Australia	23rd January 1973	Greece	11th March 1970
Austria	27th June 1969	Grenada (acc.)	19th August 1974
Bahamas (acc.)	10th July 1973	Guatemala	22nd September 1970
Bangladesh (acc.)	27th September 1979	Guinea Bissau (acc.)	20th August 1976
Barbados	21st February 1980	Haiti	2nd June 1970
Belgium	2nd May 1975	Holy See (acc.)	25th February 1971
Belize (succ.)	9th August 1985	Honduras	16th May 1973
Benin	31st October 1972	Hungary	27th May 1969
Bhutan (acc.)	23rd May 1985	Iceland	18th July 1969
Bolivia	26th May 1970	Indonesia	12th July 1979
Botswana	28th April 1969	Iran	2nd February 1970
Brunei Darussalam (acc.)	26th March 1985	Iraq	29th October 1969
Burkina Faso	3rd March 1970	Ireland	1st July 1968
Bulgaria	5th September 1969	Italy	2nd May 1975
Burundi (acc.)	19th March 1971	Ivory Coast	6th March 1973
Cameroon, United Republic of	8th January 1969	Jamaica	5th March 1970
Canada	8th January 1969	Japan	8th June 1976
Cape Verde (acc.)	24th October 1979	Jordan	11th February 1970
Central African Republic (acc.)	25th October 1970	Kenya	11th June 1970
Chad	10th March 1971	Kiribati (succ.)	18th April 1985
Colombia	8th April 1986	Korea, Republic of	23rd April 1975
Congo (acc.)	23rd October 1978	Korea, Dem. People's	
Costa Rica	3rd March 1970	Republic (acc.)	12th December 1985
Cyprus	10th February 1970	Lao People's Democratic Republic	20th February 1970
Czechoslovakia	22nd July 1969	Lebanon	15th July 1970
Democratic Kampuchea (acc.)	2nd June 1972	Lesotho	20th May 1970
Democratic Yemen	1st June 1979	Liberia	5th March 1970
Denmark	3rd January 1969	Libyan Arab Jamahiriya	26th May 1975
Dominica (succ.)	10th August 1984	Liechtenstein (acc.)	20th April 1978
Dominican Republic	24th July 1971	Luxembourg	2nd May 1975
Ecuador	7th March 1969	Madagascar	8th October 1970
Egypt	26th February 1981	Malawi (succ.)	18th February 1986
El Salvador	11th July 1971	Malaysia	5th March 1970
Equatorial Guinea (acc.)	1st November 1984	Maldives	7th April 1970
Ethiopia	5th February 1970	Mali, Republic of	10th February 1970
Fiji (acc.)	14th July 1972	Malta	6th February 1970
Finland	5th February 1969	Mauritius	25th April 1969
Gabon (acc.)	19th February 1974	Mexico	21st January 1969
Gambia	12th May 1975	Mongolia	14th May 1969
German Democratic Republic	31st October 1969	Morocco	27th November 1970

acc. = accession   succ. = succession

181

Contracting Parties	Date of Ratification Accession/Succession	Contracting Parties	Date of Ratification Accession/Succession
Nauru (acc.)	7th June 1982	Sri Lanka	5th March 1979
Nepal	5th January 1970	Sudan	31st October 1973
Netherlands	2nd May 1975	Suriname (succ.)	30th June 1976
New Zealand	10th September 1969	Swaziland	11th December 1969
Nicaragua	6th March 1973	Sweden	9th January 1970
Nigeria	27th September 1968	Switzerland	9th March 1977
Norway	5th February 1969	Syrian Arab Republic	24th September 1969
Panama	13th January 1977	Taiwan, China	27th January 1970
Papua New Guinea (acc.)	25th January 1982	Thailand (acc.)	7th December 1972
Paraguay	4th February 1970	Togo	26th February 1970
Peru	3rd March 1970	Tonga (acc.)	7th July 1971
Philippines	5th October 1972	Trinidad and Tobago	30th October 1986
Poland	12th June 1969	Tunisia	26th February 1970
Portugal (acc.)	15th December 1977	Turkey	17th April 1980
Romania	4th February 1970	Tuvalu (succ.)	19th January 1979
Rwanda (acc.)	20th May 1975	Uganda (acc.)	20th October 1982
St. Lucia (acc.)	28th December 1979	United Kingdom	27th November 1968
St. Vincent and		United States	5th March 1970
the Grenadines (succ.)	6th November 1984	Uruguay	31st August 1970
San Marino	10th August 1970	USSR	5th March 1970
Senegal	17th December 1970	Venezuela	26th September 1975
Seychelles (acc.)	12th March 1985	Viet Nam, Socialist Republic (acc.)	14th June 1982
Sierra Leone (acc.)	26th February 1975	Western Samoa (acc.)	17th March 1975
Singapore	10th March 1976	Yemen Arab Republic	14th May 1986
Solomon Islands (succ.)	17th June 1981	Yugoslavia	3rd March 1970
Somalia	5th March 1970	Zaire	4th August 1970
Spain (acc.)	5th November 1987		

acc. = accession   succ. = succession

SITUATION ON 31ST DECEMBER 1987 WITH RESPECT TO THE CONCLUSION OF SAFEGUARDS AGREEMENTS BETWEEN THE IAEA AND NON-NUCLEAR-WEAPON STATES IN CONNECTION WITH NPT*

Non-nuclear-weapon States which have signed, ratified, acceded to or succeeded to NPT[a] (1)	Date of ratification, accession or succession[a] (2)	Safeguards agreement with the IAEA (3)
Afghanistan	4th February 1970	In force: 20th February 1978
Antigua and Barbuda	1st November 1981	
Australia	23rd January 1973	In force: 10th July 1974
Austria	27th June 1969	In force: 23rd July 1972
Bahamas	10th July 1973	
Bangladesh	27th September 1979	In force: 11th June 1982
Barbados	21st February 1980	
Belgium	2nd May 1975	In force: 21st February 1977
Belize	9th August 1985	Approved by the Board, Feb. 1986
Benin	31st October 1972	
Bhutan	23rd May 1985	
Bolivia[b]	26th May 1970	Signed: 23rd August 1974
Botswana	28th April 1969	
Brunei Darussalam	25th March 1985	In force: 4th November 1987
Bulgaria	5th September 1969	In force: 29th February 1972
Burkina Faso	3rd March 1970	
Burundi	19th March 1971	
Cameroon	9th January 1969	
Canada	8th January 1969	In force: 21st February 1972
Cape Verde	24th October 1979	
Central African Republic	25th October 1970	
Chad	10th March 1971	
Colombia	8th April 1986	
Congo	23rd October 1978	
Costa Rica[b]	3rd March 1970	In force: 22nd November 1979
Côte d'Ivoire	6th March 1973	In force: 8th September 1983
Cyprus	10th February 1970	In force: 26th January 1973
Czechoslovakia	22nd July 1969	In force: 3rd March 1972
Democratic Kampuchea	2nd June 1972	
Democratic People's Republic of Korea	12th December 1985	
Democratic Yemen	1st June 1979	
Denmark[c]	3rd January 1969	In force: 21st February 1977
Dominica	10th August 1984	
Dominican Republic[b]	24th July 1971	In force: 11th October 1973
Ecuador[b]	7th March 1969	In force: 10th March 1975
Egypt	26th February 1981	In force: 30th June 1982
El Salvador[b]	11th July 1972	In force: 22nd April 1975
Equatorial Guinea	1st November 1984	Approved by the Board, June 1986
Ethiopia	5th February 1970	In force: 2nd December 1977
Fiji	14th July 1972	In force: 22nd March 1973
Finland	5th February 1969	In force: 9th February 1972
Gabon	19th February 1974	Signed: 3rd December 1979
Gambia	12th May 1975	In force: 8th August 1978
German Democratic Republic	31st October 1969	In force: 7th March 1972
Germany, Federal Republic of	2nd May 1975	In force: 21st February 1977
Ghana	5th May 1970	In force: 17th February 1975

Non-nuclear-weapon States which have signed, ratified, acceded to or succeeded to NPT[a] (1)	Date of ratification, accession or succession[a] (2)	Safeguards agreement with the IAEA (3)
Greece[d]	11th March 1970	Accession: 17th December 1981
Grenada	19th August 1974	
Guatemala[b]	22nd September 1970	In force: 1st February 1982
Guinea-Bissau	20th August 1976	
Haiti[b]	2nd June 1970	Signed: 6th January 1975
Holy See	25th February 1971	In force: 1st August 1972
Honduras[b]	16th May 1973	In force: 18th April 1975
Hungary	27th May 1969	In force: 30th March 1972
Iceland	18th July 1969	In force: 16th October 1974
Indonesia	12th July 1979	In force: 14th July 1980
Iran, Islamic Republic of	2nd February 1970	In force: 15th May 1974
Iraq	29th October 1969	In force: 29th February 1972
Ireland	1st July 1968	In force: 21st February 1977
Italy	2nd May 1975	In force: 21st February 1977
Jamaica[b]	5th March 1970	In force: 6th November 1978
Japan	8th June 1976	In force: 2nd December 1977
Jordan	11th February 1970	In force: 21st February 1978
Kenya	11th June 1970	
Kiribati	18th April 1985	
Korea, Republic of	23rd April 1975	In force: 14th November 1975
Kuwait[e]		
Lao People's Democratic Republic	20th February 1970	
Lebanon	15th July 1970	In force: 5th March 1973
Lesotho	20th May 1970	In force: 12th June 1973
Liberia	5th March 1970	
Libyan Arab Jamahiriya	26th May 1975	In force: 8th July 1980
Liechtenstein	20th April 1978	In force: 4th October 1979
Luxembourg	2nd May 1975	In force: 21st February 1977
Madagascar	8th October 1970	In force: 14th June 1973
Malawi	18th February 1986	
Malaysia	5th March 1970	In force: 29th February 1972
Maldives	7th April 1970	In force: 2nd October 1977
Mali	10th February 1970	
Malta	6th February 1970	
Mauritius	25th April 1969	In force: 31st January 1973
Mexico[b]	21st January 1969	In force: 14th September 1973
Mongolia	14th May 1969	In force: 5th September 1972
Morocco	27th November 1970	In force: 18th February 1975
Nauru	7th June 1982	In force: 13th April 1984
Nepal	5th January 1970	In force: 22nd June 1972
Netherlands[f]	2nd May 1975	In force: 21st February 1977
New Zealand	10th September 1969	In force: 29th February 1972
Nicaragua[b]	6th March 1973	In force: 29th December 1976
Nigeria	27th September 1968	
Norway	5th February 1969	In force: 1st March 1972
Panama	13th January 1977	
Papua New Guinea	25th January 1982	In force: 13th October 1983
Paraguay[b]	4th February 1970	In force: 20th March 1979
Peru[b]	3rd March 1970	In force: 1st August 1979
Philippines	5th October 1972	In force: 16th October 1974
Poland	12th June 1969	In force: 11th October 1972
Portugal[g]	15th December 1977	Accession: 1st July 1986
Romania	4th February 1970	In force: 27th October 1972
Rwanda	20th May 1975	
St. Lucia	28th December 1979	
St. Vincent and the Grenadines	6th November 1984	
Samoa	17th March 1975	In force: 22nd January 1979
San Marino	10th August 1970	Approved by the Board, Feb. 1977
Senegal	17th December 1970	In force: 14th January 1980
Seychelles	12th March 1985	

Non-nuclear-weapon States which have signed, ratified, acceded to or succeeded to NPT[a] (1)	Date of ratification, accession or succession[a] (2)	Safeguards agreement with the IAEA (3)
Sierra Leone	26th February 1975	Signed: 10th November 1977
Singapore	10th March 1976	In force: 18th October 1977
Solomon Islands	17th June 1981	
Somalia	5th March 1970	
Spain	5th November 1987	
Sri Lanka	5th March 1979	In force: 6th August 1984
Sudan	31st October 1973	In force: 7th January 1977
Suriname[b]	30th June 1976	In force: 2nd February 1979
Swaziland	11th December 1969	In force: 28th July 1975
Sweden	9th January 1970	In force: 14th April 1975
Switzerland	9th March 1977	In force: 6th September 1978
Syrian Arab Republic	24th September 1969	
Thailand	7th December 1972	In force: 16th May 1974
Togo	26th February 1970	
Tonga	7th July 1971	Approved by the Board, Feb. 1975
Trinidad and Tobago	30th October 1986	
Tunisia	26th February 1970	
Turkey	17th April 1980	In force: 1st September 1981
Tuvalu	19th January 1979	Approved by the Board, Feb. 1986
Uganda	20th October 1982	
Uruguay[b]	31st August 1970	In force: 17th September 1976
Venezuela[b]	26th September 1975	In force: 11th March 1982
Viet Nam	14th June 1982	
Yemen Arab Republic	14th May 1986	
Yugoslavia	3rd March 1970	In force: 28th December 1973
Zaire	4th August 1970	In force: 9th November 1972

a. The information reproduced in columns (1) and (2) was provided to the IAEA by depositary Governments of NPT, and an entry in column (1) does not imply the expression of any opinion on the part of the Secretariat concerning the legal status of any country or territory or of its authorities, or concerning the delimitation of its frontiers. The Table does not contain information relating to the participation of Taiwan, China in NPT.

b. The relevant safeguards agreement refers to both NPT and the Tlatelolco Treaty.

c. The NPT safeguards agreement with Denmark, in force since 1st March 1972, has been replaced by the Agreement of 5th April 1973 between the non-nuclear-weapon States of EURATOM, EURATOM and the IAEA but still applies to the Faroe Islands. Upon Greenland's secession from EURATOM as of 31st January 1985, the Agreement between the IAEA and Denmark re-entered into force as to Greenland.

d. The application of the IAEA safeguards in Greece under the agreement provisionally in force since 1st March 1972, was suspended on 17th December 1981, at which date Greece acceded to the Agreement of 5th April 1973 between the non-nuclear-weapon States of EURATOM, EURATOM and the IAEA.

e. Kuwait signed NPT on 15th August 1968 but has not yet ratified it.

f. An agreement had also been concluded in respect of the Netherlands Antilles. This agreement entered into force on 5th June 1975.

g. The NPT safeguards agreement with Portugal, in force since 14th June 1979, was suspended on 1st July 1986, on which date Portugal acceded to the Agreement between the non-nuclear weapon States of EURATOM, EURATOM and the IAEA of 5th April 1973.

*   *Source: IAEA Annual Report for 1986, updated.*

APPENDIX 3

AGREEMENTS PROVIDING FOR SAFEGUARDS, OTHER THAN THOSE IN CONNECTION WITH NPT, AS OF 31ST DECEMBER 1987*

Party(ies)[a]	Subject	Entry into force

(While the IAEA is a Party to each of the following agreements, only the State(s) Party to them is (are) listed.)

a) Project Agreements

Party(ies)[a]	Subject	Entry into force
Argentina	Siemens SUR-100	13th March 1970
	RAEP Reactor	2nd December 1964
Chile	Herald Reactor	19th December 1969
Finland[b]	FiR-1 Reactor	30th December 1960
	FINN sub-critical assembly	30th July 1963
Greece[b]	GRR-1 Reactor	1st March 1972
Indonesia	Additional core-load for TRIGA Reactor	19th December 1969
Iran, Islamic Republic of[b]	UTRR Reactor	10th May 1967
Jamaica[b]	Fuel for reasearch reactor	25th January 1984
Japan[b]	JRR-3	24th March 1959
Malaysia[b]	TRIGA-II Reactor	22nd September 1980
Mexico[b]	TRIGA-III Reactor	18th December 1963
	Siemens SUR-100	21st December 1971
	Laguna Verde Nuclear Power Plant	12th February 1974
Morocco[b]	Fuel for research reactor	2nd December 1983
Pakistan	PRR Reactor	5th March 1962
	Booster rods for KANUPP	17th June 1968
Peru[b]	Research reactor and fuel therefor	9th May 1978
Philippines[b]	PRR-1 Reactor	28th September 1966
Romania[b]	TRIGA Reactor	30th March 1973
	Experimental fuel elements	1st July 1983
Spain	Coral-I Reactor	23rd June 1967
Thailand[b]/United States	Fuel for research reactor	30th September 1986
Turkey[b]	Sub-critical assembly	17th May 1974
Uruguay[b]	URR Reactor	24th September 1965
Venezuela[b]	RV-1 Reactor	7th November 1975
Viet Nam[c]	Fuel for research reactor	1st July 1983
Yugoslavia[b]	TRIGA-II Reactor	4th October 1961
	Krsko Nuclear Power Plant	14th June 1974
Zaire[b]	TRICO Reactor	27th June 1962

b) Unilateral submissions

Party(ies)[a]	Subject	Entry into force
Albania	All nuclear material and facilities	Approved by Board, June 1986
Argentina	Atucha Power Reactor Facility	3rd October 1972
	Nuclear material	23rd October 1973
	Embalse Power Reactor Facility	6th December 1974
	Equipment and nuclear material	22nd July 1977
	Nuclear material, material, equipment and facilities	22nd July 1977
	Atucha II Nuclear Power Plant	15th July 1981
	Heavy water plant	14th October 1981
	Heavy water	14th October 1981
	Nuclear material	8th July 1982

Party(ies)[a]	Subject	Entry into force
Chile	Nuclear material	31st December 1974
	Nuclear material	22nd September 1982
	Nuclear material	18th September 1987
Cuba	Nuclear research reactor and fuel therefor	25th September 1980
	Nuclear power plant and nuclear material	5th May 1980
	Zero-power nuclear reactor and fuel therefor	7th October 1983
Democratic People's Republic of Korea	Research reactor and nuclear material for this reactor	20th July 1977
India	Nuclear material, material and facilities	17th November 1977
Pakistan	Nuclear material	2nd March 1977
Spain	Nuclear material	19th November 1974
	Nuclear material	18th June 1975
	Vandellos Nuclear Power Plant	11th May 1981
	Specified nuclear facilities	11th May 1981**
United Kingdom	Nuclear material	14th December 1972
Viet Nam	Research reactor and fuel therefor	12th June 1981

c) Tlatelolco Treaty

Colombia	All nuclear material	22nd December 1982
Mexico[d]	All nuclear material, equipment and facilities	6th September 1968
Panama	All nuclear material	23rd March 1984

d) Agreements concluded with nuclear-weapon States on the basis of voluntary offers

France	Nuclear material in facilities submitted to safeguards	12th September 1981
Union of Soviet Socialist Republics	Nuclear materials in facilities selected from list of facilities provided by the USSR	10th June 1985
United Kingdom	Nuclear material in facilities designated by the IAEA	14th August 1978
United States of America	Nuclear material in facilities designated by the IAEA	9th December 1980

e) Other agreements

Argentina/United States of America	25th July 1969
Austria[d]/United States of America	24th January 1970
Brazil/Germany, Federal Republic of[d]	26th February 1976
Brazil/United States of America	31st October 1968
Colombia/United States of America	9th December 1970
India/Canada[d]	30th September 1971
India/United States of America	27th January 1971
Iran, Islamic Republic of[d]/United States of America	20th August 1969
Israel/United States of America	4th April 1975
Japan[d]/Canada[d]	20th June 1966
Japan[d]/France	22nd September 1972
Japan/United States of America	10th July 1968
Japan[d]/United Kingdom	15th October 1968
Korea, Republic of/United States of America	5th January 1968
Korea, Republic of[d]/France	22nd September 1975
Pakistan/Canada	17th October 1969
Pakistan/France	18th March 1976
Philippines[d]/United States of America	19th July 1968
Portugal[d]/United States of America[e]	19th July 1969
South Africa/United States of America	26th July 1967
South Africa/France	5th January 1977
Spain/Germany, Federal Republic of[d]	29th September 1982
Spain/United States of America	9th December 1966
Spain/Canada[d]	10th February 1977

Party(ies)[a]	Subject	Entry into force
Sweden[d]/United States of America		1st March 1972
Switzerland[d]/United States of America[e]		28th February 1972
Turkey[d]/United States of America[e]		5th June 1969
Venezuela[d]/United States of America[e]		27th March 1968

a. An entry in this column does not imply the expression of any opinion whatsoever on the part of the Secretariat concerning the legal status of any country or territory or of its authorities or concerning the delimitation of its frontiers.

b. IAEA safeguards are being applied to the items required to be safeguarded under this (these) project agreement(s) pursuant to an agreement in connection with NPT covering the State indicated.

c. The requirement for the application of safeguards under this agreement is satisfied by the application of safeguards pursuant to the agreement of 12th June 1981.

d. Application of IAEA safeguards under this agreement has been suspended in the State indicated as the State has concluded an agreement in connection with NPT.

e. Application of IAEA safeguards under this agreement has been suspended in the United States of America in order to comply with a provision of the Agreement of 18th November 1977 between the United States and the IAEA for the application of safeguards in the United States.

* *Source:* IAEA Annual Report for 1986, updated.

** Amended in 1985 to cover specified nuclear facilities. The amendment entered into force on 8th November 1985. It is recalled that Spain ratified the NPT on 5th November 1987.

**

The Agency also applies safeguards under two agreements (Agreement of 13th October 1969 for the application of safeguards to the reactor research facility; Agreement of 6th December 1971 with IAEA and USA for the application of safeguards) to the nuclear facilities in Taiwan, China. Pursuant to the decision adopted by the IAEA Board of Governors on 9th December 1971 that the Government of the People's Republic of China is the only government which has the right to represent China in the IAEA, the relations between the Agency and the authorities in Taiwan are non-governmental. The agreements are implemented by the IAEA on that basis.

APPENDIX 4

WORLD MAP SHOWING SCOPE OF INTERNATIONAL SAFEGUARDS
AS AT 31ST DECEMBER 1987*

---

* *Source:* IAEA Bulletin, Volume 23, No. 4, December 1981, updated by the NEA Secretariat.

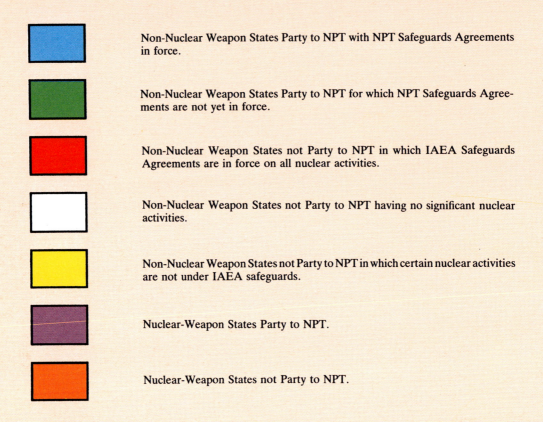

Non-Nuclear Weapon States Party to NPT with NPT Safeguards Agreements in force.

Non-Nuclear Weapon States Party to NPT for which NPT Safeguards Agreements are not yet in force.

Non-Nuclear Weapon States not Party to NPT in which IAEA Safeguards Agreements are in force on all nuclear activities.

Non-Nuclear Weapon States not Party to NPT having no significant nuclear activities.

Non-Nuclear Weapon States not Party to NPT in which certain nuclear activities are not under IAEA safeguards.

Nuclear-Weapon States Party to NPT.

Nuclear-Weapon States not Party to NPT.

The sole purpose of this map is to illustrate the scope of IAEA Safeguards and it does not imply the expression of any opinion whatsoever on the part of the Secretariat concerning the legal status of any country or territory or of its authorities, or concerning delimitation of frontiers.

SOUTH PACIFIC NUCLEAR FREE ZONE TREATY
(RAROTONGA TREATY)

Depositary :	Director of the South Pacific Bureau for Economic Co-operation, Suva, Fiji
Date of adoption :	6th August 1985
Date of entry into force :	11th December 1986

Signatories	Date of Signature	Date of Ratification
Australia	6th August 1985	11th December 1986
Cook Islands	6th August 1985	12th May 1986
Fiji	6th August 1985	4th October 1985
Kiribati	6th August 1985	28th October 1986
Nauru	18th July 1986	15th April 1987
New Zealand	6th August 1985	13th November 1986
Niue	6th August 1985	28th October 1985
Papua New Guinea	16th September 1985	
Solomon Islands	29th May 1987	
Tuvalu	6th August 1985	16th January 1986
Western Samoa	6th August 1985	20th October 1986

The Protocols to the Treaty were opened for signature on 1st December 1986. The USSR signed Protocols 2 and 3 on 15th December 1986. The People's Republic of China signed Protocols 2 and 3 on 10th February 1987.

CONVENTION ON THE ESTABLISHMENT OF A SECURITY CONTROL IN THE FIELD OF NUCLEAR ENERGY

Depositary : Organisation for Economic Co-operation
and Development, Paris
Date of adoption : 20th December 1957
Date of entry into force : 22nd July 1959

Contracting Parties	Date of Ratification	Contracting Parties	Date of Ratification
Austria	30th October 1959	Netherlands	9th July 1959
Belgium	22nd July 1959	Norway	12th February 1959
Denmark	23rd May 1959	Portugal	26th September 1959
France	23rd February 1959	Spain (acc.)	22nd July 1959
Germany, Federal Republic of	22nd July 1959	Sweden	5th January 1960
Ireland	2nd December 1958	Switzerland	21st January 1959
Italy	3rd April 1963	Turkey	20th July 1959
Luxembourg	19th May 1961	United Kingdom	10th May 1958

The application of the Security Control Regulations under the Convention was suspended on 14th October 1976.

CONVENTION ON THE PHYSICAL PROTECTION
OF NUCLEAR MATERIAL

Depositaries : International Atomic Energy Agency, Vienna ;
United Nations, New York

Date of adoption : 3rd March 1980

Date of entry into force : 8th February 1987

Signatories	Date of Signature	Date of Ratification
Argentina	28th February 1986	
Australia	22nd February 1984	22nd September 1987
Austria	3rd March 1980	
Belgium*	13th June 1980	
Brazil	15th May 1981	17th October 1985
Bulgaria	23rd June 1981	10th April 1984
Canada	23rd September 1980	21st March 1986
Czechoslovakia	14th September 1981	23rd April 1982
Denmark*	13th June 1980	
Dominican Republic	3rd March 1980	
Ecuador	26th June 1986	
Finland	25th June 1981	
France*	13th June 1980	
German Democratic Republic	21st May 1980	5th February 1981
Germany, Federal Republic of*	13th June 1980	
Greece*	3rd March 1980	
Guatemala	12th March 1980	23rd April 1985
Haiti	9th April 1980	
Hungary	17th June 1980	4th May 1984
Indonesia	3rd July 1986	5th November 1986
Ireland*	13th June 1980	
Israel	17th June 1983	
Italy*	13th June 1980	
Korea, Republic of	29th December 1981	7th April 1982
Liechtenstein	13th January 1986	25th November 1986
Luxembourg*	13th June 1980	
Mongolia	23rd January 1986	28th May 1986
Morocco	25th July 1980	
Netherlands*	13th June 1980	
Niger	7th January 1985	
Norway	26th January 1983	15th August 1985
Panama	18th March 1980	
Paraguay	21st May 1980	6th February 1985
Philippines	19th May 1980	22nd September 1981
Poland	6th August 1980	5th October 1983
Portugal*	19th September 1984	
Romania	15th January 1981	
South Africa	18th May 1981	

* Signed as a Member State of EURATOM.

Signatories	Date of Signature	Date of Ratification
Spain*	7th April 1986	
Sweden	2nd July 1980	1st August 1980
Switzerland	9th January 1987	9th January 1987
Turkey	23rd August 1983	27th February 1985
United Kingdom*	13th June 1980	
United States	3rd March 1980	13th December 1982
USSR	22nd May 1980	25th May 1983
Yugoslavia	15th July 1980	14th May 1986
European Atomic Energy Community (EURATOM)	13th June 1980	

* Signed as a Member State of EURATOM.

Appendices 9 to 15

SELECTED BILATERAL AGREEMENTS IN THE NUCLEAR FIELD

REVISED AGREEMENT FOR CO-OPERATION
BETWEEN
THE GOVERNMENT OF THE UNITED STATES OF AMERICA
AND
THE GOVERNMENT OF NORWAY
CONCERNING
PEACEFUL USES OF NUCLEAR ENERGY

(12th January 1984)

The Government of the United States of America and the Government of Norway,

Mindful that both the United States of America and Norway are Parties to the Treaty on the Non-Proliferation of Nuclear Weapons (NPT);

Re-affirming their commitment to ensuring that the international development and use of nuclear energy for peaceful purposes are carried out under arrangements which will to the maximum possible extent further the objectives of the NPT;

Affirming their support of the objectives of the International Atomic Energy Agency (IAEA) and their desire to promote universal adherence to the NPT;

Considering their close co-operation in the development, use and control of peaceful uses of nuclear energy pursuant to the Agreement for Co-operation Between the Government of the United States of America and the Government of Norway Concerning Civil Uses of Atomic Energy, signed 4th May 1967, as amended;

Desiring to continue and expand their co-operation in this field; and

Mindful that peaceful nuclear activities must be undertaken with a view to protecting the international environment from radioactive, chemical and thermal contamination;

Have agreed to revise the Agreement for Co-operation signed 4th May 1967, to read as follows:

Article 1

SCOPE OF CO-OPERATION

1.   The United States of America and Norway shall co-operate in the use of nuclear energy for peaceful purposes in accordance with the provisions of this Agreement and their applicable treaties, national laws, regulations and licence requirements.

2.   Transfers of information, material, equipment and components under this Agreement may be undertaken directly between the Parties or through authorised persons. Such transfers shall be subject to this Agreement and to such additional terms and conditions as may be agreed by the Parties.

3.   Material, equipment and components transferred from the territory of one Party to the territory of the other Party for peaceful purposes, whether directly or through a third country, will be regarded as having been transferred pursuant to the Agreement only upon confirmation, by the appropriate Government authority of the recipient Party to the appropriate Government authority of the supplier Party, that such material, equipment or components shall be subject to the Agreement.

Article 2

DEFINITIONS

For the purposes of this Agreement:

(a)   "Byproduct material" means any radioactive material (except special nuclear material yielded in or made radioactive by exposure to the radiation incident to the process of producing or utilising special nuclear material;

(b) "Component" means a component part of equipment or other item, so designated by agreement of the Parties;

(c) "Equipment" means any production or utilisation facility (including uranium enrichment and nuclear fuel reprocessing facilities), or any facility for the production of heavy water or the fabrication of nuclear fuel containing plutonium, or any other item so designated by agreement of the Parties;

(d) "High enriched uranium" means uranium enriched to twenty percent or greater in the isotope 235;

(e) "Low enriched uranium" means uranium enriched to less than twenty percent in the isotope 235;

(f) "Major critical component" means any part or group of parts essential to the operation of a sensitive nuclear facility;

(g) "Material" means source material, special nuclear material or byproduct material, radioisotopes other than byproduct material, or any other such substance so designated by agreement of the Parties;

(h) "Moderator material" means heavy water, or graphite or beryllium of a purity suitable for use in a reactor to slow down high velocity neutrons and increase the likelihood of further fission, or any other such material so designated by agreement of the Parties;

(i) "Parties" means the Government of the United States of America and the Government of Norway;

(j) "Peaceful purposes" include the use of information, material, equipment and components in such fields as research, power generation, medicine, agriculture and industry but do not include use in, research on or development of any nuclear explosive device, or any military purpose;

(k) "Person" means any individual or any entity subject to the jurisdiction of either Party but does not include the Parties to this agreement;

(l) "Previous agreement" means the Agreement for Co-operation Between the Government of the United States of America and the Government of Norway Concerning Civil Uses of Atomic Energy, signed 4th May 1967, as amended;

(m) "Production facility" means any nuclear reactor designed or used primarily for the formation of plutonium or uranium 233, any facility designed or used for the separation of the isotopes of uranium or plutonium, any facility designed or used for the processing of irradiated material containing special nuclear material, or any other item so designated by agreement of the Parties;

(n) "Reactor" means any apparatus, other than a nuclear weapon or other nuclear explosive device, in which a self-sustaining fission chain reaction is maintained by utilising uranium, plutonium or thorium, or any combination thereof;

(o) "Sensitive nuclear facility" means any facility designed or used primarily for uranium enrichment, reprocessing of nuclear fuel, heavy water production, or fabrication of nuclear fuel containing plutonium;

(p) "Sensitive nuclear technology" means any information (including information incorporated in equipment or an important component) which is not in the public domain and which is important to the design, construction, fabrication, operation or maintenance of any sensitive nuclear facility, or other such information which may be so designated by agreement of the Parties;

(q) "Source material" means *(i)* uranium, thorium, or any other material so designated by agreement of the Parties, or *(ii)* ores containing one or more of the foregoing materials, in such concentration as the Parties may agree from time to time;

(r) "Special nuclear material" means *(i)* plutonium, uranium 233, or uranium enriched in the isotope 235, or *(ii)* any other material so designated by agreement of the Parties; and

(s) "Utilisation facility" means any reactor, other than one designed or used primarily for the formation of plutonium or uranium 233, as well as pressure vessels designed to contain the core of such a facility and primary coolant pumps, fuel charging or discharging machines or control rods for such a facility, or any other item so designated by agreement of the Parties.

Article 3

TRANSFER OF INFORMATION

1.   Information concerning the use of nuclear energy for peaceful purposes may be transferred. Transfers of information may be accomplished through various means, including reports, data banks, computer programs, conferences, visits and assignments of staff to facilities. Fields which may be covered include, but shall not be limited to, the following:

(a)   development, design, construction, operation, maintenance and use of research, materials testing, experimental, demonstration power, and power reactors and reactor experiments;

(b)   the use of material in physical and biological research, medicine, agriculture and industry;

(c) fuel cycle studies of ways to meet future worldwide peaceful nuclear needs, including multilateral approaches to guaranteeing nuclear fuel supply and appropriate techniques for management of nuclear wastes;

(d) safeguards and physical security of materials, equipment and components;

(e) health, safety and environmental considerations related to the foregoing; and

(f) assessing the role nuclear power may play in national energy plans.

2. This Agreement does not require the transfer of any information which the Parties are not permitted to transfer.

3. The United States shall not transfer Restricted Data, as defined in its Atomic Energy Act of 1954, as amended, under this Agreement.

4. Sensitive nuclear technology shall not be transferred under this agreement unless provided by an amendment to this Agreement.

Article 4

TRANSFER OF MATERIAL, EQUIPMENT AND COMPONENTS

1. Material, equipment and components may be transferred for applications consistent with this Agreement. The United States shall endeavour to take such actions as necessary and feasible to ensure a reliable supply of nuclear fuel to Norway, including the export of material on a timely basis and the availability of the capacity to carry out this undertaking during the period of this Agreement.

2. Low enriched uranium may be transferred in such quantities as are required for use as fuel in reactor experiments and in reactors, for conversion or fabrication, or for such other purposes, including reasonable stockpiling, as may be agreed by the Parties.

3. Special nuclear material other than low enriched uranium and material contemplated under paragraph 6 may, if the Parties agree, be transferred for specified applications where technically and economically justified or where justified for the development and demonstration of reactor fuel cycles to meet energy security and non-proliferation objectives.

4. The quantity of special nuclear material other than low enriched uranium transferred under this Agreement shall not at any time be in excess of the quantity the Parties agree is necessary for any of the following purposes: the use in reactor experiments or loading of reactors, the efficient and continuous conduct of such reactor experiments or operation of such reactors, conversion or fabrication, and the accomplishment of other purposes as may be agreed by the Parties. If high enriched uranium in excess of the quantity required for these purposes exists in Norway, the United States shall have the right to require the return of any high enriched uranium transferred pursuant to this Agreement (including irradiated high enriched uranium) which contributes to this excess. If the United States exercises this right,

    (a) it shall, after removal of such material from the territory of Norway, reimburse Norway for the fair market value of such material; and

    (b) the Parties shall make appropriate commercial arrangements which shall not be subject to any further agreement between the Parties as otherwise contemplated under Articles 5 and 6.

5. Any high enriched uranium transferred pursuant to this agreement shall not be at a level of enrichment in the isotope 235 in excess of levels to which the Parties agree are necessary for the purposes described in paragraph 4.

6. Small quantities of special nuclear material may be transferred for use as samples, standards, detectors, targets and for such other purposes as the Parties may agree. Transfers pursuant to this paragraph shall not be subject to the quantity limitations in paragraph 4.

7. Sensitive nuclear facilities and major critical components shall not be transferred under this Agreement unless provided by an amendment to this Agreement.

Article 5

STORAGE AND RETRANSFERS

1. Material transferred pursuant to this Agreement and material used in or produced through the use of any material or equipment transferred pursuant to this Agreement may be stored by either Party, except that each Party guarantees that such plutonium or uranium 233 (except as contained in irradiated fuel elements) or high enriched uranium, over which it has jurisdiction, shall only be stored in a facility which the Parties agree meets criteria to be established in advance by the Parties.

2.   Material, equipment or components transferred pursuant to this Agreement and any special nuclear material produced through the use of any such material or equipment may be transferred by the recipient Party, except that such Party guarantees that any such material, equipment, components or special nuclear material, over which it has jurisdiction, shall not be transferred to unauthorised persons or, unless the Parties agree, beyond its territorial jurisdiction.

Article 6

REPROCESSING AND ENRICHMENT

1.   Each Party guarantees that material transferred to and under its jurisdiction pursuant to this Agreement and material used in or produced through the use of any material or equipment transferred to and under its jurisdiction pursuant to this Agreement shall not be reprocessed unless the Parties agree.

2.   Each Party guarantees that any plutonium, uranium 233, high enriched uranium or irradiated source or special nuclear material transferred to and under its jurisdiction pursuant to this Agreement, or produced through the use of any material or equipment transferred to and under its jurisdiction pursuant to this Agreement, shall not be altered in form or content, except by irradiation or further irradiation, unless the Parties agree.

3.   Each Party guarantees that uranium transferred to and under its jurisdiction pursuant to this Agreement, and uranium used in any equipment so transferred and under its jurisdiction, shall not be enriched after transfer to twenty percent or greater in the isotope 235 unless the Parties agree.

Article 7

PHYSICAL SECURITY

1.   Each Party guarantees that adequate physical security shall be maintained with respect to any material and equipment transferred to and under its jurisdiction pursuant to this agreement and with respect to any special nuclear material used in or produced through the use of any material or equipment transferred to and under its jurisdiction pursuant to this Agreement.

2.   The Parties agree to the levels for the application of physical security set forth in the Annex and, in accordance with such levels, to maintain adequate physical security measures which shall as a minimum provide protection comparable to the recommendations set forth in IAEA document INFCIRC/225/Rev.1 concerning the physical protection of nuclear material, or in any revision of that document agreed to by the Parties. In the event of any such revision, the levels in the Annex may be modified as the Parties may agree.

3.   The adequacy of physical security measures maintained pursuant to this Article shall be subject to review and consultation by the Parties periodically and whenever either Party is of the view that revised measures may be required to maintain adequate physical security.

4.   Each Party shall identify those agencies or authorities having responsibility for ensuring that levels of physical security are adequately met and having responsibility for co-ordinating response and recovery operations in the event of unauthorised use or handling of material subject to this Article. Each Party shall also designate points of contact within its national authorities to co-operate on matters of out-of-country transportation and other matters of mutual concern.

5.   This Article shall be implemented in such a manner as to avoid hampering, delay or undue interference in the Parties' nuclear activities and so as to be consistent with prudent management practices required for the economic and safe conduct of their nuclear programmes.

Article 8

NO EXPLOSIVE OR MILITARY APPLICATION

Each Party guarantees that no material, equipment or components transferred to and under its jurisdiction pursuant to this Agreement and no material used in or produced through the use of any such material, equipment or components so transferred to and under its jurisdiction shall be used for any nuclear explosive device, for research on or development of any nuclear explosive device, or for any military purpose.

Article 9

SAFEGUARDS

1.   Co-operation under this Agreement shall require the application of IAEA safeguards with respect to all nuclear activities within the territory of Norway, under its jurisdiction or carried out under its control anywhere. Implementation of the safeguards agreement pursuant to Article III(4) of the NPT, referred to in paragraph 2, shall be considered to fulfil the requirement stated in the foregoing sentence.

2.   Source and special nuclear material transferred to Norway pursuant to this Agreement and any source or special nuclear material used in or produced through the use of any material, equipment or components so transferred shall be subject to safeguards in accordance with the Agreement Between Norway and the International Atomic Energy Agency for the Application of Safeguards in Connection with the Treaty on the Non-Proliferation of Nuclear Weapons, signed on 1st March 1972.

3.   Source and special nuclear material transferred to the United States pursuant to this Agreement and any source or special nuclear material used in or produced through the use of any material, equipment or components so transferred shall be subject to the provisions of the Agreement Between the United States of America and the International Atomic Energy Agency for the Application of Safeguards in the United States, signed on 18th November 1977.

4.   If the United States of America or Norway becomes aware of circumstances which demonstrate that the IAEA for any reason is not or will not be applying safeguards in accordance with the Agreement as provided for in paragraph 2, to ensure effective continuity of safeguards the Parties shall immediately enter into arrangements which conform with IAEA safeguards principles and procedures and with the coverage required by that paragraph and which provide assurance equivalent to that intended to be secured by the system they replace.

5.   Each Party guarantees it shall take such measures as are necessary to maintain and facilitate the application of safeguards provided for under this Article.

6.   Norway shall establish and maintain a system of accounting for and control of all source and special nuclear material under its jurisdiction, the procedures of which shall be comparable to those set forth in IAEA document INFCIRC/153 (corrected), or in any revision of that document agreed to by the Parties.

7.   Upon the request of either Party, the other Party shall report or permit the IAEA to report to the requesting Party on the status of all inventories of any source and special nuclear material subject to this agreement and any other material transferred pursuant to this Agreement.

8.   The provisions of this Article shall be implemented in such a manner as to avoid hampering, delay or undue interference in the Parties' nuclear activities and so as to be consistent with prudent management practices required for the economic and safe conduct of their nuclear programmes.

Article 10

MULTIPLE SUPPLIER CONTROLS

If an agreement between either Party and another nation or group of nations provides such other nation or group of nations rights equivalent to any or all of those set forth under Articles 5, 6, or 7 with respect to material, equipment or components subject to this Agreement, the Parties may, upon the request of either of them, agree that the implementation of any such rights will be accomplished by such other nation or group of nations.

Article 11

CESSATION OF CO-OPERATION

1.   If either Party at any time following entry into force of this Agreement:

   (a)   does not comply with the provisions of Articles 5, 6, 7, 8 or 9, or

   (b)   terminates, abrogates or materially violates a safeguards agreement with the IAEA,

the other Party shall have the rights to cease further co-operation under this Agreement and to require the return of any material, equipment or components transferred under this Agreement and any special nuclear material produced through their use.

2.   If Norway at any time following entry into force of this Agreement detonates a nuclear explosive device, the United States shall have the same rights as specified in paragraph 1.

3.   If either Party exercises its rights under this Article to require the return of any material, equipment or components, it shall, after removal from the territory of the other Party, reimburse the other Party for the fair market value of such material, equipment or components. If this right is exercised, the Parties shall make such other appropriate arrangements as may be required which shall not be subject to any further agreement between the Parties as otherwise contemplated under Articles 5 and 6.

Article 12

PREVIOUS AGREEMENT TERMINATED

1.   The Agreement for Co-operation Between the Government of the United States of America and the Government of Norway Concerning Civil Uses of Atomic Energy signed on 4th May 1967, as amended, shall terminate on the date this revised Agreement enters into force.

2.   Co-operation initiated under the previous agreement shall continue in accordance with the provisions of this revised Agreement. The provisions of this Agreement shall apply to material and equipment subject to the previous Agreement.

Article 13

CONSULTATIONS, CONFIDENTIALITY, AND ENVIRONMENTAL PROTECTION

1.   The Parties undertake to consult at the request of either Party regarding the implementation of this Agreement and the development of further co-operation in the field of peaceful uses of nuclear energy.

2.   The Parties shall take every precaution to protect any information which comes to their knowledge in the implementation of this Agreement and which, at the time of transfer or receipt, is designated by the supplier Party to be proprietary or confidential.

3.   The Parties shall consult, with regard to activities under this Agreement, to identify the international environmental implications arising from such activities and shall co-operate in protecting the international environment from radioactive, chemical and thermal contamination arising from peaceful nuclear activities under this agreement and in related matters of health and safety.

Article 14

AMENDMENT

      This Agreement may be amended at any time by agreement of the Parties and in accordance with their applicable requirements.

Article 15

ENTRY INTO FORCE AND DURATION

1.   This Agreement shall enter into force on the date on which the Parties exchange diplomatic notes informing each other that they have complied with all applicable requirements for its entry into force, and shall remain in force for a period of thirty (30) years. This term may be extended for such additional periods as may be agreed between the Parties in accordance with their applicable requirements.

2.   Notwithstanding the suspension, termination or expiration of this Agreement or any co-operation hereunder for any reason, Articles 5, 6, 7, 8, 9 and 11 shall continue in effect so long as any material, equipment or components subject to these Articles remain in the territory of the Party concerned or under its jurisdiction or control anywhere, or until such time as the Parties agree that such material, equipment or components are no longer useable for any nuclear activity relevant from the point of view of safeguards.

      Done at Oslo, this 12th day of January 1984, in duplicate, in the English and Norwegian languages, both equally authentic.

Pursuant to paragraph 2 of Article 7, the agreed levels of physical security to be ensured by the competent national authorities in the use, storage and transportation of the materials listed in the attached table shall as a minimum include protection characteristics as below.

Category III

Use and storage within an area to which access is controlled.

Transportation under special precautions including prior arrangements among sender, recipient and carrier, and prior agreement between entities subject to the jurisdiction and regulation of supplier and recipient States, respectively, in case of international transport specifying time, place and procedures for transferring transport responsibility.

Category II

Use and storage within a protected area to which access is controlled, i.e., an area under constant surveillance by guards or electronic devices, surrounded by a physical barrier with a limited number of points of entry under appropriate control, or any area with an equivalent level of physical protection.

Transportation under special precautions including prior arrangements among sender, recipient and carrier, and prior agreement between entities subject to the jurisdiction and regulation of supplier and recipient States, respectively, in case of international transport, specifying time, place and procedures for transferring transport responsibility.

Category I

Material in this category shall be protected with highly reliable systems against unauthorised use as follows.

Use and storage within a highly protected area, i.e., a protected area as defined for category II above, to which, in addition, access is restricted to persons whose trustworthiness has been determined, and which is under surveillance by guards who are in close communication with appropriate response forces. Specific measures taken in this context should have as their objective the detection and prevention of any assault, unauthorised access or unauthorised removal of material.

Transportation under special precautions as identified above for transportation of categories II and III materials and, in addition under constant surveillance by escorts and under conditions which assure close communication with appropriate response forces.

TABLE

Material	Form	Category		
		I	II	III
1. Plutonium[a, f]	Unirradiated[b]	2 kg or more	Less than 2 kg but more than 500 g	500 g or less[c]
2. Uranium 235[d]	Unirradiated[b] — Uranium enriched to 20% ^{235}U or more	5 kg or more	Less than 5 kg but more than 1 kg	1 kg or less[c]
	— Uranium enriched to 10% ^{235}U but less than 20%	—	10 kg or more	Less than 10 kg[c]
	— Uranium enriched above natural, but less than 10% ^{235}U	—	—	10 kg or more
3. Uranium 233	Unirradiated[b]	2 kg or more	Less than 2 kg but more than 500 g	500 g or less[c]

a) All plutonium except that with isotopic concentration exceeding 80% in plutonium 238.
b) Material not irradiated in a reactor or material irradiated in a reactor but with a radiation level equal to or less than 100 rads/hour at one meter unshielded.
c) Less than a radiologically significant quantity should be exempted.
d) Natural uranium, depleted uranium and thorium and quantities of uranium enriched to less than 10% not falling in Category III should be protected in accordance with prudent management practice.
e) Irradiated fuel should be protected as Category I, II or III nuclear material depending on the category of the fresh fuel. However, fuel which by virtue of its original fissile material content is included as Category I or II before irradiation should only be reduced one Category level, while the radiation level from the fuel exceeds 100 rads/h at one meter unshielded.
f) The State's competent authority should determine if there is a credible threat to disperse plutonium malevolently. The State should then apply physical protection requirements for Category I, II or III of nuclear material, as it deems appropriate and without regard to the plutonium quantity specified under each category herein, to the plutonium isotopes in those quantities and forms determined by the State to fall within the scope of the credible dispersal threat.

Agreed Minute

During the negotiation of the proposed Revised Agreement for Co-operation Between the Government of the United States of America and the Government of Norway Concerning Peaceful Uses of Nuclear Energy (Agreement) signed today, the following understandings, which shall be an integral part of the Agreement, were reached.

Coverage of Agreement

For the purposes of implementing the rights specified in Articles 5, 6, and 7 with respect to special nuclear material produced through the use of material transferred pursuant to the Agreement and not used in or produced through the use of equipment transferred pursuant to the Agreement, such rights shall in practice be applied to that proportion of special nuclear material produced which represents the ratio of transferred material used in the production of the special nuclear material to the total amount of material so used, and similarly for subsequent generations.

With reference to Article 8 it is understood that "military purpose" does not include power to a military base drawn from the civil power network or production of radioisotopes to be used for diagnosis in a military hospital.

Previous Co-operation

It is noted that Article 3 does not alter the practice of mutually beneficial information sharing that has developed between the two countries and that will be continued and, as the Parties may agree, strengthened.

With respect to paragraph 2 of Article 12, in order to facilitate the application of the provisions of this agreement to material and equipment subject to the previous Agreement, the Parties shall establish a list of such material and equipment.

Storage

Concerning paragraph 1 of Article 5, the Parties note that, in practice, agreement of the Parties would be reflected in authorisation for the transfer of material under the agreement for specified applications and that this provision would otherwise be exercised with respect to such material only in the event a Party wished to store such material at facilities other than originally specified.

Spent Fuel Disposition

The Parties note their common interest in ensuring that their nuclear co-operation promotes the energy security of each Party and their mutual non-proliferation objectives. In this regard, the Parties agree that material subject to Articles 5 and 6 may be transferred by Norway to the United Kingdom or France and reprocessed at the Sellafield or La Hague reprocessing facilities, subject to the following conditions:

(1) Norway shall keep records of any such transfers and shall upon shipment notify the United States of each transfer;

(2) Prior to any such transfer, Norway shall confirm to the United States that the material to be transferred will be held within EURATOM subject to the applicable agreement for co-operation between the United States and EURATOM;

(3) Norway shall retain legal control over any plutonium separated as a result of any such transfer and shall obtain the prior agreement of the United States for the transfer of the plutonium to Norway or any other country for any use of the plutonium.

The foregoing understandings concerning fuel disposition may be terminated in whole or in part, if either Party considers that exceptional circumstances of concern from a non-proliferation or security standpoint so require; to the extent time and circumstances permit, the Parties will consult prior to any such termination. Such circumstances include, but are not limited to, a determination by either Party that the foregoing understandings cannot be continued without a significant increase of the risk of proliferation or without jeopardising its national security.

These understandings concerning spent fuel disposition do not limit the right of the Parties to agree to other activities envisaged in Articles 5 and 6.

OECD Halden Reactor Project

The Parties note the longstanding support of the United States for the OECD Halden Reactor Project, including arrangements for the test irradiation and analysis of United States fuel assemblies in connection with United States reactor safety research. The Parties confirm that in any cases where the United States has authorised material to be transferred to Norway under the Agreement for specified purposes, which in any particular case include irradiation in the Halden Reactor and the alteration in Norway of such material in form or content, no further agreement would be required pursuant to paragraph 2 of Article 6 for such analysis of the material concerned. In any other situations not covered by the preceding sentence, the United States would give favourable consideration to proposals by Norway pursuant to paragraph 2 of Article 6 that the Parties agree to alteration of material under the fuel analysis program in Norway in connection with the Halden Reactor Project.

The Parties also confirm that, notwithstanding the provisions of paragraph 2 of Article 12, moderator material transferred to Norway under the previous agreement for use in the OECD Halden Reactor Project, and special nuclear material produced through its use and not through the use of special nuclear material transferred pursuant to the Agreement, shall be subject only to Articles 7, 8, and 9 and that such moderator material shall not be transferred to unauthorised persons or, unless the Parties agree, beyond the territorial jurisdiction of Norway.

Safeguards

If either Party becomes aware of circumstances referred to in paragraph 4 of Article 9 requiring the Parties immediately to enter into arrangements referred to in that paragraph, the United States shall have the rights listed below. These rights shall be suspended if the United States agrees that the need to exercise such rights is being satisfied by the application of IAEA safeguards under arrangements pursuant to paragraph 4 of Article 9.

(1) To review in a timely fashion the design of any equipment transferred pursuant to the agreement, or of any facility which is to use, fabricate, process, or store any material so transferred or any special nuclear material used in or produced through the use of such material or equipment;

(2) to require the maintenance and production of records and of relevant reports for the purpose of assisting in ensuring accountability for material transferred pursuant to the Agreement and any source material or special nuclear material used in or produced through the use of any material, equipment or components so transferred;

(3) to designate personnel acceptable to Norway who shall have access to all places and data necessary to account for the material in paragraph(2), to inspect any equipment or facility referred to in paragraph(1), and to install any devices and make such independent measurements as may be deemed necessary to account for such material. Norway shall not unreasonably withhold its acceptance of personnel designated by the United States under this paragraph. Such personnel shall, if either Party so requests, be accompanied by personnel designated by Norway.

With reference to Article 9, it is confirmed that design information relevant to safeguards for new equipment required to be safeguarded under the Agreement shall be provided in a timely fashion to the IAEA upon its request.

The Parties confirm that paragraph 1 of Article 9 does not in any way alter Norway's commitment under the Agreement referred to in paragraph 2 of Article 9 or the implementation of that Agreement.

With further reference to Article 9, it is understood that the Agreement does not affect the rights or obligations of the United States or the IAEA pursuant to the Agreement referred to in paragraph 3 of Article 9 or the implementation of that Agreement.

Administrative Arrangements

The appropriate Government authorities of both Parties may establish administrative arrangements to ensure the effective implementation of the Agreement. Such administrative arrangements may be changed by agreement between the appropriate Government authorities of the Parties.

Consultations

The Parties shall consult periodically or at the request of either Party on ways in which their mutual non-proliferation and nuclear energy objectives can best be served. Each Party shall endeavour to avoid taking any actions that significantly alter the basis for co-operation under the Agreement. However, if either Party were to undertake any such actions, it shall endeavour to inform the other Party in advance.

ÉCHANGE DE LETTRES EN DATE DU 21 NOVEMBRE 1986 ENTRE LE GOUVERNEMENT DE LA RÉPUBLIQUE FRANÇAISE ET LE GOUVERNEMENT DU ROYAUME DE SUÈDE RELATIF AU RETRAITEMENT EN FRANCE DES COMBUSTIBLES IRRADIÉS EN SUÈDE (ENSEMBLE UNE ANNEXE)*

Le Gouvernement de la République française et le Gouvernement du Royaume de Suède, attachés au régime international de la non-prolifération, ainsi qu'aux objectifs de l'Agence Internationale de l'Energie Atomique (AIEA) tels que ceux-ci sont définis dans ses Statuts, ayant à l'esprit les divers engagements d'utilisation pacifique et de contrôle auxquels ils ont respectivement souscrit et ayant noté que les travaux de l'Agence Internationale de l'Energie Atomique sur le stockage international du plutonium (IPS) contribuent à une meilleure coopération internationale dans ce domaine, ont décidé de définir sur le plan bilatéral des dispositions relatives au plutonium issu du retraitement en France des combustibles irradiés en Suède.

A la suite des entretiens qui se sont déroulés à cette fin entre des représentants de nos deux pays, et me référant aux échanges de lettres du 10 juillet 1979, du 25 janvier et du 16 mai 1983 relatifs aux contrats de retraitement des combustibles irradiés en Suède conclus entre la Société COGEMA et la Société SKB, j'ai l'honneur de vous proposer, d'ordre de mon Gouvernement, les dispositions suivantes :

1.   Les dispositions suivantes s'appliquent aux combustibles irradiés en Suède, au plutonium, aux générations consécutives de ce plutonium ainsi qu'à d'autres matières nucléaires, issues du retraitement en France des combustibles susmentionnés.

2.   Le Gouvernement français s'engage à utiliser les matières nucléaires visées à l'Article 1er transférées en France exclusivement à des fins pacifiques et non explosives. Elles sont soumises aux garanties de l'Agence Internationale de l'Energie Atomique dans le cadre de l'Accord France-EURATOM-AIEA du 27 juillet 1978 et de ses arrangements subsidiaires.

3.   Le Gouvernement suédois s'engage à utiliser les matières nucléaires visées à l'Article 1er transférées en Suède exclusivement à des fins pacifiques et non explosives. Elles sont soumises aux garanties de l'Agence Internationale de l'Energie Atomique dans le cadre de l'Accord de garanties entre la Suède et l'Agence du 14 avril 1975 et de ses arrangements subsidiaires.

4.   Les deux Gouvernements s'engagent à soumettre les matières nucléaires visées à l'Article 1er à des mesures de protection physique conformément à leur législation nationale respective et aux arrangements internationaux auxquels ils ont souscrit. Les niveaux de protection physique sont au minimum ceux spécifiés à l'annexe B du document INFCIRC/254 AIEA. Chaque Partie se réserve le droit, conformément à sa réglementation nationale, d'appliquer sur son territoire des critères plus stricts de protection physique.

La mise en œuvre des mesures de protection physique relève de la responsabilité de chaque Partie Contractante à l'intérieur de sa juridiction.

5.   Le plutonium issu du retraitement par la COGEMA des combustibles irradiés en Suède est restitué à la Suède ou peut être vendu à un pays tiers, conformément aux dispositions de l'Article 9 ci-après, ou à la France selon les dispositions visées à l'Article 11.

6.   Le Gouvernement français s'engage à accorder les autorisations d'exportation pour les quantités de plutonium dont la destination finale en Suède est :

— l'alimentation de réacteurs électronucléaires ;
— le façonnage de combustibles pour l'alimentation de ces réacteurs ;
— l'utilisation à des fins de recherche dans des réacteurs ou laboratoires, y compris le stockage intermédiaire nécessaire au bon déroulement des opérations précitées.

7.   Les autorisations d'exportation sont accordées sur attestation du destinataire (formulaire type en annexe), spécifiant la destination finale, les quantités, l'échéancier approximatif de livraison, le calendrier d'utilisation, la forme sous laquelle la livraison aura lieu, ainsi que l'affectation de ces matières à l'une ou l'autre des utilisations susmentionnées.

Le Gouvernement suédois se porte garant auprès du Gouvernement français, à chaque livraison, des renseignements indiqués dans l'attestation mentionnée ci-dessus.

*Note: available in French only.

8. Le plutonium dont l'utilisation finale est prévue en Suède peut être transformé en éléments combustibles dans un pays tiers, si celui-ci a un accord avec la France sur le plutonium.

9. La Suède peut vendre des matières nucléaires visées à l'Article 1er à un pays tiers, soit directement à partir du territoire français, soit en les réexportant de Suède vers un pays tiers si les Gouvernements français et suédois ont donné leur accord préalable et si l'utilisation finale est identique à celles visées à l'Article 6. Dans ce cas, le Gouvernement du pays tiers acheteur doit donner aux deux Gouvernements les renseignements et garanties prévus à l'Article 7 ci-dessus.

10. Les matières nucléaires vendues par la Suède à un pays tiers conformément au paragraphe précédent peuvent être transformées en éléments combustibles dans un autre pays si celui-ci a un accord avec la France sur les matières nucléaires concernées.

11. Au cas où la France achèterait à la Suède le plutonium issu du retraitement des combustibles irradiés en Suède, les matières doivent être utilisées à des fins identiques à celles visées à l'Article 6 ci-dessus. Dans ce cas, le Gouvernement français donne au Gouvernement suédois les renseignements et garanties prévus à l'Article 7 ci-dessus.

12. Le Gouvernement suédois ainsi que le Gouvernement français notifient à l'AIEA, pour ce qui le concerne, les transferts de plutonium issu des combustibles irradiés en Suède.

13. Les deux Gouvernements se consultent pour tous les cas non prévus par cet échange de lettres ainsi que pour réexaminer celui-ci, au cas où un système de stockage international de plutonium serait mis en place dans le cadre de l'Agence Internationale de l'Energie Atomique.

14. Cet échange de lettres peut être amendé d'un commun accord par les deux Gouvernements. En tout état de cause, les dispositions des Articles 2 à 12 demeurent en vigueur aussi longtemps qu'existe l'une quelconque des matières nucléaires visées par ces Articles, à moins que les deux Parties n'en décident autrement.

Si les dispositions qui précèdent rencontrent l'agrément du Gouvernement du Royaume de Suède, j'ai l'honneur de proposer que la présente lettre et la réponse de Votre Excellence constituent un Accord entre nos deux Gouvernements qui prendra effet à la date de réponse de Votre Excellence.

Je vous prie, Monsieur l'Ambassadeur, d'agréer l'expression de ma très haute considération.

FORMULAIRE TYPE

Demande de transfert de plutonium

1. Retraiteur.

    1.1. Nom ou raison sociale ....
    1.2. Adresse ....

2. Destinataire.

    2.1. Nom ou raison sociale ....
    2.2. Adresse ....
    2.3. Activité principale ....

3. Nature de la livraison.

    3.1. Poids total du plutonium ....
    3.2. Poids du plutonium fissile ....
    3.3. Forme du plutonium ....
    3.4. Echéancier approximatif de livraison ....

4. Utilisation du plutonium.

    4.1. Fabrication de combustibles ....
        4.1.1. Nature de la fabrication ....
        4.1.2. Nom, raison sociale et adresse du fabricant ....
        4.1.3. Calendrier de fabrication ....

    4.2. Autres utilisations.

        4.2.1. Nature de l'utilisation ....
        4.2.2. Nom, raison sociale et adresse de l'utilisateur ....
        4.2.3. Calendrier d'utilisation ....

    4.3. Destination finale.

        4.3.1. Nature de l'utilisation finale ....
        4.3.2. Désignation de l'installation
        4.3.3. Nom, raison sociale et adresse de l'utilisateur final ....
        4.3.4. Calendrier d'utilisation finale ....

Je soussigné, certifie sincères et véritables les indications portées sur le présent formulaire.

        Date et lieu de signature :

        Signature :

        Nom et qualité du signataire :

AGREEMENT FOR CO-OPERATION
BETWEEN THE GOVERNMENT OF THE FEDERAL REPUBLIC OF GERMANY
AND THE GOVERNMENT OF THE PEOPLE'S REPUBLIC OF CHINA
CONCERNING PEACEFUL USES OF NUCLEAR ENERGY*

(9th May 1984)

The Government of the Federal Republic of Germany and the Government of the People's Republic of China —

On the basis of the amicable relations existing between the two States;

In view of their mutual interest in the development of the peaceful use of nuclear energy;

Supported by the Agreement on Scientific and Technological Co-operation concluded on 9th October 1978 between the Government of the Federal Republic of Germany and the Government of the People's Republic of China;

In view of the fact that the Federal Republic of Germany is Party to the Treaty on the Non-Proliferation of Nuclear Weapons, and a Member of the International Atomic Energy Agency;

In view of the fact that the People's Republic of China is a nuclear-weapon State and a Member of the International Atomic Energy Agency;

Desiring to extend and strengthen the co-operation in the field of nuclear energy;

Have agreed as follows:

Article 1

(1)   The Contracting Parties shall promote the co-operation between the Federal Republic of Germany and the People's Republic of China in the following areas of the peaceful uses of nuclear energy on the basis of non-discrimination and mutual benefit and in harmony with the laws and other regulations valid at the time in question in the national territory of each Contracting Party:

1.   Scientific research and technological development in the field of nuclear energy;

2.   Nuclear power technology;

3.   Safety of nuclear installations and radiation protection;

4.   Planning, construction and operation of nuclear power stations and research institutions;

5.   Other areas of mutual interest.

(2)   The content and extent of the co-operation as well as the concrete measures to be taken and the financial regulations to be laid down for the implementation will be the subject of special agreements to be concluded between the Contracting Parties or, with their consent, between other bodies within their territories.

---

*   Unofficial translation from *International Legal Materials*. The German text is published in Bundesgesetzblatt, II, no. 21, 23rd June 1984. The Agreement entered into force on the day of its signature.

Article 2

The co-operation between the Contracting Parties may take the following forms:

1. The exchange of scientific and technical personnel, e.g. reciprocal visits by scientists and engineers, holding of seminars, exchange of delegations and groups of experts;

2. Participation by scientists and engineers of one side in research and development activities of the other side;

3. Bilateral (or unilateral) consulting, and other technical services;

4. Joint research and joint construction;

5. Exchange of scientific information and documentation;

6. Other forms of co-operation to be agreed upon between the Parties.

Article 3

(1) The co-operation which is the subject of this present Agreement shall serve exclusively the use of nuclear energy for peaceful purposes. Nuclear material, nuclear equipment, substances and installations created specially for the purpose of the manufacture or use of nuclear material, as well as relevant technological information, which are transferred in the context of the co-operation based on this present Agreement or which are created in the course of said co-operation, shall not be utilised in any way which shall lead to the manufacture of a nuclear explosive device.

(2) Nuclear material, nuclear equipment, substances and installations created specially for the purpose of the manufacture or use of nuclear material, as well as relevant technological information, which are transferred in the context of the co-operation based on this Agreement between the Contracting Parties or which are created in the course of said co-operation shall only be transferred to a third country following prior consultations and with the mutual agreement of the Contracting Parties. In the event that the above-mentioned commodities are transferred to a third country, the Contracting Parties shall ensure that said third country meets the following requirements: solely peaceful use not leading to the creation of a nuclear explosive device, and acceptance of the safeguards measures stipulated by the International Atomic Energy Agency; the transfer of the above-mentioned commodities by the third country to another country shall not be possible without the mutual agreement of the Contracting Parties. If the third or other country is a Member State of the European Community, and if one Contracting Party has been informed beforehand about the retransfer, then the mutual consent is deemed granted. Commercial and patent regulations shall not be affected by this.

(3) Each Contracting Party shall guarantee the physical security in its own territory of the commodities mentioned in paragraph 2 in accordance with the standards stipulated in the Appendix, in order to prevent an unauthorised application or use. In the event of a transfer to a third country, each Contracting Party shall ensure, by means of an agreement concluded with said country, that a corresponding physical security is also guaranteed in said country.

Article 4

(1) The exchange of information shall take place between the Contracting Parties or between the agencies designated by them. Unless one of the Contracting Parties or an agency designated by it indicates, either prior to or during the exchange of information, that the transfer of the exchanged information is excluded or restricted, the other Contracting Party or an agency designated by it may pass on the information received to other agencies in its territory.

(2) Each Contracting Party shall ensure that the information exchanged or the information resulting from joint research or development is not publicised or divulged to third Parties without the written consent of the other Contracting Party, unless said third Party is authorised to receive said information in accordance with the present Agreement or in accordance with a special agreement concluded pursuant to Article 1 paragraph 2 of the present Agreement.

(3) The Contracting Parties shall endeavour to prevail upon the partners if the co-operation to inform each other about the degree of reliability and applicability of the information exchanged. The fact that the Contracting Parties may possibly be involved in the passing on of information in connection with this Agreement does not mean that they shall be liable for the correctness or applicability of said information.

(4) The disclosure of information with commercial value will be the subject of special agreements concluded pursuant to Article 1 paragraph 2 of the present Agreement.

(5) The provisions of this Article shall not apply to information which, on the basis of rights of or agreements with third Parties, may not be disclosed; nor shall they apply to information officially classified as secret, unless the competent authorities have given the Contracting Party in question their prior consent and an agreement on the procedure for the disclosure has been concluded.

Article 5

A Joint Committee comprising representatives of each of the Contracting Parties shall be set up to promote the co-operation on the basis of this Agreement and the conclusion of special agreements pursuant to Article 1 paragraph 2 of the present Agreement. The Committee shall meet as required at the proposal of a Contracting Party in order to examine the progress and results of the co-operation on the basis of the present Agreement, to deliberate on additional means of co-operation and, if applicable, to lay down work schedules, the duration of which shall depend on the circumstances.

Article 6

On the basis of non-discrimination and mutual benefit the Contracting Parties shall fulfil the prerequisites for the transfer of the results of their scientific and technological co-operation in the field of the peaceful use of nuclear energy.

Article 7

(1)   To the extent permitted by the laws and other regulations in force in their territories at the time in question the Contracting Parties shall grant the personnel exchanged in connection with this Agreement, as well as their next of kin living in their household, all possible assistance to facilitate their leaving and entering the respective countries, the obtaining of visas and residence permits, the import and export of their household effects, the exercise of their profession and the exemption from taxes.

(2)   Details of the above, and the treatment of material and equipment imported and exported for the purpose of the co-operation on the basis of the present Agreement, will be the subject of special agreements concluded pursuant to Article 1 paragraph 2 of this Agreement.

Article 8

Obligations incumbent upon the Contracting Parties on the basis of their respective international treaties shall remain unaffected, including obligations incumbent upon the Federal Republic of Germany on the basis of the foundation charters of the European Economic Community and the European Atomic Energy Community. The Contracting Parties should, however, avoid any effects of such obligations on the normal implementation of the present Agreement.

Article 9

The present Agreement shall also apply to Berlin (West) in harmony with the present situation.

Article 10

(1)   The present Agreement shall enter into force upon the date of signing.

(2)   The present Agreement shall apply for a period of fifteen years and shall be extended by five years respectively, unless such extension is excluded in writing by one of the Contracting Parties one year before expiry.

(3)   The duration of special agreements concluded pursuant to Article 1 paragraph 2 of the present Agreement shall remain unaffected by the expiry thereof. In the event that this Agreement shall become inoperative, the relevant clauses shall apply for as long as is necessary and to the extent necessary for the implementation of special agreements concluded pursuant to Article 1 paragraph 2 of this Agreement or for the completion of other co-operation projects already commenced in accordance with this Agreement. The provisions of this Agreement pertaining to the treatment of commodities transferred or created in the course of the co-operation shall remain unaffected by the expiry of the Agreement.

(4)   Amendments can be made to this Agreement at any time, with the consent of both Contracting Parties. They shall enter into force on the date of the relevant exchange of notes.

Done at Bonn on 9th May 1984 in duplicate, each being in German and Chinese, whereby both texts are equally authentic.

Appendix to Article 3 paragraph 3

The level of physical security to be ensured by the competent national authorities in the use, storage and transportation of the material listed in the attached table shall as a minimum meet the following requirements:

Category III

Use and storage within an area to which access is controlled. Transportation under special precautions including prior arrangements among sender, recipient and carrier, and prior agreement between entities subject to the jurisdiction and regulation of supplier and recipient States, respectively, in case of international transport, specifying time, place and procedures for transferring transport responsibility.

Category II

Use and storage within a protected area to which access is controlled, i.e., an area under constant surveillance by guards or electronic devices, surrounded by a physical barrier with a limited number of points of entry under appropriate control, or any area with an equivalent level of physical protection. Transportation under special precautions including prior arrangements among sender, recipient and carrier, and prior agreement between entities subject to the jurisdiction and regulation of supplier and recipient States, respectively, in case of international transport, specifying time, place and procedures for transferring transport responsibility.

Category I

Material in this category shall be protected with highly reliable systems against unauthorised uses as follows:

Use and storage within a highly protected area, i.e., a protected area as defined for category II above, to which, in addition, access is restricted to persons whose trustworthiness has been determined, and which is under surveillance by guards who are in close communication with appropriate response forces. Specific measures taken in this context should have as their objective the detection and prevention of any assault, unauthorised access or unauthorised removal of material.

Transportation under special safety precautions as identified above for transportation of categories II and III materials and, in addition, under constant surveillance by escorts and under conditions which assure close communication with appropriate response forces.

The Government of the Federal Republic of Germany and the Government of the People's Republic of China shall designate the agencies or authorities whose task it shall be to ensure that the level of protection is observed in an appropriate manner. Furthermore, it shall be the responsibility of said agencies or authorities to co-ordinate domestic emergency or replacement measures in the event of the unauthorised use or handling of protected material. The Government of the Federal Republic of Germany and the People's Republic of China shall designate contact offices within their respective authorities, which shall collaborate on questions of the transportation beyond national frontiers and other questions of mutual interest.

TABLE

Material	Form	Category		
		I	II	III
1. Plutonium[a]	Unirradiated[b]	2 kg or more	Less than 2 kg but more than 500 g	500 g or less[c]
2. Uranium 235	Unirradiated[b]	5 kg or more	Less than 5 kg but more than 1 kg	1 kg or less[c]
	— Uranium enriched to 20% ^{235}U or more	—		
	— Uranium enriched to 10% ^{235}U but less than 20%	—	10 kg or more	Less than 10 kg[c]
	— Uranium enriched above natural, but less than 10% ^{235}U[d]	—	—	10 kg or more
3. Uranium 233	irradiated[b]	2 kg or more	Less than 2 kg but more than 500 g	500 g or less
4. Irradiated	—	—		depleted or natural uranium, thorium or slightly enriched fuel (less than 10 % fissile content)[e, f]

a) All plutonium except that with isotopic concentration exceeding 80% in plutonium 238.
b) Material not irradiated in a reactor or material irradiated in a reactor but with a radiation level equal to or less than 100 rads/hour at one meter unshielded.
c) Less than a radiologically significant quantity should be exempted.
d) Natural uranium, depleted uranium and thorium and quantities of uranium enriched to less than 10% not falling in Category III should be protected in accordance with prudent management practice.
e) Although this level of protection is recommended, the Parties are free to use another category of physical protection, taking into account the respective circumstances.
f) Other types of fuel which, by virtue of their original fissile material content, were classed in category I or II before radiation, should only be reduced one Category level, while the radiation level from the fuel exceeds 100 rads/hour at one meter unshielded.

APPENDIX 12

AGREEMENT

**in the form of an exchange of letters between the European Atomic Energy
Community (EURATOM) and the Government of Canada intended to replace the
"Interim Arrangement concerning enrichment, reprocessing and subsequent
storage of nuclear material within the community and Canada" constituting
Annex C of the Agreement in the form of an exchange of letters of
16th January 1978 between EURATOM and the Government of Canada**

(18th December 1981)

A. Letter from the Government of Canada

Sir,

1. I have the honour to refer to the 16th January 1978 exchange of letters between the Government of Canada and the European Atomic Energy Community (EURATOM) (hereinafter referred to as the exchange of letters) amending the Agreement between the Government of Canada and the European Atomic Energy Community for co-operation in the peaceful uses of atomic energy of 6th October 1959, (hereinafter referred to as the Agreement) particularly in so far as it relates to safeguards (followed by an additional exchange of letters). I specifically refer to paragraph (e) of the exchange of letters which states that:
    "Material referred to in paragraph (c) shall be enriched beyond 20 per cent or reprocessed and plutonium or uranium enriched beyond 20 per cent shall be stored only according to conditions agreed upon in writing between the Parties (see Annex C: Interim Arrangement concerning enrichment, reprocessing and subsequent storage of nuclear material within the Community and Canada)."
    Paragraph 5 of Annex C states that the Parties would commence negotiations as soon as possible after 31st December 1979, or the termination of the INFCE study, whichever was earlier, with a view to replacing the Interim Arrangement by other arrangements that would take into account *inter alia* any results of the INFCE studies in relation to the operations in question.

2. These negotiations have now been completed and I have the honour to propose that the guidelines set forth below should cover reprocessing and plutonium storage and use:

    (a) an effective commitment to non-proliferation should have been made and should continue to be maintained by the Party envisaging reprocessing and plutonium storage and use;

    (b) all nuclear material subject to a peaceful uses commitment in facilities involved in reprocessing and the storage and use of plutonium should be subject to IAEA safeguards;

    (c) all nuclear material subject to a peaceful uses commitment in facilities involved in reprocessing and the subsequent storage and use activities, including related transport, should be subject to adequate physical protection measures;

    (d) mutually satisfactory notification and material reporting procedures should be in place between the Parties;

    (e) a description of the current and planned nuclear energy programme including in particular a detailed description of the policy, legal and regulatory elements relevant to reprocessing and plutonium storage and use should be provided by the Party envisaging such activities;

    (f) the Parties should agree to periodic and timely consultations at which *inter alia* the information provided under guideline (e) should be updated and significant changes in the nuclear energy programme would receive the fullest possible consideration;

    (g) the reprocessing and plutonium storage should only take place when the information provided on the nuclear energy programme of the Party in question has been received, when the undertakings, arrangements and other information

217

called for by the guidelines are in place or have been received and when the Parties have agreed that the reprocessing and plutonium storage are an integral part of the described nuclear energy programme; where it is proposed to carry out reprocessing or storage of plutonium when these conditions are not met, the operation should take place only when the Parties have so agreed after consultation, which should take place promptly to consider any such proposal;

(h) the reprocessing and plutonium storage envisaged should only take place so long as the commitment of the Party in question to non-proliferation does not change and so long as the commitment to periodic and timely consultations referred to in guideline (f) is honoured.

3. I note that Canada and the Community have agreed that the objectives of the above guidelines have been met.

In particular I note that Canada and the Community and its Member States, to the extent of their respective competences, have made an effective commitment to non-proliferation and have submitted all relevant material to IAEA safeguards, and to adequate physical protection measures, in paragraphs (c) and (g) of the exchange of letters completed by the letters from Member States' foreign ministers to Canadian ambassadors on physical protection. I also note that the Community has provided Canada with the description of the current and planned nuclear energy programmes of the Community and of its Member States and that the notification and material reporting procedures have been settled.

4. Finally I note that these arrangements take into account *inter alia* the results of INFCE's studies in relation to the operations in question, as envisaged in paragraph 5 of Annex C to the exchange of letters. I note that the Parties, in particular, acknowledge that the separation, storage, transportation and use of plutonium require particular measures to reduce the risk of nuclear proliferation; are determined to continue to support the development of international safeguards and other non-proliferation measures relevant to reprocessing and plutonium, including an effective and generally accepted international plutonium storage scheme; recognise the role of reprocessing in connection with the maximum use of available resources and the management of materials contained in spent fuel or other peaceful non-explosive uses, including research, in particular in the context of significant nuclear energy programmes; and desire the predictable and practical implementation of paragraph (e) of the exchange of letters taking into account both their determination to ensure the furtherance of the objective of non-proliferation and the long-term needs of the nuclear energy programmes of the Parties.

5. I have the honour to inform the Commission that pursuant to the above, in accordance with paragraph (e) of the exchange of letters, the Government of Canada agrees that material subject to the Agreement may be reprocessed and plutonium stored within the framework of the current and planned nuclear energy programme as described and up-dated from time to time by the Community and its Member States.

6. I have the honour to inform the Commission that the Agreement given by the Government of Canada in paragraph 5 will remain in force as long as the following conditions are met:

(i) that the Community maintain its commitment to non-proliferation with respect to guideline (a) which is set out in paragraph (c) of the exchange of letters, and

(ii) that the Community continue to consult with the Government of Canada, as provided for by the Agreement with a view to up-dating the described nuclear energy programmes and informing the Government of Canada of any significant changes.

7. Paragraph (e) of the exchange of letters provides that material subject to the Agreement shall be enriched beyond 20 per cent and that uranium enriched beyond 20 per cent shall be stored only according to conditions agreed upon in writing between the Parties. I have the honour to propose that the Parties agree to consult within 40 days of the receipt of a request from either Party to consider proposals for conditions to be agreed upon in writing according to which material subject to the Agreement may be enriched beyond 20 per cent or uranium enriched beyond 20 per cent may be stored.

8. I have the honour to confirm that the documents containing the descriptions of current and planned nuclear energy programmes of the Community and its Member States shall remain confidential to the Contracting Parties.

9. If the foregoing is acceptable to the European Atomic Energy Commission, I have the honour to propose that this letter, which is authentic in both English and French, together with Your Excellency's reply to that effect shall constitute the Agreement required by paragraph (e) of the exchange of letters, and replace both Annex C thereto and the 23rd December 1980 exchange of letters. This Agreement shall take effect as of the date of Your Excellency's reply to this letter.

APPENDIX 13

AGREEMENT

**in the form of an exchange of letters between the European Atomic Energy
Community (EURATOM) and the Government of Canada, amending the Agreement
between the European Atomic Energy Community (EURATOM) and the
Government of Canada of 6th October 1959 for Co-operation in the
Peaceful Uses of Atomic Energy**

(21st June 1985)

B. Letter from the Government of Canada

Sir,

I have the honour to acknowledge receipt of your letter of today's date which reads as follows:

"I refer to the Agreement between the European Atomic Energy Community (EURATOM) and the Government of
Canada for Co-operation in the Peaceful Uses of Atomic Energy, signed on 6th October 1959 and subsequently amended
by the exchange of letters of 16th January 1978 and 18th December 1981, hereinafter referred to as the "Agreement".

The nuclear relationship between EURATOM and Canada has grown significantly and undergone transformation since
1959. There is therefore some importance in updating the Agreement so that it should provide a more stable, predictable
and administratively effective legal framework for the expanded relationship between the Contracting Parties.

To this end, I have the honour to propose that the Agreement be updated and completed as follows:

1. Pursuant to Article XV(2) of the Agreement, after the intial period of 10 years, which expired on 17th November 1969,
either Contracting Party can terminate the Agreement at any time, subject to six months' notice. The Contracting
Parties hereby agree that the Agreement shall remain in force for a further period of 20 years from today's date. If
neither Contracting Party has notified the other Contracting Party of its intention to terminate the Agreement at least
six months prior to expiry of that period, the Agreement shall continue in force for additional periods of five years
each unless, at least six months before the expiration of any such additional period, a Contracting Party notifies the
other Contracting Party of its intention to terminate the Agreement.

2. Article IX(1) of the Agreement provides that the prior consent in writing of the Community or the Government of
Canada, as the case may be, is required for the transfer beyond the control of either Contracting Party of material
or equipment obtained pursuant to the Agreement or source or special nuclear material derived through the use of
such material or equipment. In order to facilitate the administration of the Agreement:

    (a) In the case of natural uranium, depleted uranium, other source material, uranium enriched to 20 per cent or
    less in the isotope U-235 and heavy water, Canada hereby provides its consent to the future retransfers of such
    items by the Community to third Parties, provided that:

        (i) such third Parties have been identified by Canada;

        (ii) procedures acceptable to both Contracting Parties relating to such retransfers shall be established;

    (b) retransfers to third Parties of material or equipment other than those referred to in (a) above, shall continue to
    require the prior written consent of Canada prior to the retransfer;

    (c) in the case of non compliance by EURATOM with the provisions in this paragraph, Canada shall have the right
    to terminate the arrangements made pursuant to this paragraph in whole or in part.

3. Further to Article IX(1) of the Agreement, Canada hereby provides its consent for the retransfer, in any given period
of 12 months, to any third Party, signatory to the NPT, of the following materials and quantities:

219

(a) special fissionable material (50 effective grams);

(b) natural uranium (500 kilograms);

(c) depleted uranium (1 000 kilograms), and

(d) thorium (1 000 kilograms).

The Joint Technical Working Group shall establish administrative arrangements for the purpose of reviewing the implementation of this provision.

4. With reference to paragraph (d) of the exchange of letters of 16th January 1978 amending the EURATOM/Canada Agreement of 1959, EURATOM agrees to waive the requirement for prior notification in cases where natural uranium, depleted uranium, other source materials, uranium enriched to 20 per cent or less in the isotope U-235 and heavy water are received by EURATOM from a third Party, identified in accordance with paragraph 2(a)(i) above, which has identified the item or the items as being subject to an Agreement with Canada. In such cases, the item or items shall become subject to the Agreement upon receipt.

5. The Contracting Parties may wish, in particular circumstances, to apply mechanisms other than those set forth in the Agreement in order to:

(a) make material subject to the Agreement, or

(b) remove material from coverage of the Agreement.

There shall be prior written agreement between the Contracting Parties in each case on the conditions under which such mechanisms are to be applied.

6. The Contracting Parties recognise that the programme provided for in Article II of the Agreement has been successfully carried out and brought to conclusion and re-affirm their commitment to mutual co-operation in nuclear research and development as laid down in Article I. They note that the list of fields of co-operation, set out in Article I, is illustrative and not exhaustive.

If the foregoing is acceptable to the Government of Canada, I have the honour to propose that this letter, which is authentic in both English and French, together with Your Excellency's reply to that effect shall constitute an agreement amending the Agreement. The present Agreement shall take effect as of the date of Your Excellency's reply to this letter."

I have the honour to inform you that the Government of Canada is in agreement with the contents of your letter, and to confirm that your letter and this reply, which is authentic in English and French, shall constitute an agreement amending the Agreement between the Government of Canada and the European Atomic Energy Community (EURATOM) of 6th October 1959, as amended, which shall enter into force on the date of this letter.

AGREED MINUTES

to the Agreement in the form of an exchange of letters between the European
Atomic Energy Community (EURATOM) and the Government of Canada, amending the
Agreement between the European Atomic Energy Community (EURATOM) and the
Government of Canada of 6th October 1959 for Co-operation in the Peaceful
Uses of Atomic Energy

1. Paragraph 2(a) of the present Agreement contemplates simplified procedures for transfers of nuclear items.

2. In implementation of such provision Canada shall provide the Community with, and keep up to date, the list of countries to which nuclear items can be transferred in accordance with the aforementioned provision. In identifying such countries Canada will take into account both the non-proliferation policy of the Canadian Government and requests made by the Community to cover its industrial and commercial interests. Canada will be prepared to consider any requests by the Community for the maintenance of any countries on the list or the inclusion of any additional countries on it.

3. During the negotiations on 19th and 20th November 1984, the Canadian delegation stated, with reference to paragraph 2(a)(ii) of the present Agreement, that Canada would use its best endeavours in discussions with other trading partners concerned progressively to simplify as far as possible, consistent with its non-proliferation policy, the notification and related procedures connected with retransfers. Canada's general aim is to establish a network of partner countries amongst which Canadian-origin nuclear material could circulate as easily as possible.

4. With reference to paragraph 5 of the present Agreement, the intention of the Contracting Parties would be, jointly and progressively, to develop a body of administrative precedents aimed at enabling individual cases to be treated expeditiously.

CO-OPERATION AGREEMENT BETWEEN THE GOVERNMENT OF THE ARGENTINE REPUBLIC AND THE GOVERNMENT OF THE FEDERATIVE REPUBLIC OF BRAZIL FOR THE DEVELOPMENT AND APPLICATION OF NUCLEAR ENERGY FOR PEACEFUL PURPOSES*

(17th May 1980)

The Government of the Argentine Republic and the Government of the Federative Republic of Brazil:

Motivated by the traditional friendship between their peoples and by the constant desire for extended co-operation that prompts their Governments;

Conscious of the right of all countries to develop and utilise nuclear energy for peaceful purposes, and likewise to possess the technology pertaining thereto;

Bearing in mind that the development of nuclear energy for peaceful purposes is a basic element in promoting the economic and social development of their peoples;

Bearing in mind the efforts that both countries have been making to enlist the service of nuclear energy for the needs of their economic and social development;

Convinced that co-operation in the use of nuclear energy for peaceful purposes will contribute to the development of Latin America;

Convinced of the need to prevent the proliferation of nuclear weapons through non-discriminatory measures that impose restrictions aimed at securing general and complete nuclear disarmament under strict international control;

Taking into consideration the objectives of the Treaty for the Prohibition of Nuclear Weapons in Latin America — the Tlatelolco Treaty;

Taking also into consideration the Scientific and Technological Co-operation Agreement signed on the same date;

Have decided to conclude this Co-operation Agreement for the development and use of nuclear energy for peaceful purposes.

Article I

The Parties shall co-operate in the development and use of nuclear energy for peaceful purposes in accordance with the requirements and priorities of their respective national nuclear programmes and with regard for the international commitments undertaken by the Parties.

Article II

The Parties shall designate the relevant competent bodies for implementing the co-operation envisaged in this Treaty.

---

* Unofficial translation by the Secretariat.

Article III

1.   The envisaged co-operation shall be applied in the following fields:

    a)   study, development and technology of experimental and power reactors, including nuclear power plants;

    b)   nuclear fuel cycle, including the search for and mining of nuclear minerals and the fabrication of fuel elements;

    c)   industrial production of materials and equipment, as well as provision of services;

    d)   production of radioisotopes and their applications;

    e)   radiological protection and nuclear safety;

    f)   physical protection of nuclear material;

    g)   basic and applied research relating to the peaceful uses of nuclear energy;

    h)   any other scientific and technological aspects of the peaceful use of nuclear energy that the Parties consider of mutual interest.

2.   Co-operation in the fields indicated in Section 1 shall be implemented in the form of:

    a)   mutual assistance in the instruction and training of scientific and technical personnel;

    b)   exchange of experts;

    c)   exchange of instructors for courses and seminars;

    d)   study fellowships;

    e)   mutual consultations on scientific and technological matters;

    f)   formation of joint working groups to carry out specific studies and projects for scientific research and technological development;

    g)   mutual supply of equipment, materials and services related to the areas indicated above;

    h)   exchange of information relating to the areas indicated above;

    i)   any other type of work that may be agreed upon under the terms of Article IV.

Article IV

In order to put into effect the collaboration envisaged in this Agreement, the competent bodies designated by each of the Parties shall conclude application agreements in which the specific conditions and procedures for co-operation, including the holding of joint technical meetings to study and evaluate programmes, shall be set forth. The competent bodies of each of the Parties shall likewise set up joint agencies for the purpose of technical and economic management of the programmes and projects agreed on, with promotion, whenever such is appropriate, of the participation of private legal entities in those agencies.

Article V

The Parties shall make free use of all information exchanged under this Agreement, except in cases in which the Party supplying the information has imposed restrictions or reservations regarding its use or dissemination. If the information exchanged is protected by patents registered with either of the Parties, the terms and conditions for its use and dissemination shall be subject to the normal legislation.

Article VI

The Parties shall facilitate mutual supply in regard to the transfer, loan, renting and sale of nuclear materials, equipment and services required for the implementation of the joint programmes and national development plans in the area of the peaceful uses of atomic energy, such operations being in all cases subject to the legal provisions in force in the Argentine Republic and the Federative Republic of Brazil.

Article VII

1.   Any material or equipment supplied by one of the Parties to the other, or any material derived from the use of the above or utilised in an item of equipment supplied under this Agreement shall be used solely for peaceful purposes. The Parties shall consult together on the application of safeguards procedures to materials and equipment supplied within the scope of this Agreement.

2.   In order to apply the safeguards procedures referred to in paragraph 1, the Parties shall, as appropriate, conclude the relevant safeguards agreements with the International Atomic Energy Agency.

Article VIII

The Parties undertake to co-operate on a reciprocal basis in the development of joint projects to be implemented in accordance with this Agreement, and to facilitate in every possible way any collaboration that may be necessary in those projects with other public and private institutions or organisations in their respective countries.

Article IX

The Parties shall consult together on situations of mutual interest that may arise on an international level in connection with the use of nuclear energy for peaceful purposes so as to co-ordinate their stand, whenever such is advisable.

Article X

The Parties shall act in such a way that any differences of opinion arising with regard to interpretation and application of this Agreement are settled through diplomatic channels.

Article XI

1.   The present Agreement shall enter into force on the date on which the exchange of instruments of ratification, which shall take place in Brazil, is effected, and shall have an initial validity of ten years, with automatic extension for successive periods of two years, unless six months before the expiry of any of those periods one Party notifies the other of its intention not to renew it.

2.   The termination of this Agreement shall not affect continuation of implementation of any application agreements that may have been concluded in conformity with the provisions of Article IV.

3.   The present Agreement shall apply provisionally from the date of its signature within the sphere of competence of the authorities responsible for its implementation.

DONE in Buenos Aires on the seventeenth of May nineteen hundred and eighty, in two original copies, in Spanish and Portuguese, both texts being equally authentic.

BETWEEN THE GOVERNMENT OF AUSTRALIA
AND THE GOVERNMENT OF THE SWISS CONFEDERATION
CONCERNING THE PEACEFUL USES OF NUCLEAR ENERGY

(28th January 1986)

The Government of Australia and the Government of the Swiss Confederation:

Reaffirming their commitment to ensuring that the international development and use of nuclear energy for peaceful purposes are carried out under arrangements which will further the objective of the non-proliferation of nuclear weapons;

Mindful that both Australia and Switzerland are non-nuclear-weapon States which are Parties to the Treaty on the Non-Proliferation of Nuclear Weapons, done at London, Moscow and Washington on 1st July 1968 (hereinafter referred to as *the Treaty*);

Recognising that Australia and Switzerland have under the Treaty undertaken not to manufacture or otherwise acquire nuclear weapons or other nuclear explosive devices and that both Governments have concluded agreements with the International Atomic Energy Agency (hereinafter referred to as *the Agency*) for the application of safeguards in their respective countries in connection with the Treaty;

Affirming their support for the objectives of the Treaty and their desire to promote universal adherence to the Treaty;

Confirming the desire of both countries to co-operate in the development and application of nuclear energy for peaceful purposes;

Desiring to establish conditions consistent with their commitment to non-proliferation under which nuclear material, material, equipment and technology can be transferred between Australia and Switzerland for peaceful non-explosive purposes;

Have agreed as follows:

Article I

For the purposes of this Agreement:

a) *appropriate authority* means, in the case of Australia, the *Australian Safeguards Office* and, in the case of Switzerland, the *Federal Office of Energy*, or such other authority as the Party concerned may from time to time notify the other Party;

b) *equipment* means the items and major components thereof specified in Part B of Annex A;

c) *material* means the non-nuclear material for reactors specified in Part A of Annex A;

d) *nuclear material* means any *source material* or *special fissionable material* as those terms are defined in Article XX of the Statute of the Agency. Any determination by the Board of Governors of the Agency under Article XX of the Agency's Statute which amends the list of materials considered to be *source material* or *special fissionable material* shall only have effect under this Agreement when both Parties to this Agreement have informed each other in writing that they accept such amendment;

e) *recommendations of the Agency* in relation to physical protection means the recommendations of document INFCIRC/225/Rev.1 entitled *The Physical Protection of Nuclear Material* as updated from time to time or any subsequent document replacing INFCIRC/225/Rev.1. Modifications of the recommendations for physical protection shall only have effect under this Agreement when both Parties have informed each other in writing that they accept such modifications;

f) *technology* means technical data in physical form including technical drawings, photographic negatives and prints, recordings, design data and technical and operating manuals designated by the supplying Party after consultations with the recipient Party prior to the transfer as important for the design, construction, operation and maintenance of enrichment, reprocessing or heavy water production facilities or major critical components thereof and any other technology as may be agreed between the Parties, but excluding data available to the public, for example in published books and periodicals, or that which has been made available internationally without restrictions upon its further dissemination.

Article II

1. The Parties shall facilitate co-operation in the development and application of nuclear energy for peaceful purposes including:

   a) the production of energy through the operation of the nuclear fuel cycle;

   b) research and its applications; and

   c) industrial co-operation.

2. The co-operation envisaged in this Article shall be effected on terms and conditions to be agreed and in accordance with this Agreement and with the laws, regulations and licensing requirements in force in Australia and in Switzerland respectively. The Parties may designate governmental authorities and natural or legal persons to undertake such co-operation.

Article III

1. This Agreement shall apply to:

   a) nuclear material, material, equipment or technology transferred between Australia and Switzerland for peaceful non-explosive purposes whether directly or through a third country;

   b) all forms of nuclear material prepared by chemical or physical processes or isotopic separation provided that the quatity of nuclear material so prepared shall only be regarded as falling within the scope of this Agreement in the same proportion as the quantity of nuclear material used in its preparation, and which is subject to this Agreement, bears to the total quantity of nuclear material so used;

   c) all generations of nuclear material produced by neutron irradiation provided that the quantity of nuclear material so produced shall only be regarded as falling within the scope of this Agreement in the same proportion as the quantity of nuclear material which is subject to this Agreement and which is used in its production, contributes to this production;

   d) equipment designed or produced by the use or by the application of technology subject to this Agreement;

   e) equipment for enrichment, reprocessing or heavy water production, the design, construction or operating processes of which are essentially of the same type as equipment subject to this Agreement and which is constructed within 20 years of the first operation of such equipment;

   f) material produced by equipment subject to this Agreement;

   g) nuclear material produced or processed in or used in connection with material or equipment subject to this Agreement.

2. The items referred to in paragraph 1 of this Article shall be transferred pursuant to this Agreement only to a natural or legal person identified by the appropriate authority of the recipient Party to the appropriate authority of the supplier Party as duly authorised to receive those items.

Article IV

1. Nuclear material referred to in Article III shall remain subject to the provisions of this Agreement until:

   a) it is determined that it is no longer usable; or

   b) it is determined that it is practicably irrecoverable for processing into a form in which it is usable for any nuclear activity relevant from the point of view of safeguards; or *c)* it has been transferred beyond the jurisdiction of Australia or beyond the jurisdiction of Switzerland in accordance with the provisions of Article IX of this Agreement; or

d) it is otherwise agreed between the Parties.

2. For the purpose of determining when nuclear material subject to this Agreement is no longer usable or is no longer practically recoverable for processing into a form in which it is usable for any nuclear activity relevant from the point of view of the safeguards referred to in Article VI both Parties shall accept a determination made by the Agency. For the purpose of this Agreement such determination shall be made by the Agency in accordance with the provisions for the termination of safeguards of the relevant safeguards agreement between the Party concerned and the Agency.

3. Material and equipment referred to in Article III shall remain subject to the provisions of this Agreement until:

a) it has been transferred beyond the jurisdiction of Australia or beyond the jurisdiction of Switzerland in accordance with the provisions of Article IX of this Agreement; or

b) it is otherwise agreed between the Parties.

4. Technology shall remain subject to this Agreement for a period jointly determined by the Parties prior to the transfer.

Article V

Nuclear material, material, equipment and technology subject to this Agreement shall not be used for, or diverted to the manufacture of nuclear weapons or other nuclear explosive devices, research on or development of nuclear weapons or other nuclear explosive devices, or be used for any *military* purpose.

Article VI

1. Where Australia is the recipient, compliance with Article V of this Agreement shall be ensured by a system of safeguards applied by the Agency in accordance with the Safeguards Agreement concluded on 10th July 1974 between Australia and the Agency in connection with the Treaty.

2. Where Switzerland is the recipient, compliance with Article V of this Agreement shall be ensured by a system of safeguards applied by the Agency in accordance with the Safeguards Agreement concluded on 6th September 1978 between Switzerland and the Agency in connection with the Treaty.

Article VII

If, notwithstanding the provisions of Article VI of this Agreement, nuclear material, material, equipment or technology subject to this Agreement is present in the territory of a Party and the Agency is not administering safeguards in the territory of that Party under the applicable Agreement concluded in accordance with Article III of the Treaty and referred to in Article VI of this Agreement, that Party shall accept safeguards under an agreement or agreements to which it and the Agency are Parties and which provide safeguards equivalent in scope and effect to those provided by the applicable Agreement referred to in Article VI of this Agreement, or, if the Agency is not administering safeguards in the territory of that Party under an agreement or agreements referred to above, the Parties shall forthwith enter into an agreement for the application of a safeguards system in the territory concerned which conforms with the principles and procedures of the Agency's safeguards system and which provides for the application of safeguards to nuclear material, material, equipment and technology subject to this Agreement.

Article VIII

1. Each Party shall take in accordance with its laws and regulations such measures as are necessary to ensure adequate physical protection of nuclear material, material, equipment and technology within its jurisdiction. In regard to nuclear material the Parties shall apply, as a minimum, measures of physical protection which satisfy the requirements of the recommendations of the Agency.

2. The Parties shall consult at the request of either Party concerning matters relating to physical protection, including the application for the purposes of this Article of recommendations which may be made from time to time by international expert groups.

Article IX

Nuclear material, material, equipment and technology subject to this Agreement shall not be transferred beyond the jurisdiction of a Party without the prior written consent of the other Party.

Article X

Nuclear material subject to this Agreement shall only be reprocessed according to conditions agreed upon in writing between the Parties, as set out in Annex B.

Article XI

Nuclear material subject to this Agreement shall not be enriched to 20 per cent or greater in the isotope U-235 without the prior written consent of the supplier Party.

Article XII

1. In applying Articles IX, X and XI of this Agreement, the supplier Party will take into account non-proliferation considerations and nuclear energy requirements of the recipient Party. The supplier Party shall not withhold its agreement for the purpose of securing commercial advantage.

2. If a Party considers that it is unable to grant consent to a matter referred to in Articles IX, X and XI of this Agreement, that Party shall provide the other Party with an immediate opportunity for full consultation on that issue.

Article XIII

1. The appropriate authorities of both Parties shall consult annually, or at any other time at the request of either Party, to ensure the effective implementation of this Agreement. The Parties may jointly invite the Agency to participate in such consultations.

2. If nuclear material subject to this Agreement is present in the territory of a Party, that Party shall, upon the request of the other Party, inform the other Party in writing of the overall conclusions of the most recent reports by the Agency on its verification activities in the territory of the requested Party for the facilities concerned.

3. The appropriate authorities of both Parties shall establish an administrative arrangement to ensure the effective fulfilment of the obligations of this Agreement. An administrative arrangement established pursuant to this paragraph may be changed with the agreement of the appropriate authorities of both Parties.

4. The cost of reports and records which either Party is required to provide pursuant to the administrative arrangement referred to in paragraph 3 of this Article shall be borne by the Party which is required to provide that report or record.

5. The Parties shall take all appropriate precautions in accordance with their laws and regulations to preserve the confidentiality of commercial and industrial secrets and other confidential information received as a result of the operation of this Agreement and designated as such by the supplier Party.

Article XIV

In the event of non-compliance by the recipient Party, with any of the provisions of Articles V to XIII inclusive or of Article XV of this Agreement, or of non-compliance with, or repudiation of, Agency safeguards arrangements by the recipient Party, the supplier Party shall have the right, subject to prior notification, to suspend or cancel further transfers of nuclear material, material, equipment and technology, and to require the recipient Party to take corrective steps. If, following consultation between the Parties, such corrective steps are not taken within a reasonable time, the supplier Party shall thereupon have the right to require the return of nuclear material, material and equipment subject to this Agreement, against payment therefor at prices then current. In the event of detonation by a Party of a nuclear explosive device, the above provisions shall also apply.

Article XV

Any dispute arising out of the interpretation or application of this Agreement which is not settled by negotiation shall, at the request of either Party, be submitted to an arbitral tribunal which shall be composed of three arbitrators appointed in accordance with the provisions of this Article. Each Party shall designate one arbitrator who may be its national and the two arbitrators so designated shall elect a third, a national of a third state, who shall be the Chairman. If within thirty days of the request for arbitration either Party has not designated an arbitrator, either Party to the dispute may request the President of the International Court of Justice to appoint an arbitrator. The same procedure shall apply if, within thirty days of the designation or appointment of the second arbitrator, the third arbitrator has not been elected. A majority of the members of the tribunal shall constitute a quorum. All decisions shall be made by majority vote of all the members of the arbitral tribunal. The arbitral procedure shall be fixed by the tribunal. The decisions of the tribunal, including all rulings concerning its constitution, procedure, jurisdiction and the division of the expenses of arbitration between the Parties, shall be binding on both Parties and shall be implemented by them.

Article XVI

1. This Agreement may be amended or revised by agreement between the Parties.

2. Any amendment or revision shall enter into force on the date the Parties, by exchange of diplomatic notes, specify for its entry into force.

Article XVII

This Agreement shall enter into force on the date the Parties, by an exchange of diplomatic notes, specify for its entry into force, and shall remain in force for an initial period of 30 years. If neither Party has notified the other Party at least 180 days prior to the expiry of such period, the present Agreement shall continue in force thereafter until 180 days after notice of termination has been given by either Party to the other Party; provided, however, that unless otherwise agreed between the Parties termination of this Agreement shall not release the Parties from obligations under this Agreement in respect of items referred to in Article III of this Agreement which remain usable or practicably recoverable for processing into a form in which they are usable for any nuclear activity relevant from the point of view of safeguards in accordance with Article IV of this Agreement.

In witness whereof the undersigned, being duly authorised thereto by their respective Governments, have signed this Agreement.

Done in duplicate in the English and French languages, both texts having equal validity, at Berne this twenty-eighth day of January 1986.

Annex A

PART A: MATERIAL

1. *Deuterium and heavy water:*

   Deuterium and any deuterium compound in which the ratio of deuterium to hydrogen exceeds 1:5000 for use in a nuclear reactor as defined in paragraph 1 of Part B below in quantities exceeding 200 kg of deuterium atoms in any period of 12 months.

2. *Nuclear grade graphite:*

   Graphite having a purity level better than 5 parts per million boron equivalent and with a density greater than 1.50 grams per cubic centimetre in quantities exceeding 30 metric tonnes in any period of 12 months.

PART B: EQUIPMENT

1. *Nuclear reactors:*

   Nuclear reactors capable of operation so as to maintain a controlled self-sustaining fission chain reaction, excluding zero energy reactors, the latter being defined as reactors with a designed maximum rate of production of plutonium not exceeding 100 grams per year.

A *nuclear reactor* basically includes the items within or attached directly to the reactor vessel, the equipment which controls the level of power in the core, and the components which normally contain or come in direct contact with or control the primary coolant of the reactor core.

It is not intended to exclude reactors which could reasonably be capable of modification to produce significantly more than 100 grams of plutonium per year. Reactors designed for sustained operation at significant power levels, regardless of their capacity for plutonium production, are not considered as *zero energy reactors*.

2. *Reactor pressure vessels*:

Metal vessels, as complete units or as major shop-fabricated parts therefor, which are especially designed or prepared to contain the core of a nuclear reactor as defined in paragraph 1 of Part B of this Annex and are capable of withstanding the operating pressure of the primary coolant.

A top plate for a reactor pressure vessel is a major shop-fabricated part of a pressure vessel.

3. *Reactor internals*:

For example, support columns and plates for the core and other vessel internals, control rod guide tubes, thermal shields, baffles, core grid plates, diffuser plates, etc.

4. *Reactor fuel charging and discharging machines*:

Manipulative equipment especially designed or prepared for inserting or removing fuel in nuclear reactors as defined in paragraph 1 of Part B of this Annex capable of on-load operation or employing technically sophisticated positioning or alignment features to allow complex off-load fuelling operations such as those in which direct viewing of or access to the fuel is not normally available.

5. *Reactor control rods*:

Rods especially designed or prepared for the control of the reaction rate in a nuclear reactor as defined in paragraph 1 of Part B of this Annex.

This item includes, in addition to the neutron absorbing part, the support or suspension structures therefor if supplied separately.

6. *Reactor pressure tubes*:

Tubes which are especially designed or prepared to contain fuel elements and the primary coolant in a reactor as defined in paragraph 1 of Part B of this Annex at an operating pressure in excess of 50 atmospheres.

7. *Zirconium tubes*:

Zirconium metal and alloys in the form of tubes or assemblies of tubes, and in quantities exceeding 500 kg per year, especially designed or prepared for use in a reactor as defined in paragraph 1 of Part B of this Annex and in which the relationship of hafnium to zirconium is less than 1:500 parts by weight.

8. *Primary coolant pumps*:

Pumps especially designed or prepared for circulating liquid metal as primary coolant for nuclear reactors as defined in paragraph 1 of Part B of this Annex.

9. *Plants for the reprocessing of irradiated fuel elements, and equipment especially designed or prepared therefor*:

A *plant for the reprocessing of irradiated fuel elements* includes the equipment and components which normally come in direct contact with and directly control the irradiated fuel and the major nuclear material and fission product processing streams. In the present state of technology, the following items of equipment are considered to fall within the meaning of the phrase *and equipment especially designed or prepared therefor*. These items are:

a) Irradiated fuel element chopping machines:

Remotely operated equipment especially designed or prepared for use in a reprocessing plant as identified above and intended to cut, chop or shear irradiated nuclear fuel assemblies, bundles or rods; and

b) Critically safe tanks (e.g. small diameter, annular or slab tanks) especially designed or prepared for use in a reprocessing plant as identified above, intended for dissolution of irradiated nuclear fuel and which are capable of withstanding hot, highly corrosive liquid, and which can be remotely loaded and maintained.

10. *Plants for the fabrication of fuel elements*:

A *plant for the fabrication of fuel elements* includes the equipment:

  a) which normally comes in direct contact with, or directly processes, or controls, the production flow of nuclear material, or

  b) which seals the nuclear material within the cladding.

The whole set of items for the foregoing operations, as well as individual items intended for any of the foregoing operations, and for other fuel fabrication operations, such as checking the integrity of the cladding or the seal, and the finish treatment to the sealed fuel.

11. *Equipment, other than analytical instruments, especially designed or prepared for the separation of isotopes of uranium*:

Equipment, other than analytical instruments, especially designed or prepared for the separation of isotopes of uranium includes each of the major items of equipment especially designed or prepared for the separation process. Such items include:

  — gaseous diffusion barriers
  — gaseous diffuser housings
  — gas centrifuge assemblies, corrosion-resistant to UF_6
  — jet nozzle separation units
  — vortex separation units
  — large UF_6 corrosion-resistant axial or centrifugal compressors
  — special compressor seals for such compressors.

12. *Plants for the production of heavy water*:

A *plant for the production of heavy water* means a plant for the production of heavy water, deuterium and deuterium compounds and equipment especially designed or prepared therefor.

Annex B

REPROCESSING

Whereas Article X of the Agreement provides that nuclear material subject to the Agreement (hereinafter referred to as NMSA) shall be reprocessed only according to conditions agreed upon in writing between the Parties:

The Parties to the Agreement:

  — acknowledge that the separation, storage, transportation and use of plutonium require particular measures to reduce the risk of nuclear proliferation;
  — recognising the role of reprocessing in connection with efficient energy use, management of materials contained in spent fuel or other peaceful non-explosive uses including research;
  — desiring predictable and practical implementation of the agreed conditions set out in this Annex, taking into account the shared non-proliferation objectives of the Parties and the long term needs of the nuclear fuel cycle programs of the recipient Party;
  — determined to continue to support the development of international institutional arrangements relevant to reprocessing and plutonium, including an effective and generally accepted international plutonium storage scheme;

Have agreed as follows:

Article 1

NMSA may be reprocessed subject to the following conditions:

  A) reprocessing shall take place under Agency safeguards for the purpose of energy use and management of materials contained in spent fuel, within the nuclear fuel cycle program as delineated and recorded in an implementing arrangement;

B) the separated plutonium shall be stored and used under Agency safeguards in accordance with the nuclear fuel cycle program as delineated and recorded in an implementing arrangement;

C) reprocessing and use of the separated plutonium for other peaceful non-explosive purposes including research shall take place only under conditions agreed upon in writing between the Parties following consultations under Article 2 of this Annex.

Article 2

Consultations shall be held within 30 days of the receipt of a request from either Party:

A) to review the operation of the provisions of this Annex;

B) to consider amendments to an implementing arrangement as provided therein;

C) to take account of improvements in international safeguards and other control techniques including the establishment of new and generally accepted international mechanisms relevant to reprocessing and plutonium;

D) to consider amendments to this Annex proposed by either Party, in particular to take account of the improvements referred to in paragraph (C) of this Article;

E) to consider proposals for reprocessing and use of the separated plutonium for other peaceful non-explosive purposes including research as referred to in Article 1(C) of this Annex.

Article 3

This Annex may be amended in accordance with Article XVI of the Agreement.

Appendices 16 to 22

URANIUM

RESSOURCES, PRODUCTION AND DEMAND

REASONABLY ASSURED RESOURCES*

(1 000 Tonnes U) — 1st January 1987

Country	Cost Ranges		Total 130/kg U or less
	$ 80/kg U or less	$ 80-130/kg U	
Algeria[2, 5]	26.00	0.00	26.00
Argentina	9.30	2.60	11.90
Australia	462.00	56.00	518.00
Brazil[1, 8]	163.05	0.00	163.05
Canada	153.00	96.00	249.00
Centr. African Rep.[1, 7]	8.00	8.00	16.00
Denmark[3]	0.00	27.00	27.00
Finland	0.00	1.50	1.50
France	53.76	11.39	65.15
Gabon[8, 10]	14.90	4.65	19.55
Germany, F.R.	0.80	4.00	4.80
Greece	0.40	0.00	0.40
India[8]	34.73	10.96	45.69
Italy	4.80	0.00	4.80
Japan	0.00	6.60	6.60
Mexico[7]	4.50	3.24	7.74
Namibia[6]	97.30	16.00	113.30
Niger	173.71	2.20	175.91
Peru[1]	0.00	1.52	1.52
Portugal	7.10	1.40	8.50
Somalia[1, 4]	0.00	6.60	6.60
South Africa[8]	247.07	102.10	349.17
Spain[1]	26.70	6.20	32.90
Sweden[9]	2.00	37.00	39.00
Turkey[1]	0.00	3.90	3.90
United States	124.00	274.00	398.00
Zaire[1, 11]	1.80	0.00	1.80
Total (rounded)	1 615.00	683.00	2 298.00
Total (adjusted)[12]	1 555.00	678.00	2 233.00

1. In situ resources.
2. Mineable resources.
3. Equivalent to recoverable resources.
4. OECD/NEA-IAEA: Uranium — Resources, Production and Demand, Paris, 1979.
5. OECD/NEA-IAEA: Uranium — Resources, Production and Demand, Paris, 1982.
6. OECD/NEA-IAEA: Uranium — Resources, Production and Demand, Paris, 1983, adjusted for production and estimated production.
7. OECD/NEA-IAEA: Uranium — Resources, Production and Demand, Paris, 1986.
8. OECD/NEA-IAEA: Uranium — Resources, Production and Demand, Paris, 1986, adjusted for production and estimated production.
9. Includes 35 000 t U in the Ranstad deposit from which no uranium production is allowed due to a veto by local authorities for environmental reasons.
10. No information on recoverability available.
11. Contains 1 000 t U in the Cu-Co-deposit Kolwezi.
12. To account for mining and milling losses, not incorporated in certain estimates.

* *Source: Uranium — Resources, Production and Demand*, a joint report by OECD/NEA and IAEA, OECD, Paris, 1988.

APPENDIX 17

ESTIMATED ADDITIONAL RESOURCES — CATEGORY I*

(1 000 Tonnes U — 1st January 1987)

Country	Cost Ranges		Total 130/kg U or less
	$ 80/kg U or less	$ 80-130/kg U	
Argentina	0.80	3.10	3.90
Austria[6]	0.70	1.00	1.70
Australia	257.00	127.00	384.00
Brazil[2, 7]	92.39	0.00	92.39
Canada	112.00	99.00	211.00
Chile[2, 4]	0.00	0.00	0.30
Denmark[5]	0.00	16.00	16.00
Finland	0.00	2.90	2.90
France	21.19	16.88	38.07
Gabon[8, 10]	1.30	8.30	9.60
Germany, F.R.	1.60	5.70	7.30
Greece	6.00	0.00	6.00
India	2.12	14.49	16.61
Indonesia[1]	0.00	7.31	7.31
Italy	0.00	1.30	1.30
Mexico[8]	0.00	2.98	2.98
Namibia[7]	30.00	23.00	53.00
Niger	283.60	16.70	300.30
Peru[2]	0.00	1.80	1.80
Portugal	1.45	0.00	1.45
Somalia[2, 5]	0.00	3.40	3.40
South Africa[8]	97.50	27.10	124.60
Spain[2]	9.00	0.00	9.00
Sweden[9]	1.00	45.30	46.30
Turkey[2]	0.00	3.20	3.20
United States[11]	0.00	0.00	0.00
Zaire[3]	1.70	0.00	1.70
Total (rounded)	919.00	426.00	1 346.00
Total (adjusted)[12]	891.00	425.00	1 316.00

1. As recoverable resources.
2. In situ resources.
3. In mineable resources.
4. No information on cost category available.
5. Equivalent to recoverable resources.
6. OECD/NEA-IAEA: Uranium — Resources, Production and Demand, Paris, 1979.
7. OECD/NEA-IAEA: Uranium — Resources, Production and Demand, Paris, 1982.
8. OECD/NEA-IAEA: Uranium — Resources, Production and Demand, Paris, 1986.
9. Includes 40 000 t U in the Ranstad deposit from which no uranium production is allowed due to a veto by local authorities for environmental reasons.
10. No information on recoverability available.
11. Estimates not separated into EAR-I and EAR-II, listed under EAR-II in OECD/NEA-IAEA: Uranium—Resources, Production and Demand, Paris 1986.
12. To account for mining and milling losses not incorporated in certain estimates.

* Source: Uranium — Resources, Production and Demand, a joint report by OECD/NEA and IAEA, OECD, Paris, 1988.

URANIUM PRODUCTION*

(Tonnes U)

Country	Pre 1981	1981	1982	1983	1984	1985	1986	Total to 1986	Expected 1987
Argentina	857	123	155	179	129	126	173	1 742	150
Australia	11 297	2 922	4 422	3 211	4 324	3 206	4 154	33 536	4 000
Belgium[1]	20	40	45	45	40	40	40	270	40
Brazil	0	4	242	189	117	115[3]	115[3]	782[3]	115[3]
Canada	138 650	7 720	8 080	7 140	11 170	10 880	11 720	195 360	11 700
Finland	30	0	0	0	0	0	0	30	0
France	34 312	2 552	2 859	3 271	3 168	3 189	3 247	52 598	3 225
Gabon	11 160	1 022	970	1 006	918	940[3]	900[3]	16 916[3]	900[3]
Germany, F.R.	385[2]	36	34	47	33	31	22	588	40
India	3 200	200	200	200	200	200[3]	200[3]	4 400[3]	200[3]
Japan	50	3	5	4	4	7	6	79	7
Mexico	42	0	0	0	0	0	0	42	0
Namibia	13 573	3 971	3 776	3 719	3 700[3]	3 400[3]	3 300[3]	35 439[3]	3 200[3]
Niger	17 545	4 363	4 259	3 426	3 276	3 181	3 110	39 160	2 960
Portugal	2 321	102	113	104	115	119	110	2 984	120
South Africa	93 586	6 131	5 816	6 060	5 721	4 900[3]	4 600[3]	126 814[3]	4 500[3]
Spain	1 390	178	150	170	196	201	215	2 500	200
Sweden	200	0	0	0	0	0	0	200	0
United States	266 700	14 800	10 300	8 100	5 700	4 400	5 200	315 200	4 200
Zaire	25 600	0	0	0	0	0	0	25 600	0
Total	620 918	44 167	41 426	36 871	38 811	34 935	37 112	854 240	35 557

1. Uranium from imported phosphates.
2. Plus 120 tonnes uranium of foreign origin.
3. Secretariat estimate.

* *Source: Uranium — Resources, Production and Demand*, a joint report by OECD/NEA and IAEA, OECD, Paris, 1988.

EVOLUTION OF URANIUM PRODUCTION *

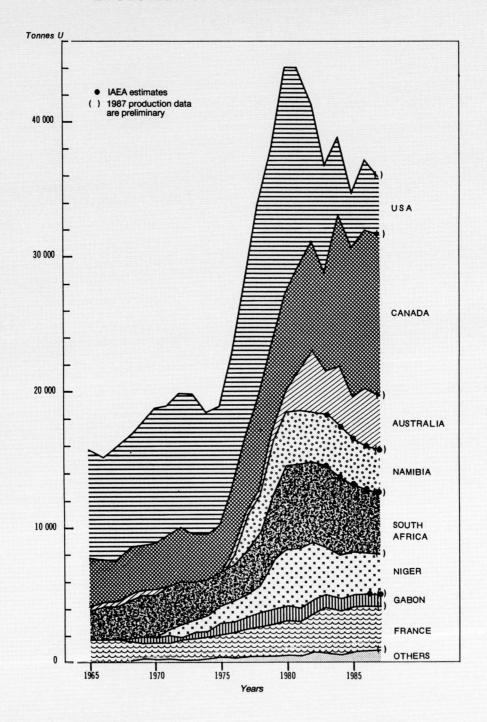

*   Does not include the Eastern European countries.
    Source: Uranium – Resources, Production and Demand, a joint report by OECD/NEA and IAEA, OECD, Paris, 1988.

APPENDIX 20

ANNUAL REACTOR RELATED URANIUM REQUIREMENTS TO 2000*

(Tonnes U)

Country	1986	1987	1988	1989	1990	1991	1992	1993	1994	1995	2000
Belgium	960	960	960	960	960	960	960	960	960	1 140	1 140
Canada	1 400	1 600	1 700	1 800	1 800	1 900	2 050	2 050	2 100	2 100	2 100
Finland	270	270	270	270	270	270	270	270	270	270	270
France	7 000	5 900	6 500	8 300	9 750	7 650	9 650	9 100	9 350	9 500	10 000
Germany, F.R.	3 300	3 300	3 300	3 500	3 500	3 500	3 500	3 500	3 500	3 500	4 000
Italy	34	765	765	235	135	534	1 050	750	500	620	n.a.
Japan	4 800	4 900	5 300[1]	5 600[1]	5 900	6 300[1]	6 600[1]	6 900[1]	7 200[1]	7 600[1]	9 200
Netherlands	100	100	100	100	100	100	100	100	100	90	90
Spain	1 019	1 301	1 393	1 059	1 400	1 088	1 515	1 713	1 410	1 536	1 670
Sweden	1 400	1 300	1 400	1 400	1 400	1 300	1 300	1 300	1 300	1 300	1 300
Switzerland	560	570	529	529	537	537	532	532	939	532	675
United Kingdom	2 500	2 400[2]	2 300[2]	2 200[2]	2 100	2 100[2]	2 000[2]	1 900[2]	1 800[2]	1 700	2 400[3]
United States	12 800	13 340	12 800	13 730	13 770	13 460	13 500	12 150	13 340	13 000	13 500
Other[5]	3 100	2 600	2 750	3 200	3 200	4 200	3 100	3 950	4 100	4 300	5 400
Total[6]	39 200	39 300	40 100	42 900	44 800	43 900	46 100	45 200	46 900	47 200	52 400[4]

1. Interpolated estimates as provided by country.
2. Interpolated by NEA Secretariat.
3. Based on lower capacity forecast.
4. Secretariat estimate.
5. Rounded, includes IAEA estimates for the following countries: Argentina; Brazil; India; Republic of Korea; Mexico; Pakistan; South Africa; Taiwan, China; Yugoslavia.
6. Rounded.
* Does not include Eastern European countries. *Source: Uranium — Resources, Production and Demand*, a joint report by OECD/NEA and IAEA, Paris, 1988.

APPENDIX 21

URANIUM STOCKS*

(Tonnes natural U-equivalent)

Country	Natural uranium stocks in concentrate and refined	Enriched uranium stocks**	Depleted uranium stocks**
Australia	2 358	—	—
Belgium	n.a.	n.a.	n.a.
Canada	n.a.	—	—
Finland	n.a.	n.a.	n.a.
France	n.a.	n.a.	n.a.
Germany, F. R.	9 000	6 000	1 000
Italy	950	2 700	500**
Japan	n.a.	n.a.	n.a.
Netherlands	n.a.	n.a.	n.a.
Portugal	609	—	—
Sweden	—	780	—
Switzerland	n.a.	n.a.	n.a.
United Kingdom	n.a.	n.a	6 000
United States	68 800	54 700	15 100

   * *Source: Uranium — Resources, Production and Demand*, Statistical Update 1986, OECD/NEA, 1986.
  ** Equivalent of natural uranium as enrichment plant feed processed at a tails assay of 0.2% U-235.
*** Calculated for a tails assay of 0.1% U-235.

INTERNATIONAL URANIUM FLOWS IN THE EARLY 1990s

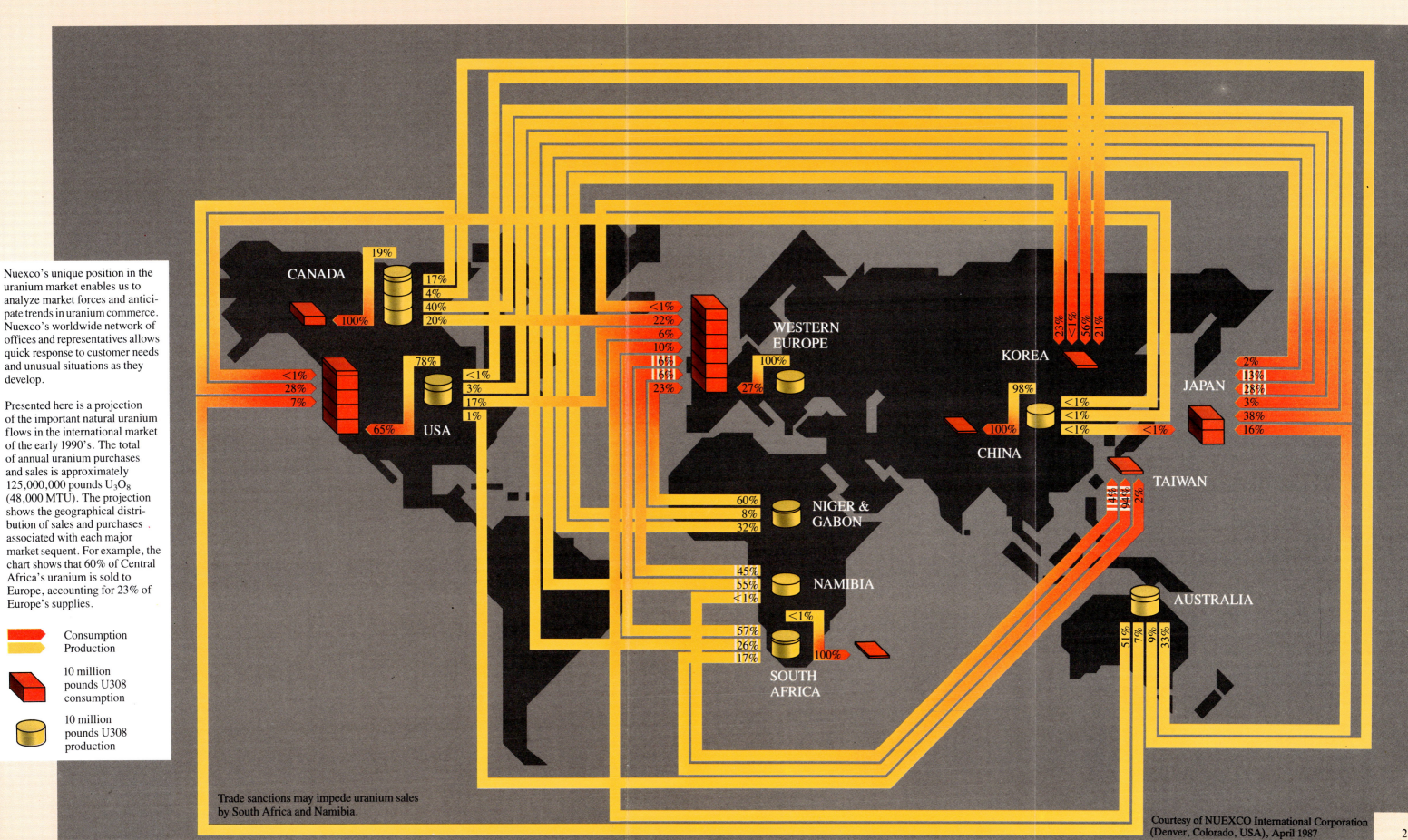

Nuexco's unique position in the uranium market enables us to analyze market forces and anticipate trends in uranium commerce. Nuexco's worldwide network of offices and representatives allows quick response to customer needs and unusual situations as they develop.

Presented here is a projection of the important natural uranium flows in the international market of the early 1990's. The total of annual uranium purchases and sales is approximately 125,000,000 pounds U_3O_8 (48,000 MTU). The projection shows the geographical distribution of sales and purchases associated with each major market sequent. For example, the chart shows that 60% of Central Africa's uranium is sold to Europe, accounting for 23% of Europe's supplies.

Consumption Production

■ 10 million pounds U308 consumption

● 10 million pounds U308 production

Trade sanctions may impede uranium sales by South Africa and Namibia.

CANADA — 19%, 17%, 4%, 40%, 20%, 100%, 78%, <1%, 28%, 7%, 65%

USA — <1%, 3%, 17%, 1%

WESTERN EUROPE — <1%, 22%, 6%, 10%, 100%, 27%, 23%

KOREA — 23%, <1%, 56%, 21%, 98%

JAPAN — 2%, 13%, 28%, 3%, 38%, 16%

CHINA — 100%, <1%, <1%, <1%

TAIWAN — 14%, 94%, 2%

NIGER & GABON — 60%, 8%, 32%

NAMIBIA — 45%, 55%, <1%, <1%

SOUTH AFRICA — 57%, 26%, 17%, 100%

AUSTRALIA — 51%, 7%, 9%, 33%

Courtesy of NUEXCO International Corporation (Denver, Colorado, USA), April 1987

241

Appendices 23 to 26

REACTORS
NUCLEAR POWER PLANT IMPORTS

APPENDIX 23

INDUSTRIALISED AND EUROPEAN CENTRALLY PLANNED COUNTRIES*

Country	Total		Domestic		Imported from													
					USA		FRG		Canada		France		Sweden		UK		USSR	
	No	MW(e)	No	MW(e)	No	MW(e)	No	MW(e)	No	MW(e)	No	MW(e)	No	MW(e)	No	MW(e)	No	MW(e)
Belgium	8	5488			5	2821					3	2667						
Bulgaria	6	3538															6	3538
Canada	23	15610	23	15610														
Czchecoslovakia	16	8327	14	7511													2	816
Finland	4	2310											2	1420			2	890
France	63	62502	63	62502														
German D. R.	11	5126															11	5126
Germany, F. R.	25	22937	25	22937														
Hungary	4	1645															4	1645
Italy	6	3272	1	35	4	3084									1	153		
Japan	45	34252	35	27090	9	7003									1	159		
Netherlands	2	507			1	55	1	452										
Poland	2	880															2	880
Romania	3	1980							3	1980								
S. Africa	2	1842									2	1842						
Spain	10	7519			8	6049	1	990			1	480						
Sweden	12	9646	9	7016	3	2630												
Switzerland	5	2932			4	2012	1	920										
UK	42	12734	42	12734														
USA	120	107874	120	107874														
USSR	83	58017	83	58017	1	632												
Yugoslavia	1	632			1	632												
Total	493	369570	415	321326	35	24286	3	2362	3	1980	6	4989	2	1420	2	312	27	12895

Note: Includes all reactors connected to the grid or under construction as of 31st December 1986.
Reactors shut down by this date are not included.

* *Source:* IAEA Power Reactor Information System.

APPENDIX 24

DEVELOPING COUNTRIES*

Country	Total		Domestic		Imported from													
					USA		FRG		Canada		France		Sweden		UK		USSR	
	No	MW(e)	No	MW(e)	No	MW(e)	No	MW(e)	No	MW(e)	No	MW(e)	No	MW(e)	No	MW(e)	No	MW(e)
Argentina	3	1627					2	1027	1	600								
Brazil	2	1871			1	626	1	1245										
China	1	288	1	288														
Cuba	2	816															2	816
India	10	2034	6	1320	2	300			2	414								
Iran	2	2400					2	2400										
Korea	9	7180			6	4751			1	629	2	1800						
Mexico	2	1308			2	1308												
Pakistan	1	125							1	125								
Taiwan	6	4918			6	4918												
Total	38	22567	7	1608	17	11903	5	4672	5	1768	2	1800					2	816

Note: Includes all reactors connected to the grid or under construction as of 31st December 1986.
Reactors shut down by this date are not included.
* *Source*: IAEA Power Reactor Information System.

APPENDIX 25

WORLD TOTAL*

	Total		Domestic		Imported from													
					USA		FRG		Canada		France		Sweden		UK		USSR	
	No	MW(e)	No	MW(e)	No	MW(e)	No	MW(e)	No	MW(e)	No	MW(e)	No	MW(e)	No	MW(e)	No	MW(e)
World total	531	392137	422	322934	52	36189	8	7034	8	3748	8	6789	2	1420	2	312	29	13711

Note: Includes all reactors connected to the grid or under construction as of 31st December 1986.
Reactors shut down by this date are not included.
* *Source*: IAEA Power Reactor Information System.

APPENDIX 26

NUCLEAR POWER PLANT IMPORTS BY REACTOR TYPE*

Reactor type	Total		Domestic		Imported from												
					USA		FRG		Canada		France		Sweden		UK		USSR	
	No	MW(e)	No	MW(e)	No	MW(e)	No	MW(e)	No	MW(e)	No	MW(e)	No	MW(e)	No	MW(e)	No	MW(e)
PHWR 100-599 MW(e)	18	6318	14	5444			1	335	3	539								
PHWR >=600 MW(e)	20	15365	14	11464			1	692	5	3209								
AGR	14	8256	14	8256														
GCR 100-599 MW(e)	25	6292	22	5500							1	480			2	312		
PWR 100-599 MW(e)	75	31029	42	17861	8	2773	1	452									24	9943
PWR >=600 MW(e)	222	221256	184	186217	22	19469	5	5555			7	6309					4	3706
BWR 100-599 MW(e)	16	7067	10	5225	6	1842												
BWR >=600 MW(e)	75	68842	59	55383	14	12039							2	1420				
LWGR >=600 MW(e)	22	22400	22	22400														
PROTOTYPES	44	5312	41	5184	2	66											1	62
Total	531	392137	422	322934	52	36189	8	7034	8	3748	8	6789	2	1420	2	312	29	13711

Note: Includes all reactors connected to the grid or under construction as of 31st December 1986.
Reactors shut down by this date are not included.
* *Source:* IAEA Power Reactor Information System.

247

Appendices 27 to 31

NUCLEAR FUEL CYCLE
MATERIAL FLOW

MATERIAL FLOW OF PWR REPROCESSING CYCLE■

The figure is an example and the numbers are approximately only. t : tonnes, m³ : cubic meters

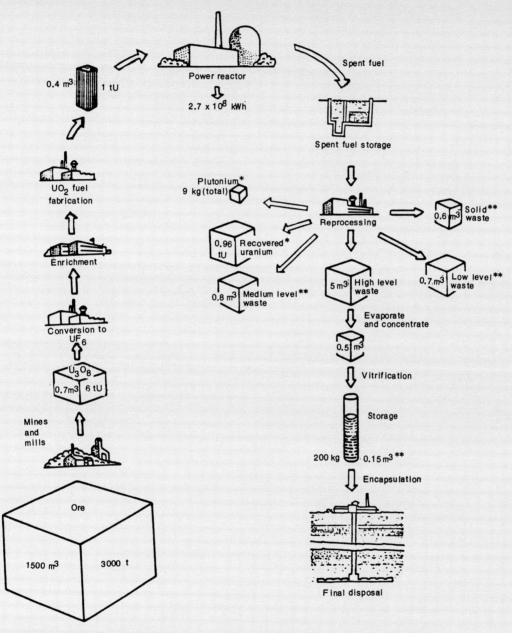

* Based on material balances in Table 14.1, Annex 14. Recovered uranium and plutonium can be recycled.
** Source : Report of INFCE Working Group 7. Solid waste includes hulls, spacers and insolubles. (3).

■ *Source : The Economics of the Nuclear Fuel Cycle*, OECD/NEA, Paris, 1985.

APPENDIX 28

MATERIAL FLOW OF PWR ONCE-THROUGH CYCLE ▪

The figure is an example and the numbers are approximately only. t : tonnes, m³ : cubic meters

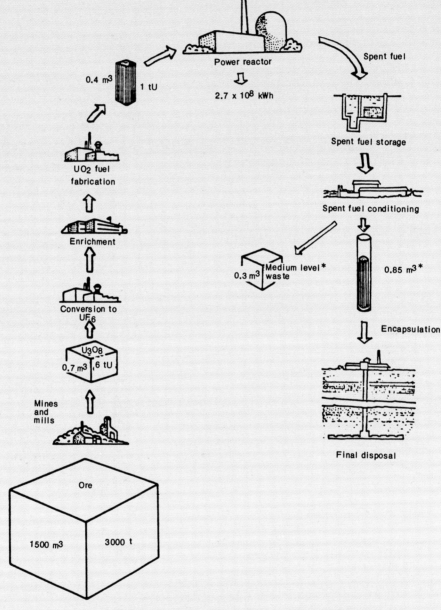

* Source : Report of INFCE Working Group 7. (3).

▪ *Source:* *The Economics of the Nuclear Fuel Cycle,* OECD/NEA, Paris, 1985.

FRONT-END AND AT REACTOR STAGE OF PWR FUEL CYCLE ▪

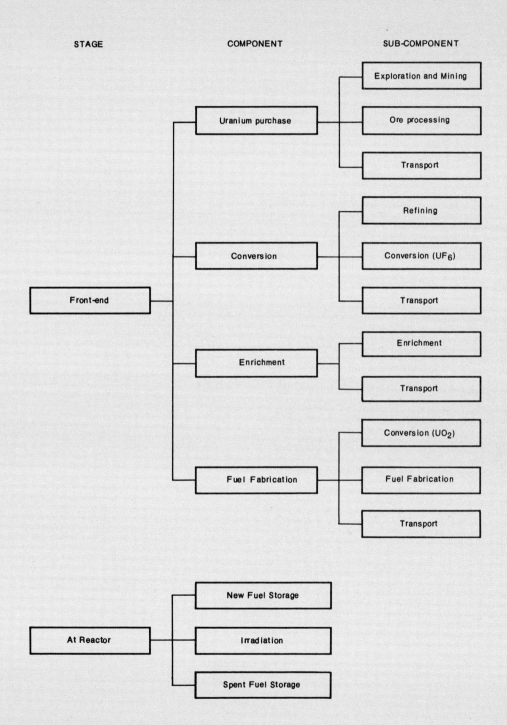

▪ Source: *The Economics of the Nuclear Fuel Cycle*, OECD/NEA, Paris, 1985.

APPENDIX 30

BACK-END STAGE OF THE PWR FUEL CYCLE ■

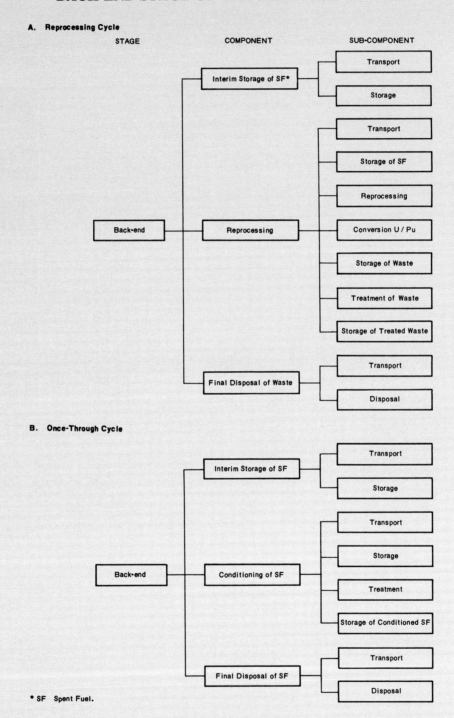

A. Reprocessing Cycle

STAGE COMPONENT SUB-COMPONENT

Interim Storage of SF*
- Transport
- Storage

Back-end — Reprocessing
- Transport
- Storage of SF
- Reprocessing
- Conversion U / Pu
- Storage of Waste
- Treatment of Waste
- Storage of Treated Waste

Final Disposal of Waste
- Transport
- Disposal

B. Once-Through Cycle

Interim Storage of SF
- Transport
- Storage

Back-end — Conditioning of SF
- Transport
- Storage
- Treatment
- Storage of Conditioned SF

Final Disposal of SF
- Transport
- Disposal

* SF   Spent Fuel.

■ *Source:*   *The Economics of the Nuclear Fuel Cycle*, OECD/NEA, Paris, 1985.

APPENDIX 31

TRANSPORT OF RADIOISOTOPES

PACKAGE SHIPMENTS*a* BY COUNTRY IN 1981
(Based on Data Supplied by Member States to the IAEA*)

Country	International package shipments	Domestic package shipments	Total package shipments
Australia	555[b]	56331[b]	56886[b]
Belgium	—[b]	78422[b]	78422[b]
Canada	433088	3947341	4380429
Finland	1[b]	24502[b]	24503[b]
France	34802	84757	119559
Germany, Fed. Rep. of	46882	414532	461414
Greece	—[b]	1338	1338
Ireland	60[b]	15500[b]	15560[b]
Italy	8166[b]	197505[b]	205671[b]
Japan	18459	705055	723514
Netherlands	—[b]	10359[b]	10359[b]
Norway	6706[c]	10298[c]	17004[c]
Spain	107[b]	48894[b]	49001[b]
Sweden	3760	119151	122911
Switzerland	774[d]	284[d]	1058[d]
United Kingdom	208168[b]	190013[b]	398181[b]
United States of America	—[b]	2819308[e]	2819308[e]

a) Including Full Load.
b) Data incomplete or not available.
c) Estimated to account for only 70% of shipments.
d) Fissile shipments only, data incomplete.
e) Known to be low by at least 500000, data incomplete, data are package combined shipments, not package shipments.
* This Table is extracted from *Estimated Annual Worldwide Shipments of Radioactive Material*, by R. B. Pope and J. D. McClure, in Packaging and Transportation of Radioactive Materials (Patram '86), Volume 1, IAEA, Vienna, 1987.

SELECTED BIBLIOGRAPHY

A. PERSONAL AUTHORS AND EDITORS

ALLEN, D. W., *The EURATOM Treaty, Chapter VI: New Hope or False Dawn*, Common Market Law Review 20, 1983.

ALONSO, G., TREVINO, J. et GONZALEZ, G., *La non-prolifération et les garanties de l'approvisionnement nucléaire (le point de vue d'un pays en développement)*, in Nuclear Inter Jura '81, Congrès de l'AIDN, Palma de Majorque, 1981, pp. 219-233.

BASSO, J. and MARSAND, J. L., *Le Club de Londres et le contrôle des transferts de technologie nucléaire*, ARES (défense et sécurité), Paris, 1980.

BERTSCH, G. K., *East West Strategic Trade — COCOM and the Atlantic Alliance*, The Atlantic Institute for International Affairs, Paris, 1983.

BETTAUER, R. J., *The Nuclear Non-Proliferation Act of 1978, Law and Policy*, in International Business, Vol. 10, No. 4, 1978, pp. 1105-1180.

BLIX, H., *Aspects juridiques des garanties de l'AIEA*, in Annuaire français de droit international, Paris, 1983.

BLIX, H., *Building Confidence in Safeguards*, in IAEA Bulletin, Vol. 26, No. 3 September 1984, pp.3-6 (also in French).

BOARDMAN, R. and KEELEY, J. F., Editors, *Nuclear Exports and World Politics — Policy and Regime*, Macmillan Press, London, 1983, 256 p.

BÖHM, P., *Die juristische Problematik des europäischen Kernbrennstoffeigentums*, NJW, 1961, p. 1553 *et seq.*

BOULANGER, W., *Das Verifikationsabkommen IAEO-EURATOM*, Atomwirtschaft 1972, pp. 510-511.

BRITO, D. L., INTRILIGATOR, M. D. and WILK, A. E, *The Strategies for Managing Nuclear Proliferation Economic and Political Issues*, Lexington, Mass., 1983, 311 p.

BROLL, Renata and SARTORELLI, C., *Les accords de swap dans le domaine nucléaire*, in Nuclear Inter Jura '87, Congrès de l'AIDN, Anvers, 1987 (à publier) (cf. SARTORELLI).

BUTTAR, P. A., *International Implications of the Australian Nuclear Fuel Cycle*, in International Harmonization in the Field of Nuclear Energy Law, Nuclear Inter Jura '85, INLA Congress, Konstanz, Nomos Verlagsgesellschaft, Baden-Baden, 1986, pp.459-483.

CAMILLERI, J. A., *The State and Nuclear Power. Conflict and Control in the Western World*, Wheatsheaf Books Ltd., Brighton, 1984, 347p.

CARBONE, S., *International Contracts - Private Autonomy and State Law: Consideration of Recent Trends with Particular Attention to their Effect on Nuclear Contracts*, in Proceedings of the CISDEN Meeting on Present Problems in Community and International Law in the Nuclear Field, Rome, June 1981, pp.22-52.

CARBONE, S., *Non-Proliferation Treaty Safeguards and Export of Nuclear Plants*, in Nuclear Inter Jura '79, INLA Congress, Buenos Aires, 1981, pp.151-160.

COURTEIX, S., *Les accords de Londres entre pays exportateurs d'équipements et de matières nucléaires*, in Annuaire français de droit international, Paris, 1976, pp.27-50.

COURTEIX, S., *Le Comité de coordination des échanges Est/Ouest (COCOM)*, Université de Strasbourg, Annuaire de l'URSS et des pays socialistes européens, Istra, Strasbourg, 1976-1977.

COURTEIX, S., *Les exportations d'équipements et de matières nucléaires et la non-prolifération des armes nucléaires*, in Nuclear Inter Jura '77, Congrès de l'AIDN, Florence, 1977, pp.214-273.

COURTEIX, S., *Exportations nucléaires et non-prolifération*, CNRS, Economica, Paris, 1978, 263p.

COURTEIX, S., *Le contrôle de la prolifération des armes nucléaires*, in McGill Law Journal, Vol. 28, July 1983, Montreal, pp.591-607.

DELAHOUSSE, J. P., *Le régime juridique des matières nucléaires dans le Traité instituant EURATOM: propriété, approvisionnement, contrôle*, in Aspects du droit de l'énergie atomique, Tome II, CNRS, Paris 1967, pp.475-503.

DELCOIGNE, G. and RUBINSTEIN, G., *Non-prolifération des armes nucléaires et systèmes de contrôle*, Editions de l'Institut de sociologie de l'Université Libre de Bruxelles, Bruxelles, 1970, 214p.

DEWITT, D., Editor, *Nuclear Non-Proliferation and Global Security*, Croom Helm Publ., London and Sydney, 1987, 283p.

DIXON, D. F., *Nuclear Proliferation and Subsequent Arrangements for Retransfer of Reprocessing*, in Virginia Journal of International Law, Vol. 20:1, 1979, pp.99-146.

DONNELLY, W., *Nuclear Weapons Proliferation and the International Atomic Energy Agency: An Analytic Report*, prepared for the United States Senate Committee on Governmental Operations, US Government Printing Office, Washington DC, 1976, p.92.

DONNELLY, W. H. and PILAT, J., *Non-Proliferation: Reassessment of US Relations with the IAEA*, Congressional Research Service, US Library of Congress, February 1983 (see PILAT).

DOREN, Ch. N. van, *Nuclear Supply and Non-Proliferation: the IAEA Committee on Assurances of Supply*, Congressional Research Service, US Library of Congress, October 1983.

DOUGHERTY, A. J., *Nuclear Co-operation Agreements and the Nuclear Non-Proliferation Act of 1978*, in Nuclear Inter Jura '79, INLA Congress, Buenos Aires, 1981, pp.161-170.

EDWARDS, D. M., *International Legal Aspects of Safeguards and the Non-Proliferation of Nuclear Weapons*, in International and Comparative Law Quarterly, London, Vol. 33, 1984, pp.1-21.

ERRERA, J., SYMON, E., VAN DER MEULEN, J. and VERNAEVE, L., *EURATOM — analyse et commentaire du Traité*, Bibliothèque de l'Institut belge de science politique, Editions de la Librairie encyclopédique, SPRL, Bruxelles, 1958, 436p.

ESTRADA OYUELA, R. A., *The International Nuclear Control of Non-Proliferation and Disarmament*, in Nuclear Inter Jura '79, INLA Congress, Buenos Aires, 1981, pp.183-194.

EVERLING, U., *Die Eigentumsproblematik bei besonderen spaltbaren Stoffen*, in Lukes, Zweites Deutsches Atomrechts Symposium, Heymans, Köln, 1974, pp.89-97.

FINNISS, G., *Les échanges d'informations techniques et le droit des brevets dans le Traité de l'EURATOM*, in Aspects du droit de l'énergie atomique, Tome II, pp.505-516, CNRS, Paris, 1967.

FIQUET, A., *Problèmes posés à l'Europe communautaire par la protection physique des matières et installations nucléaires*, in Revue trimestrielle de droit européen, Paris, 1979.

FIQUET, A., *Les problèmes de sécurité posés à l'Europe communautaire par sa politique nucléaire pacifique*, Thèse univ. Lille II, 1981, 311 p.

FISCHER, D., *International Safeguards*, Rockefeller Foundation, New York, 1979, 45 p.

FISCHER, D. and SZASZ, P., *Safeguarding the Atom: a Critical Appraisal*, SIPRI, Taylor and Francis, London, 1985, 243p.

FISCHER, G. et VIGNES, S., *L'inspection internationale*, Bruylant, Bruxelles, 1976.

FISCHER, D. and MÜLLER, H., *Non-Proliferation Beyond the 1985 Review*, Centre for European Policy Studies, No. 26, Brussels, 1985, 48 p.

FRANKO, L. G., *US Regulation on the Spread of Nuclear Technologies Through Supplier Power: Lever or Boomerang?*, in Law and Policy in International Business, United States, 1978, Vol. 10, pp.1181-1204.

FURET, M. F., *Le désarmement nucléaire*, Publications de la RGDIP, nouvelle série No. 19, Pédone, Paris, 1973, 303p.

GIJSSELS, J., *L'accord entre EURATOM et l'AIEA en application du Traité sur la non-prolifération des armes nucléaires*, in Annuaire français de droit international, Paris, 1972, pp.837-863.

GIRERD, P., *Aspects communautaires de l'approvisionnement en énergie nucléaire*, Thèse univ. Lille II, 1986, 365p.

GOHEEN, R. F., *Problems of Proliferation: US policy and the Third World*, in World Politics, Vol. XXXV, No. 2, 1983.

GOLDBLATT, J., *Non Proliferation: the Why and the Wherefore*, SIPRI, Stockholm, 1985, 400p.

GOLDSCHMIDT, B., *Le complexe atomique: histoire politique de l'énergie nucléaire*, Fayard, Paris, 1980, 493p.

GOLDSCHMIDT, B., *Les rivalités atomiques 1939-1966*, Fayard, Paris, 1967, 340p.

GOLDSCHMIDT, B. and KRATZER, M., *Peaceful Nuclear Relations: a Study of the Creation and the Erosion of Confidence*, Rockefeller Foundation, New York, 1978, 56p.

GOURRIER, J., *La notion juridique d'entreprise commune et les réalisations européennes en matière d'énergie nucléaire*, in Aspects du droit de l'énergie atomique, Tome II, CNRS, Paris, 1967, pp.517-530.

GOWING, M. and ARNOLD, L., *Independence and Deterrence: Britain and Atomic Energy, 1945-1952*, Harwell UKAEA, Authority Historian's Office, 1974, 503p.

GROUX, J. et MANIN, Ph., *Les Communautés européennes dans l'ordre international*, Commission des Communautés Européennes, Collection "Perspectives européennes", Bruxelles, 1984, 166p.

GUNNETT, Ph., *From NPT to INFCE: Developments in Thinking about Nuclear Non-Proliferation*, in International Affairs, London, 1981.

HÄCKEL, E., KAISER, K. and LELLOUCHE, P., *Nuclear Policy in Europe — France, Germany and the International Debate*, Arbeitspapiere zur Internationalen Politik 12, Forschungs institut der Deutschen Gesellschaft für Auswärtige Politik e.V. März, 1980, 133p.

HAUNSCHILD, H.-H., *The Transfer of Nuclear Technology — Necessities and Limitations*, in Symposium on International Co-operation in the Nuclear Field: Perspectives and Prospects, OECD/NEA, Paris, February 1978, pp.86-97.

HA VINH PHUONG, *Procedures for the Supply of Nuclear Materials through the IAEA*, in Experience and Trends in Nuclear Law, Legal Series No. 8, IAEA, Vienna, 1972, pp.61-65.

HA VINH PHUONG, *Nuclear Fuel Supply Arrangements through the IAEA*, in Nuclear Inter Jura '81, INLA Congress, Palma de Mallorca, 1981, pp.191-207.

HA VINH PHUONG, *The Physical Protection of Nuclear Material*, in OECD/NEA Nuclear Law Bulletin No. 35, June 1985, pp.113-119 (also in French).

HEBERT, J., *Observations sur l'arrêt rendu le 14 décembre 1971 par la Cour de Justice des Communautés Européennes relatif à l'application du Chapitre VI du Traité EURATOM*, Revue trimestrielle de droit européen, No. 2, avril-juin 1972, pp.299-314.

HERRON, L. W., *A Lawyer's View of Safeguards and Non-Proliferation*, in IAEA Bulletin, Vol. 24, No. 3, September 1982, pp.32-38 (also in French).

HERRON, L. W., *Legal Aspects of International Co-operation in the Physical Protection of Nuclear Facilities and Materials*, in Nuclear Inter Jura '81, INLA Congress, Palma de Mallorca, 1981, pp.293-303.

HEWLETT, R. and ANDERSON, O. Jr., Volume I, *A History of the United States Atomic Energy Commission, The New World, 1939/1946*, the Pennsylvania State University Press, 1962.

JONES, R. W., *The Nuclear Suppliers and Non-Proliferation*, Georgetown University, Washington DC, 1985, 272p.

JONES, R. W., MERLINI, C., PILAT, J. F. and POTTER, W. C., Editors, *The Nuclear Suppliers and Non-Proliferation*, International Policy Choices, CSIS Georgetown University, Washington DC, 1985, 272p.

JUNGK, R., *Brighter than a Thousand Suns*, Harcourt Brace Janovich Inc., New York, 1958.

KAPUR, A., *International Nuclear Proliferation, Multilateral Diplomacy and Regional Aspects*, Praeger Publishers, New York, 1979.

KENNEDY, R. T., *Nuclear Trade: Reliable Supply and Mutual Obligations*, US Department of State Bulletin No. 2091, October 1984.

KINCADE, W. H. and BERTRAM, C., *Nuclear Proliferation in the 1980s: Perspectives and Proposals*, Macmillan Publ., London, 1982, 272p.

KRAEMER, J. R. and ROWDEN, M. A., *Restrictions on the Transnational Movement of Uranium*, in Nuclear Inter Jura '87, INLA Congress, Antwerp (to be published) (see ROWDEN).

KRASS, A., *Uranium Enrichment and Nuclear Weapons Proliferation*, SIPRI, Taylor and Francis Ltd., London, 1983, 296p.

KRATZER, M. B., *New Trends in Safeguards*, speech delivered to the Institute of Nuclear Materials Management, Albuquerque N.M., July 1985.

KRAULAND, E. J., NEPA, *Nukes and Non-Proliferation: Clarifying the Transnational Impact Statement Mandate in Nuclear Export Licensing*, in Hastings International and Comparative Law Review, Vol. 4, No. 2, 1981, pp.201-272.

LANNOY, J., *L'échange des connaissances (EURATOM)*, in Les cadres juridiques de la coopération internationale en matière scientifique et le problème éuropéen, Acte du Colloque d'Aix-en-Provence, Commission des Communautés Européennes, Bruxelles, mai 1970, pp.188-203.

LELLOUCHE, P. and LESTER, R., *The Crisis of Nuclear Energy*, Washington Quarterly, Summer 1979.

LOOSCH, R., *Die Eigentumsproblematik im Hinblick auf die künftige Entwicklung der Kernenergie*, Zweites Deutsches Atomrechts-Symposium, 1974, p.83 et seq.

LUKES, R., *Die Eingentumsregelung für die besonderen spaltbaren Stoffe im Euratomvertrag*, Zweites Deutsches Atomrechts-Symposium, 1975, p.35 et seq.

LUXO, *Le transfert de technologie et l'exportation nucléaire*, Revue Générale Nucléaire, No. 4, Paris, 1978.

MANOVIL, R. M., *Nuclear International Trade: The Point of View of the Countries in Nuclear Development*, in Nuclear Inter Jura '85, INLA Congress, Konstanz, Nomos Verlagsgesellschaft, Baden-Baden, 1986, pp.386-396.

MANNING MUNTZING, L., Editor, *International Instruments for Nuclear Technology Transfers*, American Nuclear Society, La Grange Park, Illinois, 1978, 635 p.

MANNING MUNTZING, L., *International Legal and Political Issues Associated with the Export/Import of Nuclear Power Plants*, in Proceedings of the Symposium on Problems Associated with the Export of Nuclear Power Plants, IAEA, Vienna, 6th-10th March 1978, pp. 217-225.

MANNING MUNTZING, L., *Recent Developments in the United States Concerning Nuclear Non-Proliferation Policy*, in Nuclear Inter Jura '79, INLA Congress, Buenos Aires, 1981, pp.115-123.

MARSHALL, H. R., *Section 104 of the Nuclear Non-Proliferation Act of 1978: Establishment of International Nuclear Supply Assurances*, New York University Journal of International Law and Politics, 1979.

MARSHALL, H. R., *US Non-Proliferation Policy in the Decade of the Eighties: Past Dilemmas and Prospects for the Future*, in International Harmonization in the Field of Nuclear Energy Law, Nuclear Inter Jura '85, INLA Congress, Konstanz, Nomos Verlagsgesellschaft, Baden-Baden, 1986, pp.450-458.

MARWAH, O. and SCHULZ, A., *Nuclear Proliferation and the Near Nuclear Countries*, Ballinger Publ., Cambridge, Mass., 1975, 362p.

McMANUS, K. D. and MARSHALL Jr., R. H., *United States Nuclear Export Controls*, Fordham International Journal, Vol. 4, No. 2, New York.

MENNICKEN, J. B., *Die gemeinschaftliche Versorgung der EG Staaten mit Kernbrennstoffen*, Europa Archiv, 1975.

MENNICKEN, J. B., *Non-Proliferation Framework Conditions for the Supply of the European Communities with Nuclear Fuel*, in Nuclear Inter Jura '79, INLA Congress, Buenos Aires, 1981, pp. 125-138.

MEYER, S. M., *The Dynamics of Nuclear Proliferation*, University of Chicago Press, 1984, 229p.

MEYER-WOLSE, G., *Rechtsfragen des Exports von Kernanlagen in Nichtkern-waffenstaaten*, Studien zum Internationalen Wirtschaftsrecht und Atomenergierecht, Band 62, Institut für Völkerrecht der Universität Göttingen, Carl Heymann Verlag, Köln, 1979, 120p.

MORSON, S., *La politique étrangère de la France en matière d'énergie nucléaire pacifique*, Thèse univ. Lille II, 1981, 247p.

MOSS, N., *The Politics of Uranium*, Universe Books, New York, 1981, 239p.

MÜLLER, H., *Nuclear Proliferation: Facing Reality*, Centre for European Policy Studies, Nos. 14/15, Brussels, 1984, pp.3-53.

NEFF, T. C., *The International Uranium Markets*, Ballinger Publ., Cambridge, Mass., 1984, 357p.

OBOUSSIER, F., *Die Verteilung von Kernbrennstoffen: das Problem der Rohstoffe und der Anreichung*, in K. Kaiser, Kernenergie und Internationale Politik, München, Oldenburg, 1975, pp.325-355.

PANDER, J. Von, *Die Sicherheitskontrolle nach dem Nichtverbreitungsvertrag in den E. G. Staaten*, in Studien zum Internationalen Wirtschaftsrecht und Atomenergierecht, Band 58, Institut für Völkerrecht der Universität Göttingen, Carl Heymann Verlag, Köln, 1978, 622p.

PATTERSON, W. C., *The Plutonium Business and the Spread of the Bomb*, Sierra Club Books, San Francisco, 1984, 272 p.

PELZER, N., *Rechtsfragen der Sicherheitsüberwachung nach NV-Vertrag und Verifikationsabkommen*, Energiewirtschaftliche Tagesfragen 1979, 255p.

PETIT, A., *Imprévisibilité et inspections inspirées dans les garanties internationales*, in Nuclear Safeguards Technology, Proceedings of an IAEA Symposium, IAEA, Vienna, 10th-14th November 1986, Vol. 1, pp.25-30.

PILAT, J. and DONNELLY, W. H., *Nuclear Export Policies of the Reagan Administration: a Summary Analysis and Four Case Studies*, Congressional Research Service, US Library of Congress, April 1982, 42p.

PILAT, J. AND DONNELLY, W. H., *Policies for Nuclear Exports Co-operation and Non-Proliferation of Seven Nuclear Supplier States*, Congressional Research Service, US Library of Congress, May 1982, 77p.

PIROTTE, O., *L'Agence d'approvisionnement d'EURATOM: bilan et perspectives*, Mélanges Teneur, Tome II, 1977, p.573 et seq.

PIROTTE, O., *Réflexions sur quarante ans d'expérience en matière de non-prolifération horizontale*, ARES, Défense et Sécurité, Paris, Vol. 7, 1984.

POLITI, M., *Diritto internazionale et non proliferazione nucleare*, Rivista di diritto internazionale privato e processuale, Cedam Padova, 1984, 301p.

POLITI, M., *Esportazioni nucleari e politiche di non proliferazione*, Scheda per legislatione economica, Rassegna e Problemi, V, 1980-81, pp.806-821.

POTTER, W. C., *Nuclear Power and Non-Proliferation*, Oelgeschlager, Gunn and Hain, Publishers, Inc., Cambridge, Mass., 1982, 279p.

POULOSE, T. T., *Nuclear Proliferation and the Third World*, Atlantic Highlands, New Jersey, 1982, 208p.

PREUSCHEN, R. Von, *IAEO Sicherungsmassnahmen gegen die Abzweigung von Kernmaterial für Kernsprengkörper*, Thèse univ. Köln, 1984, 234p.

PUISSOCHET, J-P., *A propos d'une délibération de la Cour de Justice des Communautés Européennes — Le régime des matières nucléaires et la capacité de la Communauté de conclure des accords internationaux*, Annuaire français de droit international, Vol. XXIV, CNRS, Paris, 1978, pp.977-988.

PRINGLE, P. and SPIGELMAN, J., *The Nuclear Barons*, Holt, Rinehart and Winston Publishers, New York, 1981.

RANGARAJAN, L. N., *L'Inde, la politique nucléaire et le Traité de non-proliferation nucléaire*, in Etudes Internationales, No. 19, Tunis, juillet 1986, pp.8-33.

RANDERMANN, P. H., *Probleme im Zusammenhang mit den Verhandlungen über das Verifikationsabkommen*, Drittes Deutsches Atomrechts Symposium, 1975, 197p.

REYNERS, P., *L'Agence pour l'Energie Nucléaire de l'OCDE: ses relations avec l'AIEA et EURATOM*, Annuaire Européen, Vol. XXXII, Martinus Nijhoff, Dordrecht, 1984, pp.1-31.

ROSER, T., *Die deutsche Nuklearwirtschaft vor den Anforderungen des NV-Vertrages*, Drittes Deutsches Atomrechts Symposium, 1975, 187p.

ROTH, B., and VIROLE, J., *Guarantees in Uranium Supply Contracts: "Force majeure" and Embargo Provisions*, in Uranium and Nuclear Energy: 1981, Proceedings of the Sixth International Symposium held by the Uranium Institute, Butterworths, London, 1982, pp.350-356 (see VIROLE).

ROWDEN, M. A. and KRAEMER, J. R., *The Role of Bilateral Agreements for Co-operation in Establishing International Norms for Nuclear Exportation*, in International Harmonization in the Field of Nuclear Energy Law, Nuclear Inter Jura '85, INLA Congress, Konstanz, Nomos Verlagsgesellschaft, Baden-Baden, 1986, pp.408-433.

SAMONCHIK, O. A., *Pravovyye Problemy Ispolzovaniya Atomnoy Energii*, Institut gosudarstva i prava AN USSR, Moscow, 1985, 147p.

SANDERS, B. and RAINER, R. H., *Safeguards Agreements, their Legal and Conceptual Basis*, in Proceedings of an IAEA International Conference on Nuclear Power and its Fuel Cycle, May 1977, IAEA, Vienna, 1977, pp.395-410.

260

SARTORELLI, C., *Les contrats pour l'approvisionnement du combustible nucléaire*, in Nuclear Inter Jura '81, Congrès de l'AIDN, Palma de Majorque, 1981, pp.210-217.

SHAKER, M. I., *The Third NPT Review Conference: Issues and Prospects in Nuclear Non-Proliferation and Global Security*, Croom Helm Publ., London, 1981, pp.3-12.

SCHEAR, J. A., *Nuclear Weapons Proliferation and Nuclear Risk*, Aldershot Government Publication, UK, 1984, 185p.

SCHEINMAN, L., *The International Atomic Energy Agency and World Nuclear Order*, Resources for the Future, Washington DC., 1987, 320p.

SCHIFF, B. N., *International Nuclear Technology Transfer: Dilemmas of Dissemination and Control*, Croom Helm Publ., Canberra, 1984, 226p.

SHAW, E. N., *Europe's Nuclear Power Experiment, History of the OECD DRAGON Project*, Pergamon Press, Oxford, 1983, 338p.

SIMPSON, J. and McGREW, A., Editors, *The International Nuclear Non-Proliferation System: Challenges and Choices*, Macmillan Publ., London, 1984, 209p.

SKJOELDEBRAND, R., *World Trade International Nuclear Markets: Problems and Prospects*, in IAEA Bulletin, Vol. 26, No. 3, September 1984, pp.31-36 (also in French).

SNYDER, WELLS, SCHLESINGER, Editors, *Limiting Nuclear Proliferation*, Ballinger Publ., Cambridge, Mass., 1985, 363p.

SPECTOR, L. S., *Nuclear Proliferation Today*, Carnegie Endowment for International Peace, Random House Inc., New York, 1984, 478p.

SPECTOR, L. S., *The New Nuclear Nations*, Carnegie Endowment for International Peace, Random House Inc., New York, 1985, 367p.

STROHL, P., *La coopération internationale dans le domaine de l'énergie nucléaire — Europe et pays de l'OCDE*, SFDI, Colloque de Nancy, 21-23 mai 1981, Pédone, Paris, 1982, pp.122-158.

SZASZ, P., *The Law and Practices of the International Atomic Energy Agency*, Legal Series No. 7, IAEA, Vienna, 1970, 1176p.

TEMPUS, P., *Progress in Safeguards: 1983 Implementation*, in IAEA Bulletin, Vol. 26, No. 3, September 1984, pp.7-12 (also in French).

TREITSCHKE, W., *Vom NV-Vertrag für EURATOM aufgeworfene wesentliche Probleme und deren Lösung im Verifikationsabkommen*, Drittes Deutsches Atomrechts Symposium, 1975, 175p.

TROSTEN, L. M., *Safeguards for Nuclear Materials in Transit: Recent Developments*, in Nuclear Inter Jura '79, INLA Congress, Buenos Aires, 1981, pp.139-150.

VEDEL, V., *Le régime de propriété dans le Traité EURATOM*, in Annuaire français de droit international, 1957, p. 586 et seq.

VIROLE, J., *Contrats internationaux de droit privé concernant le cycle du combustible nucléaire et incidence des mesures d'embargo*, in Nuclear Inter Jura '79, Congrès de l'AIDN, Buenos Aires, 1981, pp.47-64.

WALKER, W. and LÖNNROTH, M., *Nuclear Power Struggles — Industrial Competition and Proliferation Control*, George Allen and Unwin, London, 1983, 204p.

WARUSFSEL, B., *Le contrôle des exportations stratégiques*, Rev. Défense Nat., Paris, 1985.

WARUSFSEL, B. et CHANTEBOUT, B., *Le contrôle des exportations de haute technologie vers les pays de l'Est*, Collection Droit-Sciences Economiques, Masson, Paris, 1987.

WAYLAND YOUNG, Editor, *Existing Mechanisms of Arms Control*, Commonwealth and International Library, 1966, 150p.

WEINBERG, ALONSO, BARKENBUS, Editors, *The Nuclear Connection. A Reassessment of Nuclear Power and Nuclear Proliferation*, New York, N.Y., Paragon House, 1985, 295p.

WILLRICH, M., *International Safeguards and Nuclear Industry*, Johns Hopkins University Press, Baltimore, 1973, 250p.

WILLRICH, M. and TAYLOR, T., *Nuclear Theft: Risks and Safeguards*, Ballinger Publ., Cambridge, Mass., 1974, 251p.

WINKLER, Th. H., *Le marché mondial du nucléaire et la prolifération dans les années 80*, in Politique Etrangère, IFRI, Paris, 1980 (47), pp.633-653.

WONDER, E., *On Comparing Nuclear Export Policies: Seeking the Determinants of International Control*, paper delivered to the Annual Meeting of the American Political Science Association, Washington DC, 1977.

ZIEGER, G., *Die rechtliche Problematik des NV-Vertrages und des Verifikationsabkommens der Europäischen Atomgemeinschaft mit der IAEO*, Drittes Deutsches Atomrechts Symposium, 1975, p.143 et seq.

B.   INTERNATIONAL ORGANISATIONS AND OTHER

International Atomic Energy Agency

The Law and Practices of the International Atomic Energy Agency (P. Szasz), Legal series No. 7, IAEA, Vienna, 1970, 1176p.

International Treaties Relating to Nuclear Control and Disarmament, Legal Series No. 9, IAEA, Vienna, 1975, 78p.

Non-Proliferation and International Safeguards, IAEA, Vienna, 1978, 75p (also in French).

Report of INFCE Working Group 3, *Assurances of Long-Term Supply of Technology, Fuel and Heavy Water and Services in the Interest of National Needs Consistent with Non-Proliferation*, IAEA, Vienna, 1980, 101p.

INFCE, *Summary Volume*, IAEA, Vienna, 1980, 72p.

Safeguards: An Introduction, IAEA, Vienna, 1981, 38p.

Agreements Registered with the International Atomic Energy Agency, Legal Series No. 3, Ninth Edition, IAEA, Vienna, 1985, 262p.

Nuclear Safeguards Technology, Proceedings of an IAEA Symposium, IAEA, Vienna 10th-14th November 1986, Vol. 1, 769p., Vol. 2, 660p.

OECD Nuclear Energy Agency

The Economics of the Nuclear Fuel Cycle, OECD/NEA, Paris, 1985, 168p. (also in French).

Nuclear Energy and Its Fuel Cycle, Prospects to 2025, 1987 Report, OECD/NEA, Paris, 188p. (also in French).

Uranium — Resources, Production and Demand, OECD/NEA-IAEA, OECD Paris, 1988, 194p. (also in French).

*
**

Comparative Study of European Nuclear Export Regulations (second edition), Volume I, Main Report, 149p. and Volume II, Appendices, Lawrence Livermore National Laboratory, Livermore, California, August 1985.

EURATOM — L'approvisionnement en question, Journée d'études organisée par l'Institut d'Etudes Européennes, Université Libre de Bruxelles le 21 juin 1980, Editions de l'Université Libre de Bruxelles, 1982.

Existe-t-il une spécificité du droit du commerce international?, Rapport de la 3ème Commission : commerce international nucléaire, in Nuclear Inter Jura '87, AIDN/INLA Congress, Antwerp (to be published).

Government Influence on International Trade in Uranium, Uranium Institute, London, 1978, 20p.

Internationalisation to Prevent the Spread of Nuclear Weapons, SIPRI, Taylor and Francis Ltd., London, 1980, 224p.

Le régime juridique du commerce international nucléaire, Rapport de la 3ème Commission: commerce international nucléaire, in International Harmonization in the Field of Nuclear Energy Law, Nuclear Inter Jura '85, AIDN/INLA Congress, Konstanz, Nomos Verlagsgesellschaft, Baden-Baden, pp.359-385.

Les entreprises de coopération technique internationale, aspects juridiques — bilan — perspectives, Table ronde, 27 avril 1985, SFDI, OCDE/AEN, AIE, ASE, 155p.

New Approaches to Non-Proliferation: A European Approach, Centre for European Policy Studies, No. 19/20, Brussels, 1985, 72p.

Nuclear Energy and Nuclear Weapon Proliferation, SIPRI, Taylor and Francis Ltd., London, 1979, 462p.

Nuclear Power and Nuclear Weapons Proliferation, Report of the Atlantic Council's Nuclear Fuel Policy Working Group, Vol. I, The Atlantic Council of the United States, 1978, 139p.

Nuclear Proliferation Fact Book, Congressional Research Service, US Library of Congress, 1985, 591p.

Nuclear Safeguards: A Reader, Congressional Research Service, US Library of Congress, December 1983, 999p.

Nuclear Weapons: US Non-Proliferation Policy in the 97th Congress, Congressional Research Service, US Library of Congress, 1983, 84p.

Prior Consent and Security of Supply in International Nuclear Trade, Uranium Institute, London, 1980, 15p.

Proliferation, Politics and the IAEA: The Issue of Nuclear Safeguards, Aspen Institute Workshop, 1985, 38p.

Review Conference of the Parties to the Treaty on the Non-Proliferation of Nuclear Weapons, Final Document, Part I, United Nations, Geneva, 1985, 62p.

Supplier Country Approaches to Nuclear Technology Transfer (extract from US publication on Technology Transfer to Middle East), pp.392-400.

The Development of Atomic Energy 1939-1984, Chronology of Events, United Kingdom Atomic Energy Authority, London, 1984, 66p.

The Reagan Administration Policy for Preventing the Further Spread of Nuclear Weapons: A Summary and Analysis of Official Statements, Congressional Research Service, US Library of Congress, May 1983.

The US and the Future of Non-Proliferation Regimes, Stanley Foundation Report, New York, 1984.

NEA Statute

Free on request

Statuts de l'AEN

Gratuit sur demande

Nuclear Law Bulletin

(Annual Subscription – two issues and supplements)
ISSN 0304-341-X
Index of the forty five issues of the Nuclear Law Bulletin (included in subscription)

£17.60   US$33.00

Bulletin de Droit Nucléaire

(Abonnement annuel – deux numéros et suppléments)
ISSN 0304-3428
Index des quarante cinq premiers numéros du Bulletin de Droit Nucléaire (compris dans l'abonnement)

F150,00   DM65.00

Licensing Systems and Inspection of Nuclear Installations (1986)
ISBN 92-64-12776-3

£12.00   US$24.00

Régime d'autorisation et d'inspection des installations nucléaires (1986)
ISBN 92-64-22776-8

F120,00   DM53

Long-term Management of Radioactive Waste – *Legal Administrative and Financial Aspects (1984)*

ISBN 92-64-12622-8

£7.00   US$14.00

Gestion à long terme des déchets radioactifs – *Aspects juridiques, administratifs et financiers (1984)*

ISBN 92-64-22622-2

F70,00   DM31.00

Nuclear Legislation, Analytical Study:
Regulatory and Institutional Framework for Nuclear Activities
Vol. 1 (1983)
Austria, Belgium, Canada, Denmark, France, Federal Republic of Germany, Greece, Iceland, Ireland, Italy, Japan, Luxembourg, Netherlands
ISBN 92-64-12534-5

£12.50   US$25.00

Législations nucléaires, étude analytique :
Réglementation générale et cadre institutionnel des activités nucléaires
Vol. 1 (1983)
Autriche, Belgique, Canada, Danemark, France, République fédérale d'Allemagne, Grèce, Islande, Irlande, Italie, Japon, Luxembourg, Pays-Bas
ISBN 92-64-22534-X

F125,00   DM56.00

Vol. 2 (1984)
New Zealand, Norway, Portugal, Spain, Sweden, Switzerland, Turkey, United Kingdom, United States
ISBN 92-64-12602-3

£15.00   US$30.00

Vol. 2 (1984)
Nouvelle-Zélande, Norvège, Portugal, Espagne, Suède, Suisse, Turquie, Royaume-Uni, États-Unis

ISBN 92-64-22602-8

F150,00   DM67.00

Nuclear Third Party Liability and Insurance – *Status and Prospects* (Proceedings of the Munich Symposium, 1984)

La responsabilité civile nucléaire et l'assurance – *Bilans et perspectives* (Compte rendu du Symposium de Munich, 1984)

ISBN 92-64-02665-7

£18.00   US$36.00

F180,00   DM80.00

WHERE TO OBTAIN OECD PUBLICATIONS
OÙ OBTENIR LES PUBLICATIONS DE L'OCDE

ARGENTINA - ARGENTINE
Carlos Hirsch S.R.L.,
Florida 165, 4º Piso,
(Galeria Guemes) 1333 Buenos Aires
Tel. 33.1787.2391 y 30.7122

AUSTRALIA - AUSTRALIE
D.A. Book (Aust.) Pty. Ltd.
11-13 Station Street (P.O. Box 163)
Mitcham, Vic. 3132   Tel. (03) 873 4411

AUSTRIA - AUTRICHE
OECD Publications and Information Centre,
4 Simrockstrasse,
5300 Bonn (Germany)   Tel. (0228) 21.60.45
Gerold & Co., Graben 31, Wien 1 Tel. 52.22.35

BELGIUM - BELGIQUE
Jean de Lannoy,
Avenue du Roi 202
B-1060 Bruxelles   Tel. (02) 538.51.69

CANADA
Renouf Publishing Company Ltd/
Éditions Renouf Ltée,
1294 Algoma Road, Ottawa, Ont. K1B 3W8
Tel: (613) 741-4333
Toll Free/Sans Frais:
Ontario, Quebec, Maritimes:
1-800-267-1805
Western Canada, Newfoundland:
1-800-267-1826
Stores/Magasins:
61 rue Sparks St., Ottawa, Ont. K1P 5A6
Tel: (613) 238-8985
211 rue Yonge St., Toronto, Ont. M5B 1M4
Tel: (416) 363-3171
Federal Publications Inc.,
301-303 King St. W.,
Toronto, Ont. M5V 1J5
Tel. (416)581-1552
Les Éditions la Liberté inc.,
3020 Chemin Sainte-Foy,
Sainte-Foy, P.Q. G1X 3V6,
Tel. (418)658-3763

DENMARK - DANEMARK
Munksgaard Export and Subscription Service
35, Nørre Søgade, DK-1370 København K
Tel. +45.1.12.85.70

FINLAND - FINLANDE
Akateeminen Kirjakauppa,
Keskuskatu 1, 00100 Helsinki 10   Tel. 0.12141

FRANCE
OCDE/OECD
Mail Orders/Commandes par correspondance :
2, rue André-Pascal,
75775 Paris Cedex 16
Tel. (1) 45.24.82.00
Bookshop/Librairie : 33, rue Octave-Feuillet
75016 Paris
Tel. (1) 45.24.81.67 or/ou (1) 45.24.81.81
Librairie de l'Université,
12a, rue Nazareth,
13602 Aix-en-Provence   Tel. 42.26.18.08

GERMANY - ALLEMAGNE
OECD Publications and Information Centre,
4 Simrockstrasse,
5300 Bonn   Tel. (0228) 21.60.45

GREECE - GRÈCE
Librairie Kauffmann,
28, rue du Stade, 105 64 Athens Tel. 322.21.60

HONG KONG
Government Information Services,
Publications (Sales) Office,
Information Services Department
No. 1, Battery Path, Central

ICELAND - ISLANDE
Snæbjörn Jónsson & Co., h.f.,
Hafnarstræti 4 & 9,
P.O.B. 1131 – Reykjavik
Tel. 13133/14281/11936

INDIA - INDE
Oxford Book and Stationery Co.,
Scindia House, New Delhi 110001
Tel. 331.5896/5308
17 Park St., Calcutta 700016   Tel. 240832

INDONESIA - INDONÉSIE
Pdii-Lipi, P.O. Box 3065/JKT.Jakarta
Tel. 583467

IRELAND - IRLANDE
TDC Publishers - Library Suppliers,
12 North Frederick Street, Dublin 1
Tel. 744835-749677

ITALY - ITALIE
Libreria Commissionaria Sansoni,
Via Lamarmora 45, 50121 Firenze
Tel. 579751/584468
Via Bartolini 29, 20155 Milano   Tel. 365083
La diffusione delle pubblicazioni OCSE viene
assicurata dalle principali librerie ed anche da :
Editrice e Libreria Herder,
Piazza Montecitorio 120, 00186 Roma
Tel. 6794628
Libreria Hœpli,
Via Hœpli 5, 20121 Milano   Tel. 865446
Libreria Scientifica
Dott. Lucio de Biasio "Aeiou"
Via Meravigli 16, 20123 Milano   Tel. 807679

JAPAN - JAPON
OECD Publications and Information Centre,
Landic Akasaka Bldg., 2-3-4 Akasaka,
Minato-ku, Tokyo 107   Tel. 586.2016

KOREA - CORÉE
Kyobo Book Centre Co. Ltd.
P.O.Box: Kwang Hwa Moon 1658,
Seoul   Tel. (REP) 730.78.91

LEBANON - LIBAN
Documenta Scientifica/Redico,
Edison Building, Bliss St.,
P.O.B. 5641, Beirut   Tel. 354429-344425

**MALAYSIA/SINGAPORE -
MALAISIE/SINGAPOUR**
University of Malaya Co-operative Bookshop
Ltd.,
7 Lrg 51A/227A, Petaling Jaya
Malaysia   Tel. 7565000/7565425
Information Publications Pte Ltd
Pei-Fu Industrial Building,
24 New Industrial Road No. 02-06
Singapore 1953   Tel. 2831786, 2831798

NETHERLANDS - PAYS-BAS
SDU Uitgeverij
Christoffel Plantijnstraat 2
Postbus 20014
2500 EA's-Gravenhage   Tel. 070-789911
Voor bestellingen: Tel. 070-789880

NEW ZEALAND - NOUVELLE-ZÉLANDE
Government Printing Office Bookshops:
Auckland: Retail Bookshop, 25 Rutland Stseet,
Mail Orders, 85 Beach Road
Private Bag C.P.O.
Hamilton: Retail: Ward Street,
Mail Orders, P.O. Box 857
Wellington: Retail, Mulgrave Street, (Head
Office)
Cubacade World Trade Centre,
Mail Orders, Private Bag
Christchurch: Retail, 159 Hereford Street,
Mail Orders, Private Bag
Dunedin: Retail, Princes Street,
Mail Orders, P.O. Box 1104

NORWAY - NORVÈGE
Narvesen Info Center – NIC,
Bertrand Narvesens vei 2,
P.O.B. 6125 Etterstad, 0602 Oslo 6
Tel. (02) 67.83.10, (02) 68.40.20

PAKISTAN
Mirza Book Agency
65 Shahrah Quaid-E-Azam, Lahore 3 Tel. 66839

PHILIPPINES
I.J. Sagun Enterprises, Inc.
P.O. Box 4322 CPO Manila
Tel. 695-1946, 922-9495

PORTUGAL
Livraria Portugal,
Rua do Carmo 70-74,
1117 Lisboa Codex   Tel. 360582/3

**SINGAPORE/MALAYSIA -
SINGAPOUR/MALAISIE**
See "Malaysia/Singapor". Voir
« Malaisie/Singapour »

SPAIN - ESPAGNE
Mundi-Prensa Libros, S.A.,
Castelló 37, Apartado 1223, Madrid-28001
Tel. 431.33.99
Libreria Bosch, Ronda Universidad 11,
Barcelona 7   Tel. 317.53.08/317.53.58

SWEDEN - SUÈDE
AB CE Fritzes Kungl. Hovbokhandel,
Box 16356, S 103 27 STH,
Regeringsgatan 12,
DS Stockholm   Tel. (08) 23.89.00
Subscription Agency/Abonnements:
Wennergren-Williams AB,
Box 30004, S104 25 Stockholm Tel. (08)54.12.00

SWITZERLAND - SUISSE
OECD Publications and Information Centre,
4 Simrockstrasse,
5300 Bonn (Germany)   Tel. (0228) 21.60.45
Librairie Payot,
6 rue Grenus, 1211 Genève 11
Tel. (022) 31.89.50
United Nations Bookshop/Librairie des Nations-
Unies
Palais des Nations,
1211 – Geneva 10
Tel. 022-34-60-11 (ext. 48 72)

TAIWAN - FORMOSE
Good Faith Worldwide Int'l Co., Ltd.
9th floor, No. 118, Sec.2
Chung Hsiao E. Road
Taipei   Tel. 391.7396/391.7397

THAILAND - THAILANDE
Suksit Siam Co., Ltd., 1715 Rama IV Rd.,
Samyam Bangkok 5   Tel. 2511630
INDEX Book Promotion & Service Ltd.
59/6 Soi Lang Suan, Ploenchit Road
Patjumamwan, Bangkok 10500
Tel. 250-1919, 252-1066

TURKEY - TURQUIE
Kültur Yayinlari Is-Türk Ltd. Sti.
Atatürk Bulvari No: 191/Kat. 21
Kavaklidere/Ankara   Tel. 25.07.60
Dolmabahce Cad. No: 29
Besiktas/Istanbul   Tel. 160.71.88

UNITED KINGDOM - ROYAUME-UNI
H.M. Stationery Office,
Postal orders only: (01)211-5656
P.O.B. 276, London SW8 5DT
Telephone orders: (01) 622.3316, or
Personal callers:
49 High Holborn, London WC1V 6HB
Branches at: Belfast, Birmingham,
Bristol, Edinburgh, Manchester

UNITED STATES - ÉTATS-UNIS
OECD Publications and Information Centre,
2001 L Street, N.W., Suite 700,
Washington, D.C. 20036 - 4095
Tel. (202) 785.6323

VENEZUELA
Libreria del Este,
Avda F. Miranda 52, Aptdo. 60337,
Edificio Galipan, Caracas 106
Tel. 951.17.05/951.23.07/951.12.97

YUGOSLAVIA - YOUGOSLAVIE
Jugoslovenska Knjiga, Knez Mihajlova 2,
P.O.B. 36, Beograd   Tel. 621.992

Orders and inquiries from countries where
Distributors have not yet been appointed should be
sent to:
OECD, Publications Service, 2, rue André-Pascal,
75775 PARIS CEDEX 16.

Les commandes provenant de pays où l'OCDE n'a
pas encore désigné de distributeur doivent être
adressées à :
OCDE, Service des Publications. 2, rue André-
Pascal, 75775 PARIS CEDEX 16.

71784-07-1988

OECD PUBLICATIONS, 2, rue André-Pascal, 75775 PARIS CEDEX 16 - No. 44171 1988
PRINTED IN FRANCE
(66 88 06 1) ISBN 92-64-13120-5